Matrix Operations for Engineers and Scientists

Alan Jeffrey

Matrix Operations for Engineers and Scientists

An Essential Guide in Linear Algebra

Springer

Prof. Dr. Alan Jeffrey[†]
16 Bruce Bldg.
University of Newcastle
NE1 7RU Newcastle upon Tyne
United Kingdom

ISBN 978-90-481-9273-1 e-ISBN 978-90-481-9274-8
DOI 10.1007/978-90-481-9274-8
Springer Dordrecht Heidelberg London New York

Library of Congress Control Number: 2010932003

Cover design: eStudio Calamar S.L., Germany

Printed on acid-free paper

Springer is part of Springer Science+Business Media (www.springer.com)

Preface

This book is based on many courses given by the author to English and American undergraduate students in engineering and the applied sciences. The book separates naturally into two distinct parts, although these are not shown as parts one and two. The first part, represented by Chapters 1–4 and a large part of Chapter 5, gives a straightforward account of topics from the theory of matrices that form part of every basic mathematics course given to undergraduate students in engineering and the applied sciences. However, the presentation of the basic material given in this book is in greater detail than is usually found in such courses. The only unusual topics appearing in the first part of the book are in Chapter 3. These are the inclusion of the technique of least-squares fitting of polynomials to experimental data, and the way matrices enter into a finite difference approximation for the numerical solution of the Laplace equation. The least-squares fitting of polynomials has been included because it is useful and provides a simple application of matrices, while the finite difference approximation for the Laplace equation shows how matrices play a vital part in the numerical solution of this important partial differential equation. This last application also demonstrates one of the ways in which very large matrix equations can be generated when seeking the numerical solution of certain types of problem.

The last part of Chapter 5 forms the start of the second part of the book, and contains various important topics which, although belonging to the subject matter of the chapter, are not discussed in courses as often as they deserve. Chapter 6 describes a matrix approach to the study of systems of ordinary differential equations and, although this approach is straightforward and found in courses for mathematics majors, it is still a relatively new topic in courses for engineers and applied scientists. In particular, the chapter shows how to use matrices when solving the homogeneous and nonhomogeneous systems of linear constant coefficient differential equations that model so many physical situations. It makes full use of the diagonalization of matrices when seeking solutions of systems of differential equations, and it also shows how the Laplace transform can be applied to matrix systems of differential equations. The chapter also provides motivation for the concept of the matrix exponential, which is then applied to differential equations.

Chapter 7 uses matrices as a typical model when explaining the notion of vector spaces that are the key to understanding many applications of mathematics. This enables the basic ideas of vector spaces to be introduced at an early stage in an undergraduate course. Chapter 8 develops the important and useful concept of a linear transformation and provides motivation by using matrices when applying linear transformations to the geometry of the plane. These applications illustrate the general ideas of linear transformations in terms of simple and familiar geometrical operations like stretching, rotating and reflecting shapes while, at the same time, relating them directly to the study of matrices. Although these applications are elementary, they are nevertheless useful, because while they can be combined to make more complicated transformations, they also serve as a foundation for the techniques used in applications as diverse as solid mechanics, crystallography and computer graphics.

This book can be used as a text for a course, to supplement an existing course, for private study, or to refresh and extend the reader's knowledge of the theory of matrices. All chapters are provided with clear and detailed illustrative examples as each new idea is introduced, so, for example, attention is drawn to the fact that a twice repeated eigenvalue does not necessarily have associated with it two linearly independent eigenvectors, and it is then shown how this influences the nature of solutions of systems of differential equations. Apart from Chapter 6 on differential equations, no systematic attempt has been made to describe the numerous applications of matrices that are possible. Nevertheless, because of the intended readership of the book, where appropriate a few relevant applications have been included. Some of these applications have already been mentioned, but others illustrate the way matrices can be used to solve linear second-order difference equations like the one that generates the Fibonacci sequence and, because of the importance of two-point boundary-value problems in applications of differential equations, it is shown how matrices enter into the numerical solution of some of these problems.

Throughout the book, worked examples are numerous and they are supplemented by exercise sets at the end of each chapter. Solutions for all of the exercises are given at the end of the book, always provided with adequate detail if a method of solution is not completely obvious. Detailed explanations of new ideas have been given throughout the book, because the author's experience has shown that an inadequate explanation when a topic is first encountered can cause unnecessary difficulties for a student at later stages of study when matrix methods need to be applied.

The ready availability of computer algebra software makes the manipulation of matrices a simple matter and, in real life applications, such software should be used whenever possible and, indeed, for complicated and large problems its use is essential. However, the use of such software tools when learning about matrices, before having first understood the underlying theory by working well-chosen examples by hand with the help of a hand-held calculator, is likely to limit the reader's ability to make full use of matrices when the time comes to apply matrix methods to new problems.

The efficient ways software manipulates matrices when performing numerical operations, like finding the rank of a matrix, its eigenvalues and eigenvectors, and accelerating computations while maintaining high accuracy, depend for their success on the use of sophisticated numerical techniques. Of necessity, the approach used in such software will differ from the way the same operations are described in this book, where only straightforward and direct methods are given, and the necessary numerical calculations in examples and exercises have been reduced to a minimum. For example, to simplify the numerical calculations involved when working with eigenvalues and eigenvectors, the worked examples and exercise sets dealing with this topic have been constructed in such a way that, whenever a characteristic equation occurs, its roots can be found by inspection. This allows the analysis to proceed without the interruption that would otherwise be caused if a numerical root-finding technique for polynomials had first to be explained and then used.

It is hoped readers will find the book helpful when working with matrices and when applying linear algebra, and that it will encourage them to apply matrix methods to the wide range of problems that are often solved less efficiently and concisely by other means.

University of Newcastle Alan Jeffrey

Contents

Chapter 1
Matrices and Linear Systems of Equations

1.1 Systems of Algebraic Equations

The practical interest in matrices arose from the need to work with linear systems of algebraic equations of the form

$$\begin{aligned} a_{11}x_1 + a_{12}x_2 + \cdots a_{1n}x_n &= b_1, \\ a_{21}x_1 + a_{22}x_2 + \cdots a_{2n}x_n &= b_2, \\ a_{31}x_1 + a_{32}x_2 + \cdots a_{3n}x_n &= b_3, \\ &\vdots \\ a_{m1}x_1 + a_{m2}x_2 + \cdots a_{mn}x_n &= b_m, \end{aligned} \tag{1.1}$$

involving the n unknowns $x_1, x_2, \ldots, x_n$, m equations with constant coefficients a_{ij}, $i = 1, 2, \ldots, m$, $j = 1, 2, \ldots n$, and m constants $b_1, b_2, \ldots, b_{mn}$ called the nonhomogeneous terms, where the coefficients a_{ij} and the b_i may be real or complex numbers. A *solution set* for system (1.1) is a set of numbers $\{x_1, x_2, \ldots x_n\}$, real or complex, that when substituted into (1.1), satisfy all m equations identically. When $m < n$ system (1.1) is said to be *underdetermined*, so as there are fewer linear equations than unknowns a unique solution set cannot be expected.

The reason for this can be seen by considering the simple underdetermined system

$$\begin{aligned} x_1 + x_2 + x_3 &= 1, \\ x_1 + 2x_2 + 3x_3 &= 2. \end{aligned}$$

Rewriting the system as

$$\begin{aligned} x_1 + x_2 &= 1 - x_3, \\ x_1 + 2x_2 &= 2 - 3x_3, \end{aligned}$$

and for the moment regarding the expressions on the right of the equality sign as known quantities, solving for x_1 and x_2 by elimination gives $x_1 = x_3$ and $x_2 = 1 - 2x_3$,

A. Jeffrey, *Matrix Operations for Engineers and Scientists*,
DOI 10.1007/978-90-481-9274-8_1,

where x_3 is unknown. Setting $x_3 = k$, where k is a parameter (an arbitrary number), the solution set $\{x_1, x_2, x_3\}$ of this underdetermined system becomes $\{k, 1 - 2k, k\}$. As k is arbitrary, the solution set of the system is not unique. It is not difficult to see that this situation generalizes to larger underdetermined systems, though then the solution set may depend on more than one unknown variable, each of which may be regarded as a parameter.

When $m > n$ system (1.1) is said to be *overdetermined*, so as n unknowns have to satisfy $m > n$ linear equations, in general no solution set will exist. That overdetermined systems may or may not have a solution set can be seen by considering the following three systems:

$$\text{(a)}\quad \begin{aligned} x_1 + x_2 &= 1 \\ x_1 + 2x_2 &= 3 \\ x_1 + 3x_2 &= 0, \end{aligned} \qquad \text{(b)}\quad \begin{aligned} x_1 + x_2 + x_3 &= 2 \\ x_1 + 2x_2 + 3x_3 &= 0 \\ x_1 - 2x_2 + x_3 &= -4 \\ 2x_1 + 3x_2 + 4x_3 &= 2, \end{aligned} \qquad \text{(c)}\quad \begin{aligned} x_1 + x_2 + x_3 &= 1 \\ x_1 + 2x_2 + 3x_3 &= 2 \\ x_2 + 2x_3 &= 1 \\ 2x_2 + 3x_3 + 4x_3 &= 3. \end{aligned}$$

System (a) can have no solution, because the left side of the third equation is the sum of the left sides of the first two equations, but this relationship is *not* true for its right side. Thus the last equation contradicts the first two equations, so the system is said to be *inconsistent*. In system (b) the last equation is seen to be the sum of the first two equations, so after discarding the last equation because it is redundant, solving the remaining three equations by elimination gives $x_1 = 2$, $x_2 = 2$ and $x_3 = -2$. Thus the overdetermined system in (b) has a unique solution set $\{2, 2, -2\}$. However, the situation in system (c) is different again, because the third equation is simply the difference between the second and first equations, while the fourth equation is the sum of the first two equations, so after discarding the last two equations which are redundant, we are left with the first two equations that have already been shown in (a) to have the nonunique solution set $\{x_1, x_2, x_3\}$of the form $\{k, 1 - 2k, k\}$, with k arbitrary (a parameter).

Finally, when $m = n$ system (1.1) is said to be *properly determined*, so as n unknowns have to satisfy n linear equations, unless one or more of the equations contradicts the other equations, a unique solution set can be expected. This is the case with the system

$$\begin{aligned} x_1 + x_2 - x_3 &= 6, \\ x_1 - x_2 + x_3 &= -4, \\ x_1 + 2x_2 - x_3 &= 8, \end{aligned}$$

which is easily seen to have the unique solution set $\{x_1, x_2, x_3\}$ given by $\{1, 2, -3\}$. Notice that when, as above, the general solution set $\{x_1, x_2, x_3\}$ is equated to $\{1, 2, -3\}$, this requires *corresponding* entries to be equal, so writing$\{x_1, x_2, x_3\} = \{1, 2, -3\}$ means that $x_1 = 1$, $x_2 = 2$ and $x_3 = -3$. This interpretation of equality between similar arrangements (arrays) of quantities, which in this case were numbers, will be seen to play an important role when matrices are introduced and their equality is defined.

1.2 Suffix and Matrix Notation

Later the solution of the system of Eq. (1.1) will be considered in detail, and it will be shown how to determine if a unique solution set exists, if a solution set exists but it is not unique, and in which case how many arbitrary parameters the solution set must contain, and if no solution set exists.

The suffix notation for the coefficients and unknowns in system (1.1) is standard, and its purpose is to show that a_{ij} is the numerical multiplier of the jth unknown x_j in the ith equation, and b_i is the corresponding nonhomogeneous term in the ith equation. With this understanding, because the numbers a_{ij} and b_j each has a sign, if the n unknowns $x_1, x_2, \ldots, x_n$ are arranged in the same order in each equation, the symbols $x_1, x_2, \ldots, x_n$ may be omitted, and the system represented instead by the array of numbers

$$\begin{array}{ccccccc} a_{11} & a_{12} & a_{13} & \cdots & a_{1n} & \vdots & b_1 \\ a_{21} & a_{22} & a_{23} & \cdots & a_{2n} & \vdots & b_2 \\ a_{31} & a_{32} & a_{33} & \cdots & a_{3n} & \vdots & b_3 \\ & & & \vdots & & & \\ a_{m1} & a_{m2} & a_{m3} & \cdots & a_{mn} & \vdots & b_m \end{array}. \tag{1.2}$$

For reasons that will appear later, the nonhomogeneous terms b_i have been separated from the array of coefficients a_{ij}, and for the time being the symbol $\vdots$ has been written in place of the equality sign. The double suffix ij serves as the "grid reference" for the position of the number a_{ij} in the array (1.2) showing that it occurs in the ith row and the jth column, while for the nonhomogeneous term b_i, the suffix i shows the row in which b_i occurs. For example, if $a_{32} = -5$, the numerical multiplier of x_2 in the third equation in (1.1) is -5, so the element in the second position of the third row in array (1.2) is -5. Similarly, if $b_3 = 4$ the nonhomogeneous term in the third equation in (1.1) is 4, so the entry b_3 in (1.2) is 4. Arrays of m rows of n numbers are called *matrices*, and a concise notation is needed if instead of algebra being performed on equations like (1.1), it is to be replaced by algebra performed on matrices. The standard notation for a **matrix** denoted by **A** that contains the entries a_{ij}, and a matrix containing the entries b_i in (1.2) is to write

$$\mathbf{A} = \begin{bmatrix} a_{11} & a_{12} & a_{13} & \cdots & a_{1n} \\ a_{21} & a_{22} & a_{23} & \cdots & a_{2n} \\ a_{31} & a_{32} & a_{33} & \cdots & a_{3n} \\ \vdots & \vdots & \vdots & \vdots & \vdots \\ a_{m1} & a_{m2} & a_{m3} & \cdots & a_{mn} \end{bmatrix}, \mathbf{b} = \begin{bmatrix} b_1 \\ b_2 \\ b_3 \\ \vdots \\ b_m \end{bmatrix}, \tag{1.3}$$

or more concisely still,

$$\mathbf{A} = [a_{ij}], \quad i = 1, 2, \ldots, m, \; j = 1, 2, \ldots, n \text{ and } \mathbf{b} = [b_i], \quad i = 1, 2, \ldots, m. \tag{1.4}$$

A different but equivalent notation that is also in use replaces the square brackets [.] by (.), in which case (1.4) become $\mathbf{A} = (a_{ij})$ and $\mathbf{b} = (b_i)$.

Expression **A** in (1.3) is called an ***m*** **×** ***n*** **matrix** to show the number of rows m and the number of columns n it contains, without specifying individual entries. The notation $m \times n$ is often called the *size* or *shape* of a matrix, as it gives a qualitative understanding of the number of rows and columns in the matrix, without specifying the individual entries a_{ij}. A matrix in which the number of rows equals the number of columns it is called a *square* matrix, so if it has n rows, it is an $n \times n$ matrix. Matrix **b** in (1.3) is called an ***m*** **element column vector**, or if the number of entries in **b** is unimportant, simply a *column vector*. A matrix with the n entries $c_1, c_2, \ldots, c_n$ of the form

$$\mathbf{c} = [c_1, \; c_2, \; c_3, \; \ldots, c_n] \tag{1.5}$$

is called an ***n*** **element row vector**, or if the number of entries in **c** is unimportant, simply a *row vector*. In what follows we use the convention that row and column vectors are denoted by bold lower case Roman characters, while other matrices are denoted by bold upper case Roman characters. The entries in matrices and vectors are called *elements*, so an $m \times n$ matrix contains mn elements, while the row vector in (1.5) is an n element row vector. As a rule, the entries in a general matrix **A** are denoted by the corresponding lower case italic letter a with a suitable double suffix, while in a row or column vector **d** the elements are denoted by the corresponding lower case italic letter d with a single suffix. The elements in each row of **A** in (1.3) form an n element row vector, and the elements in each column form an m element column vector. This interpretation of matrices as collections of row or column vectors will be needed later when the operations of matrix *transposition* and *multiplication* are defined.

1.3 Equality, Addition and Scaling of Matrices

Two matrices **A** and **B** are said to be equal, shown by writing $\mathbf{A} = \mathbf{B}$, if each matrix has the same number of rows and columns, and elements in corresponding positions in **A** and **B** are equal. For example, if

$$\mathbf{A} = \begin{bmatrix} 1 & p \\ 2 & -4 \end{bmatrix} \text{ and } \mathbf{B} = \begin{bmatrix} 1 & 3 \\ 2 & q \end{bmatrix},$$

equality is *possible* because each matrix has the same number of rows and columns, so they each have the same shape, but $\mathbf{A} = \mathbf{B}$ only if, in addition, $p = 3$ and $q = -4$.

If every element in a matrix is zero, the matrix is written $\mathbf{0}$ and called the *null* or *zero* matrix. It is not usual to indicate the number of rows and columns in a null matrix, because it will be assumed they are appropriate for whatever algebraic operations are being performed. If, for example, in the linear system of algebraic equations in (1.1) all of the nonhomogeneous terms $b_1 = b_2 = \ldots = b_m = 0$, the corresponding vector $\mathbf{b}$ in (1.3) becomes $\mathbf{b} = \mathbf{0}$, where in this case $\mathbf{0}$ in an m-dimensional column vector with every element zero. A column or row vector in which every element is zero is called a *null vector*.

Given two similar systems of equations

$$\begin{array}{c} a_{11}x_1 + a_{12}x_2 + \cdots a_{1n}x_n = b_1 \\ a_{21}x_1 + a_{22}x_2 + \cdots a_{2n}x_n = b_2 \\ a_{31}x_1 + a_{32}x_2 + \cdots a_{3n}x_n = b_3 \\ \vdots \\ a_{m1}x_1 + a_{m2}x_2 + \cdots a_{mn}x_n = b_m \end{array} \quad \text{and} \quad \begin{array}{c} \tilde{a}_{11}x_1 + \tilde{a}_{12}x_2 + \cdots \tilde{a}_{1n}x_n = \tilde{b}_1 \\ \tilde{a}_{21}x_1 + \tilde{a}_{22}x_2 + \cdots \tilde{a}_{2n}x_n = \tilde{b}_2 \\ \tilde{a}_{31}x_1 + \tilde{a}_{32}x_2 + \cdots \tilde{a}_{3n}x_n = \tilde{b}_3 \\ \vdots \\ \tilde{a}_{m1}x_1 + \tilde{a}_{m2}x_2 + \cdots \tilde{a}_{mn}x_n = \tilde{b}_m, \end{array}$$

the result of adding corresponding equations, and writing the result in matrix form, leads to the following definitions of the sum of the respective coefficient matrices and of the vectors that contain the nonhomogeneous terms

$$\mathbf{A} + \tilde{\mathbf{A}} = \begin{bmatrix} a_{11} + \tilde{a}_{11} & a_{12} + \tilde{a}_{12} & a_{13} + \tilde{a}_{13} & \cdots & a_{1n} + \tilde{a}_{1n} \\ a_{21} + \tilde{a}_{21} & a_{22} + \tilde{a}_{22} & a_{23} + \tilde{a}_{23} & \cdots & a_{2n} + \tilde{a}_{2n} \\ a_{31} + \tilde{a}_{31} & a_{32} + \tilde{a}_{32} & a_{33} + \tilde{a}_{33} & \cdots & a_{3n} + \tilde{a}_{3n} \\ \vdots & \vdots & \vdots & \vdots & \vdots \\ a_{m1} + \tilde{a}_{m1} & a_{m2} + \tilde{a}_{m2} & a_{m3} + \tilde{a}_{m3} & \cdots & a_{mn} + \tilde{a}_{mn} \end{bmatrix} \text{ and } \mathbf{b} + \tilde{\mathbf{b}}$$

$$= \begin{bmatrix} b_1 + \tilde{b}_1 \\ b_2 + \tilde{b}_2 \\ b_3 + \tilde{b}_3 \\ \vdots \\ b_m + \tilde{b}_m \end{bmatrix}.$$

This shows that if matrix algebra is to represent ordinary algebraic addition, it must be defined as follows. Matrices $\mathbf{A}$ and $\mathbf{B}$ will be said to be *conformable for addition*, or *summation*, if each matrix has the same number of rows and columns. Setting $\mathbf{A} = [a_{ij}]$, $\mathbf{B} = [b_{ij}]$, the **sum A + B** of matrices $\mathbf{A}$ and $\mathbf{B}$ is defined as the matrix

$$\mathbf{A} + \mathbf{B} = \left[a_{ij} + b_{ij}\right]. \tag{1.6}$$

It follows directly from (1.6) that

$$\mathbf{A} + \mathbf{B} = \mathbf{B} + \mathbf{A}, \tag{1.7}$$

so matrix addition is *commutative*. This means the order in which conformable matrices are added (summed) is unimportant, as it does not affect the result. It follows from (1.6) that the *difference* between matrices **A** and **B**, written $\mathbf{A} - \mathbf{B}$, is defined as

$$\mathbf{A} - \mathbf{B} = [a_{ij} - b_{ij}]. \tag{1.8}$$

The sum and difference of matrices **A** and **B** with different shapes is *not* defined.

If each equation in (1.1) is *scaled* (multiplied) by a constant k the matrices in (1.3) become

$$\begin{bmatrix} ka_{11} & ka_{12} & ka_{13} & \cdots & ka_{1n} \\ ka_{21} & ka_{22} & ka_{23} & \cdots & ka_{2n} \\ ka_{31} & ka_{32} & ka_{33} & \cdots & ka_{3n} \\ \vdots & \vdots & \vdots & \vdots & \vdots \\ ka_{m1} & ka_{m2} & ka_{m3} & \cdots & ka_{mn} \end{bmatrix} \quad \text{and} \quad \begin{bmatrix} kb_1 \\ kb_2 \\ kb_3 \\ \vdots \\ kb_m \end{bmatrix}.$$

This means that if matrix $\mathbf{A} = [a_{ij}]$ is scaled by a number k (real or complex), then the result, written $k\mathbf{A}$, is defined as $k\mathbf{A} = [ka_{ij}]$. So if $\mathbf{A} = [a_{ij}]$ and $\mathbf{B} = [b_{ij}]$ are conformable for addition and k and K are any two numbers (real or complex), then

$$k\mathbf{A} + K\mathbf{B} = [ka_{ij} + Kb_{ij}]. \tag{1.9}$$

Example 1.1. Given $\mathbf{A} = \begin{bmatrix} 4 & -1 & 3 \\ 7 & 0 & -2 \end{bmatrix}$, $\mathbf{B} = \begin{bmatrix} -4 & 2 & 2 \\ -1 & 5 & 6 \end{bmatrix}$, find $\mathbf{A} + \mathbf{B}$, $\mathbf{A}$ $\mathbf{B}$ and $2\mathbf{A} + 3\mathbf{B}$.

Solution. The matrices are conformable for addition because each has two rows and three columns (they have the same shape). Thus from (1.6), (1.7) and (1.8)

$$\mathbf{A} + \mathbf{B} = \begin{bmatrix} 0 & 1 & 5 \\ 6 & 5 & 4 \end{bmatrix}, \ \mathbf{A} - \mathbf{B} = \begin{bmatrix} 8 & -3 & 1 \\ 8 & -5 & -8 \end{bmatrix} \text{ and } 2\mathbf{A} + 3\mathbf{B} = \begin{bmatrix} -4 & 4 & 12 \\ 11 & 15 & 14 \end{bmatrix}.$$

1.4 Some Special Matrices and the Transpose Operation

Some square matrices exhibit certain types of symmetry in the pattern of their coefficients. Consider the $n \times n$ matrix

$$\mathbf{A} = \begin{bmatrix} a_{11} & a_{12} & a_{13} & \cdots & a_{1n} \\ a_{21} & a_{22} & a_{23} & \cdots & a_{2n} \\ a_{31} & a_{32} & a_{33} & \cdots & a_{3n} \\ \vdots & \vdots & \vdots & \vdots & \vdots \\ a_{n1} & a_{n2} & a_{n3} & \cdots & a_{nn} \end{bmatrix},$$

then the diagonal drawn from top left to bottom right containing the elements a_{11}, a_{22}, a_{33}, . . . , a_{nn} is called the *leading diagonal* of the matrix.

A square matrix **A** is said to be *symmetric* if its numerical entries appear symmetrically about the leading diagonal. That is, the elements of an $n \times n$ symmetric matrix **A** are such that

$$a_{ij} = a_{ji}, i, j = 1, 2, ..., n \; (\textit{condition for symmetry}). \tag{1.10}$$

Another way of defining a symmetric matrix is to say that if a new matrix **B** is constructed such that row 1 of **A** is written as column 1 of **B**, row 2 of **A** is written as column 2 of **B**, . . . , and row n of **A** is written as column n of **B**, then the matrices **A** and **B** are identical if **B** = **A**. For example, if

$$\mathbf{A} = \begin{bmatrix} 1 & 4 & 3 \\ 4 & 2 & 6 \\ 3 & 6 & 4 \end{bmatrix} \quad \text{and} \quad \mathbf{B} = \begin{bmatrix} 1 & 5 & 7 \\ 9 & 4 & 5 \\ 1 & 0 & 1 \end{bmatrix},$$

then **A** is seen to be a symmetric matrix, but **B** is not symmetric.

Belonging to the class of symmetric matrices are the $n \times n$ **diagonal matrices,** all of whose elements are zero away from the leading diagonal. A diagonal matrix **A** with entries λ_1, λ_2, . . . , λ_n on its leading diagonal, some of which may be zero, is often written $\mathbf{A} = \text{diag}\{\lambda_1, \lambda_2, \ldots, \lambda_n\}$. An important special case of diagonal matrices are the *identity matrices*, also called *unit matrices*, which are denoted collectively by the symbol **I**. These are diagonal matrices in which each element on the leading diagonal is 1 (and all remaining entries are zeros). When written out in full, if $\mathbf{A} = \text{diag}\{2, -3, 1\}$, and **I** is the 3×3 identity matrix, then

$$\mathbf{A} = \text{diag}\{2, -3, 1\} = \begin{bmatrix} 2 & 0 & 0 \\ 0 & -3 & 0 \\ 0 & 0 & 1 \end{bmatrix} \quad \text{and} \quad \mathbf{I} = \begin{bmatrix} 1 & 0 & 0 \\ 0 & 1 & 0 \\ 0 & 0 & 1 \end{bmatrix}.$$

As with the null matrix, it is not usual to specify the number of rows in an identity matrix, because the number is assumed to be appropriate for whatever algebraic operation is to be performed that involves **I**. If, for any reason, it is necessary to show the precise shape of an identity matrix, it is sufficient to write $\mathbf{I}_n$ to show an $n \times n$ identity matrix is involved. In terms of this notation, the 3×3 identity matrix shown above becomes $\mathbf{I}_3$.

A different form of symmetry occurs when the $n \times n$ matrix $\mathbf{A} = [a_{ij}]$ is *skew symmetric*, in which case its entries a_{ij} are such that

$$a_{ij} = -a_{ji} \text{ for } i, j = 1, 2, \ldots, n \ (\textit{condition for skew symmetry}). \tag{1.11}$$

Notice that elements on the leading diagonal of a skew symmetric matrix must all be zero, because by definition $a_{ii} = -a_{ii}$, and this is only possible if $a_{ii} = 0$ for $i = 1, 2, \ldots, n$.

A typical example of a skew symmetric matrix is

$$\mathbf{A} = \begin{bmatrix} 0 & 1 & 3 & -2 \\ -1 & 0 & 4 & 6 \\ -3 & -4 & 0 & -1 \\ 2 & -6 & 1 & 0 \end{bmatrix}.$$

Other square matrices that are important are *upper* and *lower triangular matrices*, denoted respectively by **U** and **L**. In **U** all elements below the leading diagonal are zero, while in **L** all elements above the leading diagonal are zero. Typical examples of upper and lower triangular matrices are

$$\mathbf{U} = \begin{bmatrix} 2 & 0 & 8 \\ 0 & 1 & 6 \\ 0 & 0 & -3 \end{bmatrix} \text{ and } \mathbf{L} = \begin{bmatrix} 3 & 0 & 0 \\ 5 & 1 & 0 \\ -9 & 7 & 0 \end{bmatrix}.$$

The need to construct matrices in which rows and columns have been interchanged (not necessarily square matrices) leads to the introduction of the *transpose operation*. The *transpose* of an $m \times n$ matrix **A**, denoted by $\mathbf{A}^{\mathrm{T}}$, is the $n \times m$ matrix derived from **A** by writing row 1 of **A** as column 1 of $\mathbf{A}^{\mathrm{T}}$, row 2 of **A** as column 2 of $\mathbf{A}^{\mathrm{T}}, \ldots,$ and row m of **A** as column m of $\mathbf{A}^{\mathrm{T}}$. Obviously, the transpose of a transposed matrix is the original matrix, so $(\mathbf{A}^{\mathrm{T}})^{\mathrm{T}} = \mathbf{A}$. Typical examples of transposed matrices are

$$[1 \quad -4 \quad 7]^{\mathrm{T}} = \begin{bmatrix} 1 \\ -4 \\ 7 \end{bmatrix}, \begin{bmatrix} 1 \\ -4 \\ 7 \end{bmatrix}^{\mathrm{T}} = [1, -4, 7] \text{ and } \begin{bmatrix} 2 & 0 & 5 \\ 1 & -1 & 4 \end{bmatrix}^{\mathrm{T}} = \begin{bmatrix} 2 & 1 \\ 0 & -1 \\ 5 & 4 \end{bmatrix}.$$

Clearly, a square matrix **A** is symmetric if $\mathbf{A}^{\mathrm{T}} = \mathbf{A}$, and it is skew symmetric if $\mathbf{A}^{\mathrm{T}} = -\mathbf{A}$. The matrix transpose operation has many uses, some of which will be encountered later.

A useful property of the transpose operation when applied to the sum of two $m \times n$ matrices **A** and **B** is that

$$[\mathbf{A} + \mathbf{B}]^{\mathrm{T}} = \mathbf{A}^{\mathrm{T}} + \mathbf{B}^{\mathrm{T}}. \tag{1.12}$$

This proof of this result is almost immediate. If $\mathbf{A} = [a_{ij}]$ and $\mathbf{B} = [b_{ij}]$, by definition

$$\mathbf{A} + \mathbf{B} = \begin{bmatrix} a_{11} + b_{11} & a_{12} + b_{12} & \cdots & a_{1n} + b_{1n} \\ a_{21} + b_{21} & a_{22} + b_{22} & \cdots & a_{2n} + b_{2n} \\ \vdots & \vdots & \vdots & \vdots \\ a_{m1} + b_{m1} & a_{m2} + b_{m2} & \cdots & a_{mn} + b_{mn} \end{bmatrix}.$$

Taking the transpose of this result, and then using the rule for matrix addition, we have

$$[\mathbf{A} + \mathbf{B}]^{\mathrm{T}} = \begin{bmatrix} a_{11} + b_{11} & a_{21} + b_{21} & \cdots & a_{m1} + b_{m1} \\ a_{12} + b_{12} & a_{22} + b_{22} & \cdots & a_{m2} + b_{m2} \\ \vdots & \vdots & \vdots & \vdots \\ a_{1n} + b_{1n} & a_{2n} + b_{2n} & \cdots & a_{nm} + b_{nm} \end{bmatrix} = \mathbf{A}^{\mathrm{T}} + \mathbf{B}^{\mathrm{T}},$$

and the result is established.

An important use of matrices occurs in the study of properly determined systems of n linear first order differential equations in the n unknown differentiable functions $x_1(t), x_2(t), \ldots, x_n(t)$ of the independent variable t:

$$\begin{aligned} \frac{dx_1(t)}{dt} &= a_{11}x_1(t) + a_{12}x_2(t) + \cdots + a_{1n}x_n(t), \\ \frac{dx_2(t)}{dt} &= a_{21}x_1(t) + a_{22}x_2(t) + \cdots + a_{2n}x_n(t), \\ &\vdots \\ \frac{dx_n(t)}{dt} &= a_{n1}x_1(t) + a_{n2}x_2(t) + \cdots + a_{nn}x_n(t). \end{aligned} \tag{1.13}$$

In the next chapter matrix multiplication will be defined, and in anticipation of this we define the coefficient matrix of system (1.13) as $\mathbf{A} = [a_{ij}]$, and the column vectors $\mathbf{x}(t)$, and $d\mathbf{x}(t)/dt$ as

$$\begin{aligned} \mathbf{x}(t) &= [x_1(t), x_2(t), \ldots, x_n(t)]^{\mathrm{T}} \text{ and} \\ \frac{d\mathbf{x}(t)}{dt} &= \left[\frac{dx_1(t)}{dt}, \frac{dx_2(t)}{dt}, \ldots, \frac{dx_n(t)}{dt}\right]^{\mathrm{T}}, \end{aligned} \tag{1.14}$$

where the transpose operation has been used to write a column vector as the transpose of a row vector to save space on the printed page. System (1.13) can be written more concisely as

$$\frac{d\mathbf{x}(t)}{dt} = \mathbf{A}\mathbf{x}(t), \tag{1.15}$$

where $\mathbf{Ax}(t)$ denotes the product of matrix $\mathbf{A}$ and vector $\mathbf{x}(t)$, in this order, which will be defined in Chapter 2. Notice how the use of the transpose operation in Eq. (1.14) saves space on a printed page, because had it not been used, column vectors like $\mathbf{x}(t)$ and $d\mathbf{x}(t)/dt$ when written out in full would have become

$$\mathbf{x}(t) = \begin{bmatrix} x_1(t) \\ x_2(t) \\ \vdots \\ x_n(t) \end{bmatrix} \quad \text{and} \quad \frac{d\mathbf{x}(t)}{dt} = \begin{bmatrix} \frac{dx_1(t)}{dt} \\ \frac{dx_2(t)}{dt} \\ \vdots \\ \frac{dx_n(t)}{dt} \end{bmatrix}.$$

Exercises

1. Write down the coefficient matrix $\mathbf{A}$ and nonhomogeneous term matrix $\mathbf{b}$ for the linear nonhomogeneous system of equations in the variables x_1, x_2, x_3 and x_4:

$$\begin{aligned} 3x_1 + 2x_2 - 4x_3 + 5x_4 &= 4, \\ 3x_1 + 2x_2 - x_4 + 4x_3 &= 3, \\ 4x_2 - 2x_1 + x_3 + 5x_4 &= 2, \\ 6x_3 + 3x_1 + 2x_2 &= 1. \end{aligned}$$

2. If $\mathbf{A} = \begin{bmatrix} 2 & 0 & 5 \\ 1 & 3 & 1 \end{bmatrix}$, $\mathbf{B} = \begin{bmatrix} -1 & 2 & 3 \\ -2 & 4 & 6 \end{bmatrix}$, find $\mathbf{A} + 2\mathbf{B}$ and $3\mathbf{A} - 4\mathbf{B}$.

3. If $\mathbf{A} = \begin{bmatrix} 1 & 3 & a \\ 2 & b & -1 \\ -2 & c & 3 \end{bmatrix}$ and $\mathbf{B} = \begin{bmatrix} 1 & 2 & -2 \\ 3 & 6 & 4 \\ 0 & -1 & 3 \end{bmatrix}$, find a, b and c if $\mathbf{A} = \mathbf{B}^{\mathrm{T}}$.

4. If $\mathbf{A} = \begin{bmatrix} 2 & 4 \\ 6 & 1 \\ 0 & 3 \end{bmatrix}$ and $\mathbf{B} = \begin{bmatrix} 4 & 1 & -3 \\ 2 & -3 & 1 \end{bmatrix}$, find $3\mathbf{A} - \mathbf{B}^{\mathrm{T}}$ and $2\mathbf{A}^{\mathrm{T}} + 4\mathbf{B}$.

5. If $\mathbf{A} = \begin{bmatrix} 3 & 0 & 1 \\ 1 & 4 & 3 \\ 5 & 1 & 2 \end{bmatrix}$ and $\mathbf{B} = \begin{bmatrix} 0 & 4 & 1 \\ 2 & 5 & 1 \\ 3 & -2 & 2 \end{bmatrix}$, find $\mathbf{A}^{\mathrm{T}} + \mathbf{B}$ and $2\mathbf{A} + 3(\mathbf{B}^{\mathrm{T}})^{\mathrm{T}}$.

6. If matrices $\mathbf{A}$ and $\mathbf{B}$ are conformable for addition, prove that $(\mathbf{A} + \mathbf{B})^{\mathrm{T}} = \mathbf{A}^{\mathrm{T}} + \mathbf{B}^{\mathrm{T}}$.

7. Given

$$\mathbf{A} = \begin{bmatrix} a_{11} & 4 & -3 & a_{14} \\ a_{21} & a_{22} & a_{23} & a_{24} \\ a_{31} & 6 & a_{33} & 7 \\ 1 & a_{42} & a_{43} & a_{44} \end{bmatrix},$$

what conditions, if any, must be placed on the undefined coefficients a_{ij} if (a) matrix $\mathbf{A}$ is to be symmetric, and (b) matrix $\mathbf{A}$ is to be skew symmetric?

8. Prove that every $n \times n$ matrix $\mathbf{A}$ can be written as the sum of a symmetric matrix $\mathbf{M}$ and a skew symmetric matrix $\mathbf{S}$. Write down an arbitrary 4×4 matrix and use your result to find the matrices $\mathbf{M}$ and $\mathbf{S}$.
9. Consider the underdetermined system

$$\begin{aligned} x_1 + x_2 + x_3 &= 1, \\ x_1 + 2x_2 + 3x_3 &= 2, \end{aligned}$$

solved in the text. Rewrite it as the two equivalent systems

$$\text{(a)} \quad \begin{aligned} x_1 + x_3 &= 1 - x_2 \\ x_1 + 3x_3 &= 2 - 2x_2 \end{aligned} \quad \text{and} \quad \text{(b)} \quad \begin{aligned} x_2 + x_3 &= 1 - x_1 \\ 2x_2 + 3x_3 &= 2 - x_1. \end{aligned}$$

Find the solution set of system (a) in terms of an arbitrary parameter $p = x_2$, and the solution set of system (b) in terms of an arbitrary parameter $q = x_1$. By comparing solution sets, what can you deduce about the solution set found in the text in terms of the arbitrary parameter $k = x_3$, and the solution sets for systems (a) and (b) found, respectively, in terms of the arbitrary parameters p and q?
10. Consider the two overdetermined systems

$$\text{(a)} \quad \begin{aligned} x_1 - 2x_2 + 2x_3 &= 6 \\ x_1 + x_2 - x_3 &= 0 \\ x_1 + 3x_2 - 3x_3 &= -4 \\ x_1 + x_2 + x_3 &= 3 \end{aligned} \quad \text{and} \quad \text{(b)} \quad \begin{aligned} 2x_1 + 3x_2 - x_3 &= 2 \\ x_1 - x_2 + 2x_3 &= 1 \\ 4x_1 + x_2 + 3x_3 &= 4 \\ x_1 + 4x_2 - 3x_3 &= 1. \end{aligned}$$

In each case try to find a solution set, and comment on the result.

Chapter 2
Determinants, and Linear Independence

2.1 Introduction to Determinants and Systems of Equations

Determinants can be defined and studied independently of matrices, though when square matrices occur they play a fundamental role in the study of linear systems of algebraic equations, in the formal definition of an inverse matrix, and in the study of the eigenvalues of a matrix. So, in anticipation of what is to follow in later chapters, and before developing the properties of determinants in general, we will introduce and motivate their study by examining the solution a very simple system of equations.

The theory of determinants predates the theory of matrices, their having been introduced by Leibniz (1646–1716) independently of his work on the calculus, and subsequently their theory was developed as part of algebra, until Cayley (1821–1895) first introduced matrices and established the connection between determinants and matrices. Determinants are associated with square matrices and they arise in many contexts, with two of the most important being their connection with systems of linear algebraic equations, and systems of linear differential equations like those in (1.12).

To see how determinants arise from the study of linear systems of equations we will consider the simplest linear nonhomogeneous system of algebraic equations

$$\begin{aligned} a_{11}x_1 + a_{12}x_2 &= b_1, \\ a_{21}x_1 + a_{22}x_2 &= b_2. \end{aligned} \tag{2.1}$$

These equations can be solved by elimination as follows. Multiply the first equation by a_{22}, the second by a_{12}, and subtract the results to obtain an equation for x_1 from which the variable x_2 has been *eliminated* . Next, multiply the first equation by a_{21}, the second by a_{11}, and subtract the results to obtain an equation for x_2, where this time the variable x_1 has been *eliminated*. The result is the solution set $\{x_1, x_2\}$ with its elements given by given by

$$x_1 = \frac{b_1a_{22} - b_2a_{12}}{a_{11}a_{22} - a_{12}a_{21}}, \quad x_2 = \frac{b_2a_{11} - b_1a_{21}}{a_{11}a_{22} - a_{12}a_{21}}. \tag{2.2}$$

A. Jeffrey, *Matrix Operations for Engineers and Scientists*,
DOI 10.1007/978-90-481-9274-8_2, © Springer Science+Business Media B.V. 2010

For this solution set to exist it is necessary that the denominator $a_{11}a_{22} - a_{12}a_{21}$ in the expressions for x_1 and x_2 does not vanish. So setting $\Delta = a_{11}a_{22} - a_{12}a_{21}$, the condition for the existence of the solution set $\{x_1, x_2\}$ becomes $\Delta \neq 0$.

In terms of a square matrix of coefficients whose elements are the *coefficients* associated with (2.1), namely

$$\mathbf{A} = \begin{bmatrix} a_{11} & a_{12} \\ a_{21} & a_{22} \end{bmatrix}, \tag{2.3}$$

the *second-order* determinant associated with **A**, written either as det **A** or as $|\mathbf{A}|$, is defined as the *number*

$$\det \mathbf{A} = |\mathbf{A}| = \begin{vmatrix} a_{11} & a_{12} \\ a_{21} & a_{22} \end{vmatrix} = a_{11}a_{22} - a_{12}a_{21}, \tag{2.4}$$

so the denominator in (2.2) is $\Delta = \det \mathbf{A}$.

Notice how the *value* of the determinant in (2.4) is obtained from the elements of **A**. The expression on the right of (2.4), called the *expansion* of the determinant, is the product of elements on the leading diagonal of **A**, from which is subtracted the product of the elements on the cross-diagonal that runs from the bottom left to the top right of the array **A**. The classification of the type of determinant involved is described by specifying its *order*, which is the number of rows (equivalently columns) in the square matrix **A** from which the determinant is derived. Thus the determinant in (2.4) is a *second-order* determinant. Specifying the *order* of a determinant gives some indication of the magnitude of the calculation involved when expanding it, while giving *no* indication of the value of the determinant. If the elements of **A** are numbers, det **A** is seen to be a number, but if the elements are functions of a variable, say t, then det **A** becomes a function of t. In general determinants whose elements are functions, often of several variables, are called *functional determinants*. Two important examples of these determinants called *Jacobian determinants*, or simply *Jacobians*, will be found in Exercises 14 and 15 at the end of this chapter.

Notice that in the conventions used in this book, when a matrix is written out in full, the elements of the matrix are enclosed within square brackets, thus [. . .], whereas the notation for its determinant, which is only associated with a square matrix, encloses its elements between vertical rules, thus |...|, and these notations should not be confused

Example 2.1. Given (a) $\mathbf{A} = \begin{bmatrix} 1 & 3 \\ -4 & 6 \end{bmatrix}$ and (b) $\mathbf{B} = \begin{bmatrix} e^t & e^t \\ \cos t & \sin t \end{bmatrix}$, find det **A** and det **B**.

Solution. By definition (a)det$\mathbf{A} = \begin{vmatrix} 1 & 3 \\ -4 & 6 \end{vmatrix} = (1 \times 6) - (3) \times (-4) = 18.$
(b) $\det \mathbf{B} = \begin{vmatrix} e^t & e^t \\ \cos t & \sin t \end{vmatrix} = (e^t) \times (\sin t) - (e^t) \times (\cos t) = e^t(\sin t - \cos t).$

It is possible to express the solution set $\{x_1, x_2\}$ in (2.2) entirely in terms of determinants by defining the three second-order determinants

$$\Delta = \det \mathbf{A} = \begin{vmatrix} a_{11} & a_{12} \\ a_{21} & a_{22} \end{vmatrix}, \quad \Delta_1 = \begin{vmatrix} b_1 & a_{12} \\ b_2 & a_{22} \end{vmatrix}, \quad \Delta_2 = \begin{vmatrix} a_{11} & b_1 \\ a_{21} & b_2 \end{vmatrix}, \tag{2.5}$$

because then the solutions in (2.2) become

$$x_1 = \frac{\Delta_1}{\Delta}, \quad x_2 = \frac{\Delta_2}{\Delta}. \tag{2.6}$$

Here Δ is the determinant of the coefficient matrix in system (2.1), while the determinant Δ_1 in the numerator of the expression for x_1 is obtained from Δ by replacing its *first column* by the nonhomogeneous terms b_1 and b_2 in the system, and the determinant Δ_2 in the numerator of the expression for x_2 is obtained from Δ by replacing its *second column* by the nonhomogeneous terms b_1 and b_2. This is the simplest form of a result known as *Cramer's rule* for solving the two simultaneous first-order algebraic equations in (2.1), in terms of determinants, and its generalization to n nonhomogeneous equations in n unknowns will be given later, along with its proof.

2.2 A First Look at Linear Dependence and Independence

Before developing the general properties of determinants, the simple system (2.1) will be used introduce the important concepts of the *linear dependence* and *independence* of equations. Suppose the second equation in (2.1) is proportional to the first equation, then for some constant of proportionality $\lambda \neq 0$ it will follow that $a_{21} = \lambda a_{11}$, $a_{22} = \lambda a_{12}$ and $b_2 = \lambda b_1$. If this happens the equations are said to be *linearly dependent*, though when they are not proportional, the equations are said to be *linearly independent*. Linear dependence and independence between systems of linear algebraic equations is important, irrespective of the number of equations and unknowns that are involved. Later, when the most important properties of determinants have been established, a determinant test for the linear independence of n homogeneous linear equations in n unknowns will be derived.

When the equations in system (2.1) are linearly dependent, the system only contains one equation relating x_1 and x_2, so one of the equations can be discarded, say the second equation. This means that one of the variables, say x_1, can only be determined in terms of the other variable x_2, so in this sense the values of x_1 and x_2, although related, become indeterminate because then x_2 is arbitrary. To discover the effect this has on the solutions in (2.2), suppose the second equation is λ times the first equation, so that $a_{21} = \lambda a_{11}$, $a_{22} = \lambda a_{12}$ and $b_2 = \lambda b_1$.

Substituting these results into (2.2), and canceling the nonzero scale factor λ, gives

$$x_1 = \frac{b_1 a_{12} - b_1 a_{12}}{a_{11} a_{12} - a_{12} a_{11}} \text{ and } x_2 = \frac{b_1 a_{11} - b_1 a_{11}}{a_{11} a_{12} - a_{12} a_{11}},$$

showing that both the numerators and the denominator in the expressions for x_1 and x_2 vanish, confirming that x_1 and x_2 are indeterminate. A comparison of this result with (2.6) shows that when two rows of a determinant are proportional, its value is zero. This is, in fact, a result that is true for all determinants and not just for second-order determinants.

The indeterminacy of the solution set is hardly surprising, because one of the equations in system (2.1) is redundant, and assigning x_2 an arbitrary value $x_2 = k$, say, will determine x_1 in terms of k as $x_1 = (b_1 - a_{12}k)/a_{11}$, so the solution set $\{x_1, x_2\}$ then takes the form $\{(b_1 - a_{12}k)/a_{11}, k\}$, where k is a parameter. Thus, when the two equations are linearly dependent, that is when $\Delta = 0$, a solution set will exist but it will not be unique, because the solution set will depends on the parameter k, which may be assigned any nonzero value. If, however, $\Delta \neq 0$ the equations will be linearly independent, and the solution set in (2.2) will exist and be unique.

A different situation arises if the left sides of the equations in (2.1) are proportional, but the constants on the right do not share the same proportionality constant, because then the equations imply a contradiction, and no solution set exists. When this happens the equations are said to be *inconsistent*. A final, very important result follows from the solution set (2.6) when the system of Eq. (2.1) is *homogeneous*; which occurs when $b_1 = b_2 = 0$. The consequence of this is most easily seen from (2.2), which is equivalent to (2.6). When the equations are homogeneous, the numerators in (2.2) both vanish because each term in the expansion of the determinant contains a zero factor, so if $\Delta = \det \mathbf{A} \neq 0$, it follows that the solution $x_1 = x_2 = 0$ is unique. This zero solution is called the *null solution*, or the *trivial solution*. Thus the only solution of a *linearly independent* set of homogeneous equations in system (2.1) is the null solution. However, if $\Delta = 0$ the equations will be *linearly dependent* (proportional), and then a solution will exist but, as has been shown, it will be such that x_1 will depend on the variable x_2, which may be assigned arbitrarily. These results will be encountered again when general systems of equations are considered that may be homogeneous or nonhomogeneous.

2.3 Properties of Determinants and the Laplace Expansion Theorem

Having seen something of the way determinants enter into the solution of the system of Eq. (2.1), it is time to return to the study of determinants. The definition of det **A** in (2.4) can be used to establish the following general properties of second-order determinants which, it turns out, are also properties common to determinants of all orders, though determinants of order greater than two have still to be defined.

Theorem 2.1 *Properties of det A.*

1. *Multiplication of the elements of any one row (column) of det* **A** *by a constant k changes the value of the determinant to k det* **A**. *Equivalently, multiplication of det* **A** *by k can be replaced by multiplying the elements of any one row (column) of det* **A** *by k.*
2. *If every element in a row (column) of det* **A** *is zero, then det* $\mathbf{A} = 0$.
3. *If two rows (columns) of det* **A** *are the identical, or proportional, then det* $\mathbf{A} = 0$.
4. *The value of a determinant is unchanged if a constant multiple of each element in a row (column) is added the corresponding element in another row (column).*
5. *If two rows (columns) in det* **A** *are interchanged, the sign of det* **A** *is changed.*
6. *det* $\mathbf{A} =$ *det* $\mathbf{A}^T$.
7. *If det* **A** *and det* **B** *are determinants of equal order, then det(***AB***) = det* **A***det* **B**.

Proof. Result 1 follows directly from definition (2.4), because each product in the definition of det **A** is multiplied by k. Result 2 also follows directly from definition (2.4), because then a coefficient in each of the products in the definition of det **A** is zero. Result 3 is an extension of the result considered previously where a row was proportional to another row. The result follows from the fact that if two rows (columns) in det **A** are equal, or proportional, the two products in the definition of det **A** cancel. To prove result 4 suppose, for example, that k times each element in the first row of det **A** is added to the corresponding element in the second row, to give det **B** where

$$\det \mathbf{B} = \begin{vmatrix} a_{11} & a_{12} \\ ka_{11} + a_{21} & ka_{12} + a_{22} \end{vmatrix}.$$

Expanding det **B** and canceling terms gives

$$\det \mathbf{B} = a_{11}(ka_{12} + a_{22}) - a_{12}(ka_{11} + a_{21}) = a_{11}a_{22} - a_{12}a_{21} = \det \mathbf{A}.$$

Similar reasoning establishes the equivalent results concerning the other row of the determinant, and also its two columns. Result 5 follows because interchanging two rows (or columns) in det **A** reverses the order of the products in the definition of det **A** in (2.4), and so changes the sign of det **A**. The proof of result 6 is left as Exercise 2.3, and the proof of result 7 will be postponed until Chapter 3, where it is given in Section 3.4 for second-order determinants, using an argument that extends directly to determinates of any order.

◆

In this account of determinants the ***n*th order determinant** associated with an $n \times n$ coefficient matrix $\mathbf{A} = [a_{ij}]$ will be defined in terms of determinants of order $n - 1$ and then, after stepping down recursively to still lower-order determinants, to a definition in terms of a sum of second-order determinants. To proceed to a definition of an nth order determinant, the definition is first extended to a third-order determinant

$$\det \mathbf{A} = \begin{vmatrix} a_{11} & a_{12} & a_{13} \\ a_1 & a_{22} & a_{23} \\ a_{31} & a_{32} & a_{33} \end{vmatrix}. \tag{2.7}$$

The third-order determinant in (2.7) is defined in terms of second-order determinants as

$$\det \mathbf{A} = a_{11}\begin{vmatrix} a_{22} & a_{23} \\ a_{32} & a_{33} \end{vmatrix} - a_{12}\begin{vmatrix} a_{21} & a_{23} \\ a_{31} & a_{33} \end{vmatrix} + a_{13}\begin{vmatrix} a_{21} & a_{22} \\ a_{31} & a_{32} \end{vmatrix}. \tag{2.8}$$

To remember this definition, notice how the terms are obtained. The first term is the product of a_{11} times the second-order determinant obtained from **A** by omitting the row and column containing a_{11}, the second term is $(-1) \times a_{12}$ times the determinant obtained from **A** by omitting the row and column containing a_{12} and, finally, the third term is a_{13} times the determinant obtained from **A** by omitting the row and column containing a_{13}.

Reasoning as in the proof of Theorem 2.1 and using the fact that a third-order determinant is expressible as a sum of multiples of second-order determinants, it is a straightforward though slightly tedious matter to show that the properties of second-order determinants listed in Theorem 2.1 also apply to third-order determinants, though the proofs of these results are left as exercises.

Determinants of order greater than three will be defined after the cofactors of a determinant have been defined. As already mentioned, the statements in Theorem 2.1 are true for determinants of all orders, though their proof for higher-order determinants will be omitted.

The three determinants

$$\begin{vmatrix} a_{22} & a_{23} \\ a_{32} & a_{33} \end{vmatrix}, \quad \begin{vmatrix} a_{21} & a_{23} \\ a_{31} & a_{33} \end{vmatrix}, \quad \begin{vmatrix} a_{21} & a_{22} \\ a_{31} & a_{32} \end{vmatrix}$$

that occurred in (2.8) are called, respectively, the minors associated with the elements a_{11}, a_{12} and a_{13} in the first row of det **A**. These minors will be denoted by M_{11}, M_{12} and M_{13}, using the same suffixes as the elements a_{11}, a_{12} and a_{13} to which they correspond, so that

$$M_{11} = \begin{vmatrix} a_{22} & a_{23} \\ a_{32} & a_{33} \end{vmatrix}, \quad M_{12} = \begin{vmatrix} a_{21} & a_{23} \\ a_{31} & a_{33} \end{vmatrix}, \quad M_{13} = \begin{vmatrix} a_{21} & a_{22} \\ a_{31} & a_{32} \end{vmatrix}. \tag{2.9}$$

Remember, that the elements of the minors M_{1i} for $i = 1$, 2 and 3, are obtained from the elements of **A** by omitting the elements in row 1 and column i.

Corresponding to the minors M_{11}, M_{12} and M_{13}, are what are called the cofactors C_{11}, C_{12} and C_{13} associated with the elements a_{11}, a_{12} and a_{13}, and these are defined in terms of the minors as

$$C_{11} = (-)^{1+1}M_{11}, \; C_{12} = (-1)^{1+2}M_{12} \text{ and } C_{13} = (-1)^{1+3}. \tag{2.10}$$

The effect of the factors $(-1)^{1+i}$ for $i = 1, 2, 3$ in the definitions of the cofactors C_{11}, C_{12} and C_{13} is to introduce an alternation of sign in the pattern of the minors. Using (2.6) and (2.9) allows us to write $\det \mathbf{A} = a_{11}M_{11} - a_{12}M_{12} + a_{13}M_{13}$, so from (2.10) this becomes

$$\det \mathbf{A} = a_{11}C_{11} + a_{12}C_{12} + a_{13}C_{13}. \tag{2.11}$$

This result is called the expansion of det **A** in terms of the cofactors of the elements of its first row.

There is a minor is associated with every element of a determinant, and not only the elements of its first row. The minor associated with the general element a_{ij}, for $i, j = 1, 2, 3$ is denoted by M_{ij}, and for a third-order determinant it is the numerical value of the 2×2 determinant derived from det **A** by deleting the elements in its ith row and jth column.

Example 2.2. Find the minors and cofactors of the elements of the first row of det **A**, and also the value of det **A**, given that

$$\mathbf{A} = \begin{bmatrix} -4 & 3 & -1 \\ -2 & 4 & 2 \\ 1 & 10 & 1 \end{bmatrix}.$$

Solution. We have

$$M_{11} = \begin{vmatrix} 4 & 2 \\ 10 & 1 \end{vmatrix} = -16, \quad M_{12} = \begin{vmatrix} -2 & 2 \\ 1 & 1 \end{vmatrix} = -4, \quad M_{13} = \begin{vmatrix} -2 & 4 \\ 1 & 10 \end{vmatrix} = -24,$$

so the corresponding cofactors are

$$C_{11} = (-1)^{1+1}(-16) = -16, \; C_{12} = (-1)^{1+2}(-4) = 4, \; C_{13} = (-1)^{1+3}(-24) = -24.$$

From (2.11), when the determinant is expanded in terms of the elements of the first row,

$$\begin{aligned} \det \mathbf{A} &= a_{11}C_{11} + a_{12}C_{12} + a_{13}C_{13} \\ &= (-4) \times (-16) + 3 \times 4 + (-1) \times (-24) = 100. \end{aligned}$$

◆

To extend the role of the cofactor associated with the minor of any element of det **A** we start by expanding the expression for a third-order determinant in (2.8), to obtain

$$\det \mathbf{A} = \mathrm{D} = a_{11}a_{22}a_{33} - a_{11}a_{23}a_{32} + a_{12}a_{23}a_{31} - a_{12}a_{21}a_{33} + a_{13}a_{21}a_{32} - a_{13}a_{22}a_{31}. \tag{2.12}$$

Next, we define the cofactor C_{ij} associated with the general element a_{ij} in $\mathbf{A}$ to be

$$C_{ij} = (-1)^{i+j} M_{ij}, \tag{2.13}$$

where for this third-order determinant M_{ij} is the 2×2 minor obtained from det $\mathbf{A}$ by deleting the elements in its ith row and jth column. Using this definition of a general cofactor, and rearranging the terms in (2.12) to give results similar to (2.8), but this time with terms a_{i1}, a_{i2} and a_{i3} multiplying the determinants, it is easily shown that

$$\det \mathbf{A} = a_{i1}C_{i1} + a_{i2}C_{i2} + a_{i3}C_{i3}, \quad \text{for } i = 1,\ 2 \text{ or } 3. \tag{2.14}$$

This result provides three different, but equivalent, ways of calculating det $\mathbf{A}$, the first of which was encountered in (2.11). Expressed in words, result (2.14) says that det $\mathbf{A}$ is equal to the sum of the products of the elements and their respective cofactors in any row of the determinant. The result is important, and it is called the expansion of det $\mathbf{A}$ in terms of the elements and cofactors of the ith row of the determinant. So (2.11) is seen to be the expansion of det $\mathbf{A}$ in terms of the elements and cofactors of its first row.

A different rearrangement of the terms in (2.12) shows that

$$\det \mathbf{A} = a_{1j}C_{1j} + a_{2j}C_{2j} + a_{3j}C_{3j}, \quad \text{for } j = 1,\ 2 \text{ or } 3, \tag{2.15}$$

providing three more ways of expanding det $\mathbf{A}$. When expressed in words, this expansion says that det $\mathbf{A}$ can be calculated as the sum of the products of the elements and their respective cofactors in any column of the determinant. Result (2.15) is called the expansion of det $\mathbf{A}$ in terms of the elements and cofactors of the jth column of the determinant.

It remains for us to determine the effect of forming the sum of the products of the elements of a row, or column, with the corresponding cofactors of a different row, or column. To resolve this, let δ be the sum of the products of the elements of row i with the cofactors of row s, so that $\delta = \sum_{j=1}^{3} a_{ij}C_{sj}$ for $s \neq j$. Now δ can be interpreted as a third-order determinant with the elements a_{ij} forming its ith row, and the remaining elements taken to be the cofactors C_{sj}. As $s \neq j$, it follows that each cofactor will contain elements from row i, so when the third-order determinant is reconstructed, it will contain another row equal to the ith row except, possibly, for a change of sign throughout the row. Thus the determinant δ will either have two identical rows, or two rows which are identical apart from a change of sign. So by an extension of the results of Theorem 2.1 (see Example 2.3), the determinant must vanish. A similar argument shows that the sum of products formed by multiplying the elements of a column with the corresponding cofactors of a different column is also zero, so that $\sum_{i=1}^{3} a_{ij}C_{ik} = 0$ for $k \neq j$.

The extension of these expansions to include nth-order determinants follows from (2.13) and (2.14) by defining the nth order determinant as either

$$\det \mathbf{A} = \sum_{j=1}^{n} a_{ij}C_{ij} \quad \text{for } i = 1, 2, \ldots, n \tag{2.16}$$

or as

$$\det \mathbf{A} = \sum_{i=1}^{n} a_{ij}C_{ij} \quad \text{for } j = 1, 2, ..., n. \tag{2.17}$$

Notice that now the cofactors C_{ij} are determinants of order $n - 1$. These expressions provide equivalent recursive definitions for an nth-order determinant in terms of second-order determinants, because any determinant of order $n \geq 3$ can always be reduced to a sum of products involving second-order determinants. A determinant $\det \mathbf{A}$ is said to be singular if $\det \mathbf{A} = 0$, and nonsingular if $\det \mathbf{A} \neq 0$. Chapter 3 will show it is necessary that $\det \mathbf{A} \neq 0$ when defining an important matrix $\mathbf{A}^{-1}$ called the inverse matrix associated with a square matrix $\mathbf{A}$, or more simply the inverse of $\mathbf{A}$.

To avoid the tedious algebraic manipulations involved when extending the results of Theorem 2.1 to determinants of order n, we again mention that the properties listed in the theorem apply to determinants of all orders. However, some of the properties in Theorem 2.1 are almost self-evident for determinants of all orders, as for example the properties 1, 2 and 3.

The extension of the previous results to an nth-order determinant yields the following fundamental expansion theorem due to Laplace.

Theorem 2.2 *The Laplace Expansion of a Determinant.*
Let $\mathbf{A} = [a_{ij}]$ *be an* $n \times n$ *matrix, and let the cofactor associated with* a_{ij} *be* C_{ij}*. Then, for any* i,

$$\det \mathbf{A} = a_{i1}C_{i1} + a_{i2}C_{i2} + \ldots + a_{in}C_{in} \quad (\textit{expansion by elements of the ith row}),$$

and for any j,

$$\det \mathbf{A} = a_{1j}C_{1j} + a_{2j}C_{2j} + \cdots + a_{nj}C_{nj} \quad (\textit{expansion by elements of the jth column})$$

while for any i *with* $s \neq i$

$$a_{i1}C_{s1} + a_{i2}C_{s2} + \cdots + a_{in}C_{sn} = 0 \quad (\textit{expansion using different rows})$$

or for any j *with* $k \neq j$

$$a_{1j}C_{1k} + a_{2j}C_{2k} + \cdots + a_{nj}C_{nk} = 0 \quad (\textit{expansion using different columns}).$$

◆

Example 2.3. (a) Expand the determinant in Example 2.2 in terms of elements and cofactors of the third column. (b) Compute the sum of the products of the elements in the first row and the corresponding cofactors of the second row, and hence confirm that the result is zero. (c) Reconstruct the determinant corresponding to the calculation in (b), and hence show why the result is zero.

Solution.

(a) To expand the determinant using elements and cofactors of the third column it is necessary to compute C_{13}, C_{23} and C_{33}. We have

$$\mathbf{A} = \begin{bmatrix} -4 & 3 & -1 \\ -2 & 4 & 2 \\ 1 & 10 & 1 \end{bmatrix}, \text{ so } C_{13} = (-)^{1+3}\begin{vmatrix} -2 & 4 \\ 1 & 10 \end{vmatrix} = -24,$$

$$C_{23} = (-)^{2+3}\begin{vmatrix} -4 & 3 \\ 1 & 10 \end{vmatrix} = 43, \; C_{33} = (-)^{3+3}\begin{vmatrix} -4 & 3 \\ -2 & 4 \end{vmatrix} = -10.$$

Expanding det **A** in terms of the elements and cofactors of the third column gives det **A** $= (-1) \times (-24) + 2 \times 43 + 1 \times (-10) = 100$, in agreement with Example 2.2.

(b) To form the sum of the products of the elements of the first row with the corresponding cofactors of the second row it is necessary to compute C_{21}, C_{22} and C_{23}. We have

$$C_{21} = (-1)^{2+1}\begin{vmatrix} 3 & -1 \\ 10 & 1 \end{vmatrix} = -13\,, \quad C_{22} = (-1)^{2+2}\begin{vmatrix} -4 & -1 \\ 1 & 1 \end{vmatrix} = -3\,,$$

$$C_{23} = (-1)^{2+3}\begin{vmatrix} -4 & 3 \\ 1 & 10 \end{vmatrix} = 43\,.$$

So the required expansion in terms of elements of the first row and the corresponding cofactors in the second row becomes

$$(-4) \times (-13) + 3 \times (-3) + (-1) \times 43 = 0\,,$$

confirming the third property in Theorem 2.2.

(c) To reconstruct the third-order determinant δ corresponding to the sum of products of the elements in the first row and the cofactors in the second row used in (b) we first write δ as

$$\delta = (-4) \times C_{21} + 3 \times C_{22} + (-1) \times C_{23}.$$

Substituting for the cofactors this becomes

$$\delta = 4 \times \begin{vmatrix} 3 & -1 \\ 10 & 1 \end{vmatrix} + 3 \times \begin{vmatrix} -4 & -1 \\ 1 & 1 \end{vmatrix} + 1 \times \begin{vmatrix} -4 & 3 \\ 1 & 10 \end{vmatrix}.$$

To express this result as the appropriate expansion of a determinant it is necessary restore the correct signs to the multipliers 4, 3 and 1 in the above expression to make them equal to the elements in the first row of **A**, namely −4, 3 and −1. To do this we use result 1 from Theorem 2.1 which shows that when a determinant is multiplied by −1, this multiplier can be taken inside the determinant and used as a multiplier for any one of its rows. To be consistent, we will change the signs of the terms in the last rows of the determinants, so that δ becomes

$$\delta = -\left\{(-4) \times \begin{vmatrix} 3 & -1 \\ -10 & -1 \end{vmatrix} + (3) \times \begin{vmatrix} -4 & -1 \\ -1 & -1 \end{vmatrix} + (-1) \times \begin{vmatrix} -4 & -3 \\ -1 & -1 \end{vmatrix}\right\}.$$

Recognizing that these three determinants are now the cofactors of the elements −4, 3 and −1 in the first row of the determinant that is to be reconstructed, allows the result can be written

$$\delta = -\begin{vmatrix} -4 & 3 & -1 \\ -4 & 3 & -1 \\ -1 & -10 & -1 \end{vmatrix}.$$

This determinant has two identical rows, and so vanishes, showing why result (b) yields the value zero.

◆

The equivalent definitions of an nth order determinant in Theorem 2.2 permit the immediate evaluation of some important and frequently occurring types of determinants. The first case to be considered occurs when det **A** is the nth-order diagonal determinant

$$\det \mathbf{A} = \begin{vmatrix} a_{11} & 0 & 0 & \cdots & 0 \\ 0 & a_{22} & 0 & \cdots & 0 \\ 0 & 0 & a_{33} & \cdots & 0 \\ \vdots & \vdots & \vdots & \vdots & \vdots \\ 0 & 0 & 0 & \cdots & a_{nn} \end{vmatrix} = a_{11}a_{22}a_{33}\cdots a_{nn}. \tag{2.18}$$

This follows because expanding the determinant in terms of elements of the first row, gives det $\mathbf{A} = a_{11}C_{11}$, where the cofactor C_{11} is the determinant of order $n - 1$ with the same diagonal structure as det **A**. Expanding C_{11} in terms of the elements of its first row gives det $\mathbf{A} = a_{11}a_{22}C_{11}^{(1)}$, where $C_{11}^{(1)}$ is now the cofactor belonging to determinant C_{11} corresponding to the first element a_{22} in its first row. Continuing this process n times gives the stated result det $\mathbf{A} = a_{11}a_{22}a_{33}\cdots a_{nn}$.

Two other determinants whose values can be written down at sight are the determinants det **L** and det **U** associated, respectively, with the upper and lower triangular $n \times n$ matrices **L** and **U**. We have

$$\det \mathbf{L} = \begin{vmatrix} a_{11} & 0 & 0 & \cdots & 0 \\ a_{21} & a_{22} & 0 & \cdots & 0 \\ a_{31} & a_{32} & a_{33} & \cdots & 0 \\ \vdots & \vdots & \vdots & \vdots & \vdots \\ a_{n1} & a_{n2} & a_{n3} & \cdots & a_{nn} \end{vmatrix} = a_{11}a_{22}a_{33}\cdots a_{nn} \tag{2.19}$$

and

$$\det \mathbf{U} = \begin{vmatrix} a_{11} & a_{12} & a_{13} & \cdots & a_{1n} \\ 0 & a_{22} & a_{23} & \cdots & a_{2n} \\ 0 & 0 & a_{33} & \cdots & a_{3n} \\ \vdots & \vdots & \vdots & \vdots & \vdots \\ 0 & 0 & 0 & \cdots & a_{nn} \end{vmatrix} = a_{11}a_{22}a_{33}\cdots a_{nn}. \tag{2.20}$$

Result (2.18) is obtained in a manner similar to the derivation of (2.18), by repeated expansion of det **L** in terms of the elements of its first row, while result (2.20) follows by a similar repeated expansion of det **U** in terms of elements of its first column.

The next example illustrates how the properties of Theorem 2.1 can sometimes be used to evaluate a determinant without first expanding it with respect to either the elements in its rows or the elements in its columns. The determinant involved has a special form, and it is called an alternant, also known as a Vandermonde determinant.

Example 2.4. Show without direct expansion that

$$\begin{vmatrix} 1 & 1 & 1 \\ a & b & c \\ a^2 & b^2 & c^2 \end{vmatrix} = (b-a)(c-a)(c-b)\,.$$

Solution. Using property 4 of Theorem 2.1, which leaves the value of a determinant unchanged, we subtract column 1 from columns 2 and 3 to obtain

$$\begin{vmatrix} 1 & 1 & 1 \\ a & b & c \\ a^2 & b^2 & c^2 \end{vmatrix} = \begin{vmatrix} 1 & 0 & 0 \\ a & (b-a) & (c-a) \\ a^2 & (b^2-a^2) & (c^2-a^2) \end{vmatrix} = \begin{vmatrix} 1 & 0 & 0 \\ a & (b-a) & (c-a) \\ a^2 & (b+a)(b-a) & (c+a)(c-a) \end{vmatrix}.$$

Next we use property 1 of Theorem 2.1 to remove factors $(b-a)$ and $(c-a)$ from the second and third columns to obtain

$$\begin{vmatrix} 1 & 1 & 1 \\ a & b & c \\ a^2 & b^2 & c^2 \end{vmatrix} = (b-a)(c-a)\begin{vmatrix} 1 & 0 & 0 \\ a & 1 & 1 \\ a^2 & (b+a) & (c+a) \end{vmatrix}.$$

Finally, subtracting column two from column three we find that

$$\begin{vmatrix} 1 & 1 & 1 \\ a & b & c \\ a^2 & b^2 & c^2 \end{vmatrix} = (b-a)(c-a)\begin{vmatrix} 1 & 0 & 0 \\ a & 1 & 0 \\ a^2 & (b+a) & (c-b) \end{vmatrix}.$$

The determinant is now of lower triangular form, so from (2.19) its value is $(c - b)$. So, as required, we have shown the value of this alternant to be

$$\begin{vmatrix} 1 & 1 & 1 \\ a & b & c \\ a^2 & b^2 & c^2 \end{vmatrix} = (b-a)(c-a)(c-b).$$

♦

2.4 Gaussian Elimination and Determinants

The expansion of a determinant using Theorem 2.2 is mainly of theoretical interest, because to evaluate a determinant of order n requires $n!$ multiplications. So, evaluating a determinant of order 8 requires 40,320 multiplications, while evaluating a determinant of order 15 requires approximately 1.31×10^9 multiplications. If, for example, this method of evaluating a determinant were to be performed on a computer where one multiplication takes 1/1,000 s, the evaluation of a determinant of order 15 would take approximately 41.5 years. Clearly, when the order is large, some other way must be found by which to evaluate determinants if this prohibitive number of multiplications is to be avoided, not to mention the buildup of round-off errors that would result. A better method is essential, because many applications of mathematics lead to determinants with orders far larger than 15.

The way around this difficulty is found in property 4 of Theorem 2.1. Subtracting a_{21}/a_{11} times the first row of the determinant from the second row reduces to zero the element immediately below a_{11}. Similarly, subtracting a_{31}/a_{11} times the first row of the determinant from the third row reduces to zero the element in row three below a_{11}, while neither of these operations changes the value of the determinant. So, proceeding down the first column in this manner leads to a new determinant in which the only nonzero entry in its first column is a_{11}. If this procedure is now applied to the second column of the modified determinant, starting with the new coefficient $\tilde{a}_{22}$ that is now in row 2 and column 2, it will reduce to zero all entries below the element $\tilde{a}_{22}$. Proceeding in this way, column by column, the determinant will eventually be replaced by an equivalent nth-order determinant of upper triangular form, the value of which follows, as in (2.20), by forming the product of all the elements in its leading diagonal. This way of evaluating a determinant, called the Gaussian elimination method, or sometimes the Gaussian reduction method, converts a determinant to upper triangular form, whose value is simply the products of the elements on its leading diagonal. This method requires significantly fewer multiplications than the direct expansion used in the definition, and so is efficient when applied to determinants of large order. Software programs are based on a refinement of this method, and even on a relatively slow PC the evaluation of a determinant of order 50 may take only a few seconds.

It can happen that at the ith stage of this reduction process a zero element occurs on the leading diagonal, thereby preventing further reduction of the determinant. This difficulty is easily overcome by interchanging the ith row with a row below it in which the ith element is not zero, after which the reduction continues as before. However, after such an interchange of rows, the sign of the determinant must be changed as required by property 5 of Theorem 2.1. If, on the other hand, at some stage of the reduction process a complete row of zeros is produced, further simplification is impossible, and this shows the value of the determinant is zero or, in other words, that the determinant is singular. The following Example shows how such a reduction proceeds in a typical case when a row interchange becomes necessary.

Remember that an interchange of rows changes the sign of a determinant, so if p interchanges become necessary during the Gaussian elimination process used to calculate the value of determinant, then the sign of the upper triangular determinant that is obtained must be multiplied by $(-1)^p$ in order to arrive at the value of the original determinant.

Example 2.5. Evaluate the following determinant by reducing it to upper triangular form:

$$\det \mathbf{A} = \begin{vmatrix} 1 & 3 & 2 & 1 \\ 1 & 3 & 6 & 3 \\ 0 & 2 & 1 & 5 \\ 0 & 2 & 1 & 1 \end{vmatrix}.$$

Solution. Subtracting row 1 from row 2 gives

$$\det \mathbf{A} = \begin{vmatrix} 1 & 3 & 2 & 1 \\ 1 & 3 & 6 & 3 \\ 0 & 2 & 1 & 5 \\ 0 & 2 & 1 & 1 \end{vmatrix} = \begin{vmatrix} 1 & 3 & 2 & 1 \\ 0 & 0 & 4 & 2 \\ 0 & 2 & 1 & 5 \\ 0 & 2 & 1 & 1 \end{vmatrix}.$$

The second element in row 2 is zero, so subtracting multiples of row 2 from rows 3 and 4 cannot reduce to zero the elements in the column below this zero element. To overcome this difficulty we interchange rows 2 and 3, because row 3 has a nonzero element in its second position, and compensate for the row interchange by changing the sign of the determinant, to obtain

$$\det \mathbf{A} = \begin{vmatrix} 1 & 3 & 2 & 1 \\ 0 & 0 & 4 & 2 \\ 0 & 2 & 1 & 5 \\ 0 & 2 & 1 & 1 \end{vmatrix} = -\begin{vmatrix} 1 & 3 & 2 & 1 \\ 0 & 2 & 1 & 5 \\ 0 & 0 & 4 & 2 \\ 0 & 2 & 1 & 1 \end{vmatrix}.$$

Finally, subtracting the new row 2 from row 4 produces the required upper triangular form

$$\det \mathbf{A} = -\begin{vmatrix} 1 & 3 & 2 & 1 \\ 0 & 2 & 1 & 5 \\ 0 & 0 & 4 & 2 \\ 0 & 2 & 1 & 1 \end{vmatrix} = -\begin{vmatrix} 1 & 3 & 2 & 1 \\ 0 & 2 & 1 & 5 \\ 0 & 0 & 4 & 2 \\ 0 & 0 & 0 & -4 \end{vmatrix},$$

so from (2.20),

$$\det \mathbf{A} = -(1) \times (2) \times (4) \times (-4) = 32 .$$

◆

Once the inverse matrix has been introduced, matrix algebra will be used to prove the following generalization of Cramer's rule to a nonhomogeneous system of n linear equations in the n unknowns $x_1, x_2, \ldots, x_n$. However, it will be useful to state this generalization in advance of its proof.

Theorem 2.3 *The Generalized Cramer's Rule.*
The system of n *nonhomogeneous linear equations in the variables* $x_1, x_2, \ldots, x_n$

$$\begin{aligned} a_{11}x_1 + a_{12}x_2 + \cdots + a_{1n}x_n &= b_1, \\ a_{21}x_1 + a_{22}x_2 + \cdots + a_{2n}x_n &= b_2, \\ &\cdots \\ a_{n1}x_1 + a_{n2}x_2 + \cdots + a_{nn}x_b &= b_n \end{aligned} \tag{2.21}$$

has the solution set $\{x_1, x_2, \ldots, x_n\}$ *given by*

$$x_1 = \frac{\Delta_1}{\Delta}, \; x_2 = \frac{\Delta_2}{\Delta}, \cdots, \; x_n = \frac{\Delta_n}{\Delta}, \tag{2.22}$$

provided $\Delta \neq 0$, *where*

$$\Delta = \begin{vmatrix} a_{11} & a_{12} & \cdots & a_{1n} \\ a_{21} & a_{22} & \cdots & a_{2n} \\ \vdots & \vdots & \vdots & \vdots \\ a_{n1} & a_{n2} & \cdots & a_{nn} \end{vmatrix}, \; \Delta_1 = \begin{vmatrix} b_1 & a_{12} & \cdots & a_{1n} \\ b_2 & a_{22} & \cdots & a_{2n} \\ \vdots & \vdots & \vdots & \vdots \\ b_n & a_{n1} & a_{n2} & a_{nn} \end{vmatrix}, \ldots, \tag{2.23}$$

$$\Delta_n = \begin{vmatrix} a_{11} & a_{12} & \cdots & b_1 \\ a_{21} & a_{22} & \cdots & b_2 \\ \vdots & \vdots & \vdots & \vdots \\ a_{n1} & a_{n2} & \cdots & b_n \end{vmatrix}.$$

◆

Notice that in (2.23) $\Delta = \det \mathbf{A}$ is the determinant of the coefficient matrix $\mathbf{A}$, and the determinant Δ_i for $i = 1, 2, \ldots, n$ is derived from Δ by replacing its ith column by the column vector containing the nonhomogeneous terms $b_1, b_2, \ldots, b_n$.

2.5 Homogeneous Systems of Equations and a Test for Linear Independence

Consider the system of n homogeneous linear equations in the n independent variables $x_1, x_2, \ldots, x_n$:

$$\begin{aligned} a_{11}x_1 + a_{12}x_2 + \cdots + a_{1n}x_n &= 0, \\ a_{21}x_1 + a_{22}x_2 + \cdots + a_{2n}x_n &= 0, \\ &\ldots\ldots\ldots\ldots \\ a_{n1}x_1 + a_{n2}x_2 + \cdots + a_{nn}x_n &= 0. \end{aligned} \tag{2.24}$$

Accepting the validity of this generalization of Cramer's rule, it follows that if the determinant of the coefficients $\det \mathbf{A} \neq 0$, the only possible solution of (2.24) is the null solution $x_1 = x_2 = \cdots = x_n = 0$. This means that no equation in (2.24) can be expressed as the sum of multiples of other equations belonging to the system, so the equations in the system are linearly independent. Suppose, however, that one of the equations is formed by the addition of multiples of some of the remaining equations, making it linearly dependent on other equations in the system. Subtracting these same multiples of equations from the linearly dependent equation will reduce it to an equation of the form $0x_1 + 0x_2 + \cdots + 0x_n = 0$, leading to a row of zeros in the equivalent coefficient matrix. It then follows immediately that $\det \mathbf{A} = 0$, and the same conclusion follows if more than one of the equations in (2.24) is linearly dependent on the other equations. We have established the following useful result.

Theorem 2.4 *Determinant Test for Linear Independence.*
A necessary and sufficient condition that the n homogeneous equations in (2.24) with the coefficient matrix $\mathbf{A}$ *are linearly independent is that* $\det \mathbf{A} \neq 0$. *Conversely, if* $\det \mathbf{A} = 0$, *the equations are linearly dependent.*

♦

It follows from Theorem 2.4 that if $m < n$ of the equations in (2.24) are linearly independent, it is only possible to solve for m of the unknown variables $x_1, x_2, \ldots, x_n$ in terms of the of the remaining $n - m$ variables that can then be regarded as arbitrary parameters.

The next example shows how this situation arises when dealing with a system of four equations, only two of which are linearly independent.

Example 2.6. Show only two of the following four linear homogeneous equations are linearly independent, and find the solution set if two of the unknowns are assigned arbitrary values.

$$\begin{aligned}
2x_1 - 3x_2 + x_3 + 2x_4 &= 0,\\
3x_1 + 2x_2 - 3x_3 - x_4 &= 0,\\
x_1 + 5x_2 - 4x_3 - 3x_4 &= 0,\\
5x_1 - x_2 - 2x_3 + x_4 &= 0.
\end{aligned}$$

Solution. Later a simple way will be found of determining which equations may be taken to be linearly independent. However, for the moment, it will suffice to notice that the third equation is obtained by subtracting the first equation from the second equation, and the fourth equation is obtained by adding the first and second equations. So we may take the first and second equations as being linearly independent, and the last two equations as being redundant because of their linear dependence on the first two equations. The linear dependence of this system of equations is easily checked by using the determinant test in Theorem 2.4, because

$$\det \mathbf{A} = \begin{vmatrix} 2 & -3 & 1 & 2 \\ 3 & 2 & -3 & -1 \\ 1 & 5 & -4 & -3 \\ 5 & -1 & -2 & 1 \end{vmatrix} = 0.$$

While the determinant test establishes the existence of linear dependence amongst the equations in system (2.24), it does not show how many of the equations are linearly independent.

As we know by inspection that the first two equations contain all of the information in this system, the last two equations can be disregarded, and we can work with the first two equations

$$\begin{aligned}
2x_1 - 3x_2 + x_3 + 2x_4 &= 0,\\
3x_1 + 2x_2 - 3x_3 - x_4 &= 0.
\end{aligned}$$

If we set $x_3 = k_1$ and $x_4 = k_2$, each of which is arbitrary, the system reduces to the two equations for x_1 and x_2,

$$\begin{aligned}
2x_1 - 3x_2 &= -k_1 - 2k_2,\\
3x_1 + 2x_2 &= 3k_1 + k_2.
\end{aligned}$$

Solving these equations for x_1 and x_2 shows the solution set $\{x_1, x_2, x_3, x_4\}$ for the system has for its elements

$$x_1 = \tfrac{7}{13}k_1 - \tfrac{1}{13}k_2\,, \quad x_2 = \tfrac{9}{13}k_1 + \tfrac{8}{13}k_2\,, \quad x_3 = k_1\,, \quad x_4 = k_2\,,$$

where the quantities k_1 and k_2 are to be regarded as arbitrary parameters.

♦

Corollary 2.5. *Linear Dependence of the Columns of a Determinant.*
If in system (2.24) det $\mathbf{A} = 0$*, then the columns of the determinant are linearly dependent.*

Proof. The result is almost immediate, and it follows from the fact that the rows of det $\mathbf{A}^{\mathrm{T}}$ are the columns of det $\mathbf{A}$. The vanishing of det $\mathbf{A}$ implies linear dependence between the rows of det $\mathbf{A}$, but det $\mathbf{A} =$ det $\mathbf{A}^{\mathrm{T}}$, so the vanishing of det $\mathbf{A}$ implies linear dependence between the columns of det $\mathbf{A}$.

♦

2.6 Determinants and Eigenvalues: A First Look

An important type of determinant associated with an $n \times n$ matrix $\mathbf{A} = [a_{ij}]$ has the form $\det[\mathbf{A} - \lambda\mathbf{I}]$, where λ is a scalar parameter. To interpret the matrix expression $\mathbf{A} - \lambda\mathbf{I}$ we need to anticipate the definition of the multiplication of a matrix by a scalar. This is accomplished by defining the matrix $\lambda\mathbf{I}$ to be the matrix obtained from the unit matrix $\mathbf{I}$ by multiplying each of its elements by λ, so if $\mathbf{I}$ is the 3×3 unit matrix,

$$\lambda\begin{bmatrix} 1 & 0 & 0 \\ 0 & 1 & 0 \\ 0 & 0 & 1 \end{bmatrix} = \begin{bmatrix} \lambda & 0 & 0 \\ 0 & \lambda & 0 \\ 0 & 0 & \lambda \end{bmatrix}.$$

Example 2.7. Given

$$\mathbf{A} = \begin{bmatrix} 1 & 2 & 0 \\ 2 & -1 & -2 \\ 0 & -2 & 1 \end{bmatrix},$$

find $\mathbf{A} - \lambda\mathbf{I}$ and write down $\det[\mathbf{A} - \lambda\mathbf{I}]$.

Solution. We have

$$\mathbf{A} - \lambda\mathbf{I} = \begin{bmatrix} 1 & 2 & 0 \\ 2 & -1 & -2 \\ 0 & -2 & 1 \end{bmatrix} - \lambda\begin{bmatrix} 1 & 0 & 0 \\ 0 & 1 & 0 \\ 0 & 0 & 1 \end{bmatrix} = \begin{bmatrix} 1 & 2 & 0 \\ 2 & -1 & -2 \\ 0 & -2 & 1 \end{bmatrix} - \begin{bmatrix} \lambda & 0 & 0 \\ 0 & \lambda & 0 \\ 0 & 0 & \lambda \end{bmatrix},$$

from which it follows that

$$\mathbf{A} - \lambda\mathbf{I} = \begin{bmatrix} 1-\lambda & 2 & 0 \\ 2 & -1-\lambda & -2 \\ 0 & -2 & 1-\lambda \end{bmatrix}, \text{ and so } \det[\mathbf{A} - \lambda\mathbf{I}] = \begin{vmatrix} 1-\lambda & 2 & 0 \\ 2 & -1-\lambda & -2 \\ 0 & -2 & 1-\lambda \end{vmatrix}.$$

♦

If $\mathbf{A}$ is an $n \times n$ matrix, when expanded det $[\mathbf{A} - \lambda\mathbf{I}]$ yields a polynomial $p(\lambda)$ of degree n in λ, where n is the order of det $\mathbf{A}$. In Example 2.7 the polynomial $p(\lambda)$ given by

$$p(\lambda) = \det[\mathbf{A} - \lambda\mathbf{I}] = \begin{vmatrix} 1-\lambda & 2 & 0 \\ 2 & -1-\lambda & -2 \\ 0 & -2 & 1-\lambda \end{vmatrix} = -\lambda^3. + \lambda^2 + 9\lambda - 9.$$

The roots of det $[\mathbf{A} - \lambda\mathbf{I}] = 0$, that is the zeros of $p(\lambda)$, are called the eigenvalues of the matrix $\mathbf{A}$, so in Example 2.7 the polynomial $p(\lambda) = 0$ becomes the cubic equation $\lambda^3 - \lambda^2 - 9\lambda + 9 = 0$. This has the roots $\lambda = 1$, $\lambda = -3$ and $\lambda = 3$, so these are the eigenvalues of matrix $\mathbf{A}$. The expression $p(\lambda)$ is called the characteristic polynomial of matrix $\mathbf{A}$, and $p(\lambda) = 0$ is called the characteristic equation of matrix $\mathbf{A}$. As the eigenvalues of a square matrix $\mathbf{A}$ are the roots of a polynomial it is possible for the eigenvalues of $\mathbf{A}$ to be complex numbers, even when all of the elements of $\mathbf{A}$ are real. It is also important to recognize that only square matrices have eigenvalues, because when $\mathbf{A}$ is an $m \times n$ matrix with $m \neq n$, det $\mathbf{A}$ has no meaning.

Theorem 2.5 *The eigenvalues of A and A^T.*
The matrix $\mathbf{A}$ *and its transpose* $\mathbf{A}^T$ *have the same characteristic polynomial, and the same eigenvalues.*
Proof. The results follow directly from Property 6 of Theorem 2.1, because $\mathbf{A}$ and $\mathbf{A}^T$ have the same characteristic polynomial, and hence the same eigenvalues. ◆

Example 2.8. If $\mathbf{A} = \begin{bmatrix} 1 & 3 & 2 \\ -1 & 2 & 4 \\ 1 & 0 & -1 \end{bmatrix}$, then $\mathbf{A}^T = \begin{bmatrix} 1 & -1 & 1 \\ 3 & 2 & 0 \\ 2 & 4 & -1 \end{bmatrix}$, and routine calculations confirm that

$$p(\lambda) = \det[\mathbf{A} - \lambda\mathbf{I}] = \det[\mathbf{A}^T - \lambda\mathbf{I}] = \lambda^3 - 2\lambda^2 - 3,$$

so the characteristic polynomials are identical. The eigenvalues determined by $p(\lambda) = 0$ are

$$\lambda_1 = 2.48558,\ \ \lambda_2 = 0.24279 - 1.07145i \ \text{ and } \ \lambda_3 = \bar{\lambda}_2 = 0.24279 + 1.07145i,$$

so in this case one eigenvalue is real and the other two are complex conjugates. ◆

Exercises

1. Evaluate the determinants

$$\text{(a) } \det\mathbf{A} = \begin{vmatrix} 7 & 3 & 4 \\ 1 & 2 & 1 \\ 3 & 0 & 2 \end{vmatrix},\ \text{(b) } \det\mathbf{B} = \begin{vmatrix} 1 & -3 & 2 \\ 4 & 5 & 6 \\ 5 & 2 & 8 \end{vmatrix},\ \text{(c) } \det\mathbf{C} = \begin{vmatrix} 0 & 1 & 0 \\ 1 & 0 & 0 \\ 0 & 0 & 1 \end{vmatrix}.$$

2. Evaluate the determinants

$$\text{(a)}\quad \det \mathbf{A} = \begin{vmatrix} \sin t & \cos t & 1 \\ -\cos t & \sin t & 0 \\ e^t & 0 & 0 \end{vmatrix},\ \text{(b)}\ \det \mathbf{B} = \begin{vmatrix} e^{-t}\sin t & e^{-t}\cos t & 0 \\ -e^{-t}\cos t & e^{-t}\sin t & 1 \\ e^t & 0 & 1 \end{vmatrix}.$$

3. Construct a 3×3 matrix $\mathbf{A}$ of your own choice, and by expanding the determinants $\det \mathbf{A}$ and $\det \mathbf{A}^{\mathrm{T}}$ show that $\det \mathbf{A} = \det \mathbf{A}^{\mathrm{T}}$. Prove that if $\mathbf{A}$ is any $n \times n$ matrix, then it is always true that $\det \mathbf{A} = \det \mathbf{A}^{\mathrm{T}}$.
4. Evaluate the determinant

$$\det \mathbf{A} = \begin{vmatrix} 2 & 0 & -1 & 3 \\ 1 & 4 & 9 & 0 \\ -2 & 1 & 3 & -1 \\ 4 & 0 & 3 & 2 \end{vmatrix}.$$

5. Show without expanding the determinant that

$$\begin{vmatrix} 1+a & a & a \\ b & 1+b & b \\ b & b & 1+b \end{vmatrix} = (1+a+2b)\,.$$

6. Show without expanding the determinant that

$$\begin{vmatrix} x^3+1 & 1 & 1 \\ 1 & x^3+1 & 1 \\ 1 & 1 & x^3+1 \end{vmatrix} = x^6(x^3+3)\,.$$

7. Evaluate the following determinant by reducing it to upper triangular form

$$\Delta = \begin{vmatrix} 2 & 1 & 0 & 1 \\ 3 & 2 & 4 & 2 \\ 1 & 2 & 1 & 3 \\ 0 & 3 & 1 & 1 \end{vmatrix}.$$

8. Use Cramer's rule to solve the system of equations

$$\begin{aligned} x_1 + 2x_2 - x_3 &= 9, \\ 2x_1 - 3x_2 + 5x_3 &= -2, \\ 4x_1 - 2x_2 - 3x_3 &= 7\,. \end{aligned}$$

9. Are the equations in the following two systems linearly dependent?

$$\text{(a)}\quad \begin{aligned} x_1 - 2x_2 + 4x_3 &= 0 \\ 3x_1 + 6x_2 + 2x_3 &= 0 \\ 7x_1 + 22x_2 - 2x_3 &= 0, \end{aligned} \qquad \text{(b)}\quad \begin{aligned} 3x_1 - x_2 + 2x_3 &= 0 \\ x_1 + 4x_2 + 6x_3 &= 0 \\ 3x_1 - x_2 + 4x_3 &= 0\ . \end{aligned}$$

10. Are the equations in the following system linearly independent? Give a reason for your answer.

$$\begin{aligned} x_1 + 2x_2 - x_3 - x_4 &= 0, \\ 2x_1 - x_2 + 2x_3 + 2x_4 &= 0, \\ 4x_1 - 7x_2 + 8x_3 + 8x_4 &= 0, \\ 3x_1 - x_2 + 3x_3 - 2x_4 &= 0\ . \end{aligned}$$

11. Given that

$$\mathbf{A} = \begin{bmatrix} 2 & 0 & -1 \\ -1 & 1 & 1 \\ 0 & 0 & 3 \end{bmatrix},$$

confirm by direct computation that if a constant k is subtracted from each element on the leading diagonal of matrix $\mathbf{A}$, the eigenvalues of the modified matrix are the eigenvalues of matrix $\mathbf{A}$ from each of which is subtracted the constant k. Could this result have been deduced without direct computation, and if so how? Is this result only true for this matrix $\mathbf{A}$, or is it a general property of the eigenvalues of $n \times n$ matrices?

12. Construct a square matrix of your choice, and verify by direct expansion that the characteristic polynomials of $\mathbf{A}$ and $\mathbf{A}^{\mathrm{T}}$ are identical.

The calculation of integrals over areas and volumes is often simplified by changing the variables involved to ones that are more natural for the geometry of the problem. When an integral is expressed in terms of the Cartesian coordinates x, y and z, a change of the coordinates to u_1, u_2 and u_3 involves making a transformation of the form

$$x = f(u_1,\ u_2,\ u_3),\quad y = g(u_1,\ u_2,\ u_3),\quad z = h(u_1,\ u_2,\ u_3),$$

and when this is done a scale factor J enters the transformed integrand to compensate for the change of scales. The factor J is a functional determinant denoted by $\frac{\partial(x,y,z)}{\partial(u_1,u_2,u_3)}$, where

$$J = \frac{\partial(x,y,z)}{\partial(u_1,u_2,u_3)} = \begin{vmatrix} \frac{\partial x}{\partial u_1} & \frac{\partial x}{\partial u_2} & \frac{\partial x}{\partial u_3} \\ \frac{\partial y}{\partial u_1} & \frac{\partial y}{\partial u_2} & \frac{\partial y}{\partial u_3} \\ \frac{\partial z}{\partial u_1} & \frac{\partial z}{\partial u_2} & \frac{\partial z}{\partial u_3} \end{vmatrix},$$

and J is called the Jacobian of the transformation or, more simply, just the Jacobian. If the Jacobian vanishes at any point P, the transformation fails to establish a unique correspondence at that point between the point (x_P, y_P, z_P) and the transformed point

$$(u_{1\mathrm{P}}, u_{2P}, u_{3\mathrm{P}}).$$

In Exercises 13 and 14, find the Jacobian of the given transformation, and determine when $J = 0$. Give a geometrical reason why the transformation fails when $J = 0$.

13. Find the Jacobian for the cylindrical polar coordinates $x = r\cos\phi$, $y = r\sin\phi$, $z = z$ where the coordinate system is shown in Fig. 2.1.

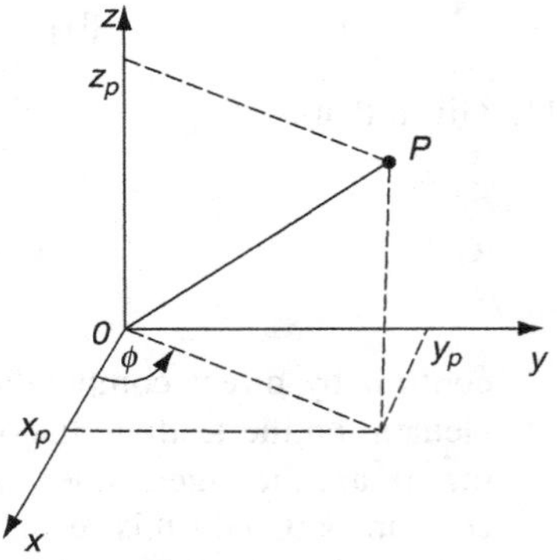

Fig. 2.1 The cylindrical polar coordinate system

14. Find the Jacobian for the *spherical polar coordinates* $x = r\sin\theta\cos\phi$, $y = r\sin\theta\sin\phi$, $z = r\cos\theta$ where the coordinate system is shown in Fig. 2.2.

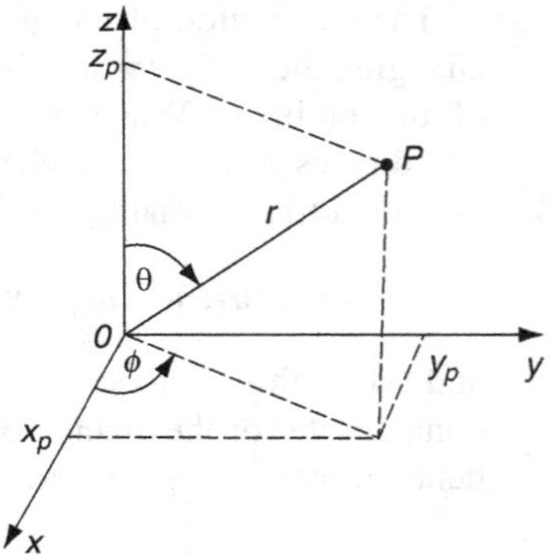

Fig. 2.2 The spherical polar coordinate system

Chapter 3
Matrix Multiplication, the Inverse Matrix and Partitioning

3.1 The Inner Product, Orthogonality and the Norm

Matrix multiplication is based on the product **ab** of an n element row vector $\mathbf{a} = [a_1, a_2, \ldots, a_n]$ and an n element column vector $\mathbf{b} = [b_1, b_2, \ldots, b_n]^{\mathrm{T}}$. This product of vectors written **ab**, and called the *inner product* or *scalar product* of the matrix row vector **a** and the matrix column vector **b**, is defined as

$$\mathbf{ab} = a_1b_1 + a_2b_2 + \cdots + a_nb_n = \sum_{i=1}^{n} a_ib_i \qquad (3.1)$$

The inner product is *only* defined if the vectors **a** and **b** each has the same number of elements.

The name *scalar product* is used because although it is the product of a row vector and a column vector, each with n elements, the result is a single scalar quantity (a number when the elements of **a** and **b** are numbers). For example, the scalar product of the two four element vectors $\mathbf{a} = [1, -2, 4, 3]$ and $\mathbf{b} = [2, 1, 0, 5]^{\mathrm{T}}$ is

$$\mathbf{ab} = (1) \times (2) + (-2) \times (1) + (4) \times (0) + (3) \times (5) = 15.$$

If **a** is not a null vector, the scalar product of the matrix vectors **a** and $\mathbf{a}^{\mathrm{T}}$ is such that $\mathbf{aa}^{\mathrm{T}} = a_1^2 + a_2^2 + \cdots + a_n^2 = \sum_{i=1}^{n} a_i^2 > 0$, and the quantity denoted by $\|\mathbf{a}\|$, where $\|\mathbf{a}\| = \sqrt{\mathbf{aa}^{\mathrm{T}}} = \left(a_1^2 + a_2^2 + \cdots + a_n^2\right)^{1/2}$, is called the *Euclidean norm* of vector **a**, also known as the *Frobenius norm*. The more familiar name *Euclidean norm* is used here because of its use with space vectors. To understand why this is, let $\mathbf{a} = a_1\mathbf{i} + a_2\mathbf{j} + a_3\mathbf{k}$ be a vector in three-dimensional Euclidean space with **i**, **j**, **k** unit vectors in the x, y and z directions. Then $\|\mathbf{a}\| = \sqrt{a_1^2 + a_2^2 + a_3^2}$ is the magnitude (length) of the space vector **a**, though in vector analysis the magnitude of a vector is usually denoted by $|\mathbf{a}|$.

The use of the term *vector* for a row or column matrix is deliberate, because Chapter 7 will show that matrices are an important example of what is called a

A. Jeffrey, *Matrix Operations for Engineers and Scientists*,
DOI 10.1007/978-90-481-9274-8_3, © Springer Science+Business Media B.V. 2010

linear vector space. In a vector space composed of matrix row and column vectors there are special vectors that play the part of the three-dimensional unit space vectors **i**, **j** and **k** that are used in the calculus and vector analysis when constructing general space vectors by scaling and vector addition. Two n element matrix vectors **a** and **b** are said to be *orthogonal* if $\mathbf{ab} = 0$, and to be *orthonormal* if in addition to $\mathbf{ab} = 0$ it is also true that $\|\mathbf{a}\| = 1$ and $\|\mathbf{b}\| = 1$. The last two conditions are equivalent to requiring the matrix vectors **a** and **b** to be such that $\mathbf{aa}^{\mathrm{T}} = 1$ and $\mathbf{b}^{\mathrm{T}}\mathbf{b} = 1$. Here the requirement that $\|\mathbf{a}\| = 1$ is a generalization to matrix vectors of the concept of unit space vector like **i**, **j** or **k**, while **ab** is a generalization to matrix vectors of the scalar product **u.v** of space vectors **u** and **v** which are orthogonal (perpendicular) if their scalar product $\mathbf{u.v} = 0$, while the vectors are orthonormal if in addition to $\mathbf{u.v} = 0$ it is also true that **u** and **v** are both *unit space vectors* (each has the Euclidean norm 1).

3.1.1 A Digression on Norms

The essential features of the norm of a matrix vector **a** are that:

(i) $\|\mathbf{a}\| > 0$ when $\mathbf{a} \neq \mathbf{0}$, and $\|\mathbf{a}\| = 0$ if and only if $\mathbf{a} = \mathbf{0}$,
(ii) $\|\mathbf{a} + \mathbf{b}\| \leq \|\mathbf{a}\| + \|\mathbf{b}\|$ (the *triangle inequality*),
(iii) $\|\lambda\mathbf{a}\| = |\lambda|\|\mathbf{a}\|$ when λ is any scalar multiplier.

Properties (i) to (iii) will be familiar from the study of three-dimensional space vectors.

The norm of a vector serves many purposes, one of the most important examples of which occurs when only a finite number of linearly independent matrix vectors can be found. This will be seen to be the case when eigenvectors are introduced in Chapter 5. If, say, these linearly independent matrix vectors are $\mathbf{v}_1, \mathbf{v}_2, \ldots, \mathbf{v}_n$, then their norms $\hat{\mathbf{v}}_1, \hat{\mathbf{v}}_2, \ldots, \hat{\mathbf{v}}_n$ play the part of the unit vectors **i**, **j** and **k** in three space dimensions when constructing more general vectors. Chapter 7 will show how by using an inner product and normed vectors, a linear transformation described by a matrix can *project* an n element matrix vector in a space S onto what is called a sub-space S of S. This is analogous to projecting a three-dimensional space vector onto a plane, which is a two-dimensional sub-space of three-dimensional space. Yet another application of normed vectors occurs in numerical analysis when matrix vectors are iterated, because working with vectors scaled according to their norm prevents them from either growing without bound, or from becoming vanishingly small, as the number of iterations increases. Although they will not be used here, we mention that other norms are possible, like the *infinity norm* and the p-norms for matrix vectors. If the elements of a vector **a** are denoted by $a_1, a_2, \ldots, a_n$, these norms are defined as:

(i) The *infinity norm* $\|\mathbf{a}\|_\infty = \max\{|a_1|, |a_2|, \ldots, |a_n|\}$

and

(ii) The *p-norm* $\|\mathbf{a}\|_p = \|\mathbf{a}\|_p = [|a_1|^p + |a_2|^p + \cdots + |a_n|p]^{1/p}$, where p is a positive integer.

It can be seen from (ii) that the Euclidean, or Frobenius norm, is the 2-norm $\|\mathbf{a}\|_2$. There are also norms for general matrices, one of the most useful being the *Frobenius norm* for an $m \times n$ matrix $\mathbf{A}$ defined as $\|\mathbf{A}\|_F = \left[\sum_{i=1}^{m}\sum_{j=1}^{n}|a_{ij}|^2\right]^{1/2}$, A simpler norm for an $n \times n$ matrix $\mathbf{A}$ that will be needed later when discussing the matrix exponential $e^{\mathbf{A}}$ is $\|\mathbf{A}\|_M = \max\{|a_{ij}| : i,j = 1, \ldots, n\}$.

3.2 Matrix Multiplication

Let $\mathbf{A}$ be an $m \times n$ matrix and $\mathbf{B}$ be an $n \times r$ matrix, with $\mathbf{a}_i$ the ith row of $\mathbf{A}$ and $\mathbf{b}_j$ the jth column of $\mathbf{B}$, so in abbreviated form $\mathbf{A}$ and $\mathbf{B}$ can be written

$$\mathbf{A} = \begin{bmatrix} \mathbf{a}_1 \\ \mathbf{a}_2 \\ \vdots \\ \mathbf{a}_m \end{bmatrix} \text{ and } \mathbf{B} = [\mathbf{b}_1 \quad \mathbf{b}_2 \quad \cdots \quad \mathbf{b}_r], \tag{3.2}$$

where it will be shown later that the matrices $\mathbf{a}_i$ and $\mathbf{b}_j$ can be considered to be special types of matrices called *block matrices*.

The two matrices $\mathbf{A}$ and $\mathbf{B}$ are said to be *conformable for multiplication* if the number of columns in $\mathbf{A}$ is equal to the number of rows in $\mathbf{B}$. The matrices $\mathbf{A}$ and $\mathbf{B}$ above satisfy this condition, because $\mathbf{A}$ is an $m \times n$ matrix and $\mathbf{B}$ is an $n \times r$ matrix so that $\mathbf{A}$ has n columns and $\mathbf{B}$ has n rows. The matrix product $\mathbf{M} = \mathbf{AB}$, with the matrices arranged in *this* order, is an $m \times r$ matrix whose element m_{ij} in the ith row and jth column is defined as the inner (scalar) product $\mathbf{a}_i\mathbf{b}_j$. In terms of $\mathbf{a}_i$, $\mathbf{b}_j$ and the inner products $\mathbf{a}_i\mathbf{b}_j$, the matrix product $\mathbf{AB}$ can be written

$$\mathbf{AB} = \begin{bmatrix} \mathbf{a}_1\mathbf{b}_1 & \mathbf{a}_1\mathbf{b}_2 & \cdots & \mathbf{a}_1\mathbf{b}_r \\ \mathbf{a}_2\mathbf{b}_1 & \mathbf{a}_2\mathbf{b}_2 & \cdots & \mathbf{a}_2\mathbf{b}_r \\ \vdots & \vdots & \vdots & \vdots \\ \mathbf{a}_n\mathbf{b}_1 & \mathbf{a}_n\mathbf{b}_2 & \cdots & \mathbf{a}_n\mathbf{b}_r \end{bmatrix}. \tag{3.3}$$

Remember that for the product of an $m \times n$ matrix $\mathbf{A}$ and a $p \times r$ matrix $\mathbf{B}$ to be conformable for the product $\mathbf{AB}$, it is necessary that $n = p$, when the result of the product will be an $m \times r$ matrix. It is because of the way the sum of the products of the elements of rows of $\mathbf{A}$ with the elements of columns of $\mathbf{B}$ is combined, that the formation of the matrix product $\mathbf{AB}$ is most easily remembered as "the product of the rows of $\mathbf{A}$ with the columns of $\mathbf{B}$". If the number of columns in $\mathbf{A}$ is not equal to the number of rows in **B**, the scalar product $\mathbf{a}_i\mathbf{b}_j$ is *not* defined, and then the matrix product **AB** will *not* exist. By convention, a matrix $\mathbf{A} = [a]$ containing a single element a may be regarded either as the simplest possible matrix, that is as a 1×1 matrix, or as the scalar quantity a, depending on the context in which it occurs.

It is clear from the definition of a matrix product that the order in which matrices are multiplied is important. If the product **AB** is defined, it is not necessary that the product **BA** exists, and even when it does exist, in general $\mathbf{AB} \neq \mathbf{BA}$. This situation is described by saying that matrix multiplication is *noncommutative*, meaning that, in general, when a matrix product is defined, the order in which the matrices appear in the product *cannot* be changed. Before examining more complicated examples of matrix multiplication, let us first apply result (3.3) to determine the following simple matrix products.

Example 3.1. Form the matrix products **AB** and **BA**, given that

$$\mathbf{A} = \begin{bmatrix} 1 & 2 & 3 \\ 2 & 1 & 1 \end{bmatrix}, \quad \mathbf{B} = \begin{bmatrix} 1 & 0 \\ 2 & 1 \\ 1 & 1 \end{bmatrix}.$$

Solution.

$$\mathbf{AB} = \begin{bmatrix} 1 & 2 & 3 \\ 2 & 1 & 1 \end{bmatrix} \begin{bmatrix} 1 & 0 \\ 2 & 1 \\ 1 & 1 \end{bmatrix}$$

$$= \begin{bmatrix} 1\times1+2\times2+3\times1 & 1\times0+2\times1+3\times1 \\ 2\times1+1\times2+1\times1 & 2\times0+1\times1+1\times1 \end{bmatrix} = \begin{bmatrix} 8 & 5 \\ 5 & 2 \end{bmatrix}.$$

Similarly

$$\mathbf{BA} = \begin{bmatrix} 1 & 0 \\ 2 & 1 \\ 1 & 1 \end{bmatrix} \begin{bmatrix} 1 & 2 & 3 \\ 2 & 1 & 1 \end{bmatrix} = \begin{bmatrix} 1 & 2 & 3 \\ 1 & 5 & 7 \\ 3 & 3 & 4 \end{bmatrix}.$$

So, although the products **AB** and **BA** are both defined, $\mathbf{AB} \neq \mathbf{BA}$.

◊

When performing matrix multiplication it is necessary to use some terminology that makes clear the order in which matrices occur in a matrix product. In a matrix product **AB**, this order is made clear by saying matrix **B** is *pre-multiplied* by matrix **A**, or that matrix **A** is *post-multiplied* by matrix **B**. So to *pre-multiply* means to "multiply from the left", while to *post-multiply* means to "multiply from the right".

Another feature of matrix multiplication that differs from ordinary algebraic multiplication is that, in general, the cancellation of matrix factors in a matrix equation is *not* permissible. It is also the case that the matrix product $\mathbf{AB} = \mathbf{0}$ does not necessarily imply either that $\mathbf{A} = \mathbf{0}$ or that $\mathbf{B} = \mathbf{0}$ nor, when the product exists, does it necessarily imply that $\mathbf{BA} = \mathbf{0}$. This can be illustrated by considering the products **AB** and **BA**, where

$$\mathbf{A} = \begin{bmatrix} 2 & 2 \\ 3 & 3 \end{bmatrix} \text{ and } \mathbf{B} = \begin{bmatrix} -1 & 1 \\ 1 & -1 \end{bmatrix}.$$

Here, although $\mathbf{A} \neq \mathbf{0}$, $\mathbf{B} \neq \mathbf{0}$, the product $\mathbf{AB} = \begin{bmatrix} 0 & 0 \\ 0 & 0 \end{bmatrix}$, while $\mathbf{BA} = \begin{bmatrix} 1 & 1 \\ -1 & -1 \end{bmatrix}$.

Similarly, cancellation of the matrix factor **A** from the equation $\mathbf{AB} = \mathbf{AC}$ is *not* permissible, because this matrix equation does not necessarily imply that $\mathbf{B} = \mathbf{C}$. This can be illustrated by considering the matrix equation $\mathbf{AB} = \mathbf{AC}$, with

$$\mathbf{A} = \begin{bmatrix} 2 & 2 \\ 3 & 3 \end{bmatrix}, \ \mathbf{B} = \begin{bmatrix} 1 & 0 \\ 2 & 2 \end{bmatrix}, \ \mathbf{C} = \begin{bmatrix} 2 & 0 \\ 1 & 2 \end{bmatrix},$$

because $\mathbf{AB} = \mathbf{AC} = \begin{bmatrix} 6 & 4 \\ 9 & 6 \end{bmatrix}$, but $\mathbf{B} \neq \mathbf{C}$.

Example 3.2. Form the matrix products **AB**, **BA**, **AC**, **AI** and **IA**, and explain why $\mathbf{AC}^{\mathrm{T}}$ does not exist, given that

$$\mathbf{A} = \begin{bmatrix} 1 & 3 & -2 \\ 0 & 4 & 1 \end{bmatrix}, \ \mathbf{B} = \begin{bmatrix} 2 & 1 \\ 4 & 5 \\ 1 & 2 \end{bmatrix}, \ \mathbf{C} = \begin{bmatrix} 4 \\ 1 \\ -3 \end{bmatrix},$$ and **I** is a conformable identity (unit) matrix.

Solution. The matrix product **AB** is defined because **A** is a 2×3 matrix and **B** is a 3×2 matrix, so the product **AB** is a 2×2 matrix. Let $\mathbf{a}_i$ be the ith row of **A** and $\mathbf{b}_j$ be the jth row of **B**, then $\mathbf{a}_1\mathbf{b}_1 = (1)\times(2) + (3)\times(4) + (-2)\times(1) = 12$, $\mathbf{a}_1\mathbf{b}_2 = (1)\times(1) + (3)\times(5) + (-2)\times(2) = 12$, $\mathbf{a}_2\mathbf{b}_1 = (0)\times(2) + (4)\times(4) + (1)\times(1) = 17$, $\mathbf{a}_2\mathbf{b}_2 = (0)\times(1) + (4)\times(5) + (1)\times(2) = 22$. Thus the matrix product

$$\mathbf{AB} = \begin{bmatrix} 12 & 12 \\ 17 & 22 \end{bmatrix}.$$

The matrix product **BA** is also defined, though it is a 3×3 matrix. The calculation of **BA** proceeds as with the product **AB**, but this time the rows of **B** contain only two elements, as do the columns of **A**, so **BA** is a (3×3) matrix. The calculation is routine, so by way of example only the details of the calculation for the element in row one and column two of the product **BA** are given. The calculation of this element involves the scalar product of the two element vectors $[2, 1]$ and $[3, 4]^{\mathrm{T}}$, given by $[2, 1]\,[3, 4]^{\mathrm{T}} = (2)\times(3) + (1)\times(4) = 10$. Completing the calculations gives

$$\mathbf{BA} = \begin{bmatrix} 2 & 10 & -3 \\ 4 & 32 & -3 \\ 1 & 11 & 0 \end{bmatrix}.$$

As **A** is a (2×3) matrix, and **C** is a (3×1) matrix, the matrix product **AC** is the (2×1) matrix

$$\mathbf{AC} = \begin{bmatrix} 1 & 3 & -2 \\ 0 & 4 & 1 \end{bmatrix} \begin{bmatrix} 4 \\ 1 \\ -3 \end{bmatrix} = \begin{bmatrix} (1) \times (4) + (3) \times (1) + (-2) \times (-3) \\ (0) \times (4) + (4) \times (1) + (1) \times (-3) \end{bmatrix} = \begin{bmatrix} 13 \\ 1 \end{bmatrix}.$$

As **A** is a 2×3 matrix, for compatibility the product **AI** will be defined if the identity matrix **I** is taken to be a 3×3 matrix, in which case a simple calculation confirms that $\mathbf{AI} = \mathbf{A}$. However, for the product **IA** to be conformable it is necessary for **I** to be the 2×2 identity matrix, from which it then follows that $\mathbf{IA} = \mathbf{A}$.

This illustrates the fact that in multiplications the identity matrix **I** acts like the number 1 (unity) in ordinary multiplication. If the shape of the identity matrices involved must be made clear, in the first of these calculations we could write $\mathbf{AI}_3 = \mathbf{A}$, where $\mathbf{I}_3$ is a 3×3 identity matrix, while in the second calculation we could write $\mathbf{I}_2\mathbf{A} = \mathbf{A}$, where $\mathbf{I}_2$ is a 2×2 identity matrix. However, the identification of a unit matrix in this way is seldom necessary, since it is always understood that the symbol **I** represents whatever identity matrix is appropriate for the algebraic operation that is to be performed.

Finally, the matrix product $\mathbf{AC}^{\mathrm{T}}$ is not defined, because **A** is a 2×3 matrix, and $\mathbf{C}^{\mathrm{T}}$ is a 1×3 matrix.

◇

By definition, if a general matrix **A** is multiplied (scaled) by a constant k, then *each* element of matrix **A** is multiplied by k. This definition of scaling a general matrix by a constant k is in agreement with the definition of the meaning of $\lambda\mathbf{I}$ introduced at the end of Chapter 2 when considering $\det[\mathbf{A} - \lambda\mathbf{I}]$. So, for example,

$$k \begin{bmatrix} a_{11} & a_{12} & \cdots & a_{1n} \\ a_{21} & a_{22} & \cdots & a_{2n} \\ \vdots & \vdots & \vdots & \vdots \\ a_{m1} & a_{m2} & \cdots & a_{mn} \end{bmatrix} = \begin{bmatrix} ka_{11} & ka_{12} & \cdots & ka_{1n} \\ ka_{21} & ka_{22} & \cdots & ka_{2n} \\ \vdots & \vdots & \vdots & \vdots \\ ka_{m1} & ka_{m2} & \cdots & ka_{mn} \end{bmatrix}. \tag{3.4}$$

To return to the study of linear systems of algebraic equations, we now examine the relationship between nonhomogeneous first-order algebraic systems and matrix multiplication, by considering the system

$$\begin{aligned} a_{11}x_1 + a_{12}x_2 + \cdots + a_{1n}x_n &= b_1, \\ a_{21}x_1 + a_{22}x_2 + \cdots + a_{2n}x_n &= b_2, \\ &\vdots \\ a_{m1}x_1 + a_{m2}x_2 + \cdots + a_{mn}x_n &= b_m. \end{aligned} \tag{3.5}$$

Defining the matrices

$$\mathbf{A} = \begin{bmatrix} a_{11} & a_{12} & \cdots & a_{1n} \\ a_{21} & a_{22} & \cdots & a_{2n} \\ \vdots & \vdots & \vdots & \vdots \\ a_{m1} & a_{m2} & \cdots & a_{mn} \end{bmatrix}, \quad \mathbf{x} = \begin{bmatrix} x_1 \\ x_2 \\ \vdots \\ x_n \end{bmatrix}, \quad \mathbf{b} = \begin{bmatrix} b_1 \\ b_2 \\ \vdots \\ b_m \end{bmatrix}, \tag{3.6}$$

allows system (3.6) to be written in the concise form

$$\mathbf{Ax} = \mathbf{b}. \tag{3.7}$$

Division by a matrix is *not* defined, so the matrix Eq. (3.7) cannot be divided by **A** to find **x**. However, if $m = n$, and det $\mathbf{A} \neq 0$, it will be shown later that a new matrix denoted by $\mathbf{A}^{-1}$can be defined with the property that $\mathbf{A}^{-1}\mathbf{A} = \mathbf{AA}^{-1} = \mathbf{I}$, where the matrix $\mathbf{A}^{-1}$ is called the *inverse* of matrix **A**.

Using this property, and pre-multiplying (3.7) by $\mathbf{A}^{-1}$, that is multiplying it from the left by $\mathbf{A}^{-1}$, it becomes $\mathbf{A}^{-1}\mathbf{Ax} = \mathbf{A}^{-1}\mathbf{b}$ but $\mathbf{A}^{-1}\mathbf{A} = \mathbf{I}$, and $\mathbf{Ix} = \mathbf{x}$, so the solution of (3.7) is seen to be given by $\mathbf{x} = \mathbf{A}^{-1}\mathbf{b}$, whenever $\mathbf{A}^{-1}$ exists. This reasoning raises the important question of how to find $\mathbf{A}^{-1}$ for any given square matrix **A**, though this matter will be postponed until later in this chapter. However, if **A** is not a square matrix the inverse of **A** does *not* exist.

It is a consequence of the definition of matrix multiplication that when **A**, **B** and **C** are conformable for the product **ABC**, pre-multiplying **C** by the product **AB** is the same as post-multiplying **A** by the product **BC**, so that

$$(\mathbf{AB})\mathbf{C} = \mathbf{A}(\mathbf{BC}). \tag{3.8}$$

An immediate consequence of (3.8) is that for any integer n we may use exponent notation and write

$$\underbrace{\mathbf{AA}\cdots\mathbf{A}}_{n \text{ times}} = \mathbf{A}^n, \tag{3.9}$$

where, for consistency we define $\mathbf{A}^0 = \mathbf{I}$.

Example 3.3. Verify property (3.8) given that

$$\mathbf{A} = \begin{bmatrix} 1 & 4 \\ 3 & 2 \end{bmatrix}, \quad \mathbf{B} = \begin{bmatrix} 2 & 6 \\ -1 & 0 \end{bmatrix}, \quad \mathbf{C} = \begin{bmatrix} 3 & 2 \\ 1 & 4 \end{bmatrix},$$

and show that **ABC** $\neq$ **CBA**.

Solution.

$$\mathbf{AB} = \begin{bmatrix} -2 & 6 \\ 4 & 18 \end{bmatrix}, \quad (\mathbf{AB})\mathbf{C} = \begin{bmatrix} -2 & 6 \\ 4 & 18 \end{bmatrix}\begin{bmatrix} 3 & 2 \\ 1 & 4 \end{bmatrix} = \begin{bmatrix} 0 & 20 \\ 30 & 80 \end{bmatrix}$$

and

$$\mathbf{BC} = \begin{bmatrix} 12 & 28 \\ -3 & -2 \end{bmatrix}, \quad \mathbf{A}(\mathbf{BC}) = \begin{bmatrix} 1 & 4 \\ 3 & 2 \end{bmatrix}\begin{bmatrix} 12 & 28 \\ -3 & -2 \end{bmatrix} = \begin{bmatrix} 0 & 20 \\ 30 & 80 \end{bmatrix}.$$

A routine calculation shows that

$$\mathbf{CBA} = \begin{bmatrix} 58 & 52 \\ 16 & 4 \end{bmatrix},$$

so in this case **ABC** $\neq$ **CBA**.

◇

3.3 Quadratic Forms

An important connection exists between $n \times n$ matrices with real elements, and *quadratic forms* $Q(x_1, x_2, \ldots, x_n)$ in the n real variables $x_1, x_2, \ldots, x_n$, where by definition the quadratic form

$$Q(x_1, x_2, \ldots, x_n) = \sum_{i,j=1}^{n} \alpha_{ij} x_i x_j. \tag{3.10}$$

The coefficients α_{ij} can be represented in matrix form by defining an $n \times n$ matrix

$$\tilde{\mathbf{A}} = \begin{bmatrix} \alpha_{11} & \alpha_{12} & \cdots & \alpha_{1n} \\ \alpha_{21} & \alpha_{22} & \cdots & \alpha_{2n} \\ \vdots & \vdots & \vdots & \vdots \\ \alpha_{n1} & \alpha_{n2} & \cdots & \alpha_{nn} \end{bmatrix}, \tag{3.11}$$

which then enables (3.10) to be written

$$Q(\mathbf{x}) = \mathbf{x}^{\mathrm{T}} \tilde{\mathbf{A}} x, \tag{3.12}$$

where $\mathbf{x} = [x_1, x_2, \ldots, x_n]^{\mathrm{T}}$ is an n element column vector, and $\mathbf{x}^{\mathrm{T}} = [x_1, x_2, \ldots, x_n]^{\mathrm{T}}$ is its transpose (an n element row vector).

Quadratic forms have many uses. They are introduced here because the process of simplifying (3.10) to an equivalent sum involving only the squares of n new variables, say $y_1, y_2, \ldots, y_n$, involves finding a linear change of the variables in $\mathbf{x}$ to the variables in $\mathbf{y}$ that has the effect of reducing $\tilde{\mathbf{A}}$ to a diagonal matrix. Later we will see how this same process, called the *diagonalization* of a matrix, plays an important role when working with systems of linear differential equations.

A typical quadratic form involving the two real variables x and y is

$$Q(x, y) = \alpha_{11}x^2 + (\alpha_{12} + \alpha_{21})xy + \alpha_{22}y^2. \tag{3.13}$$

This can be written in the matrix form

$$Q(x, y) = [x \quad y]\begin{bmatrix} \alpha_{11} & \alpha_{12} \\ \alpha_{21} & \alpha_{22} \end{bmatrix}\begin{bmatrix} x \\ y \end{bmatrix},$$

because

$$[x \quad y]\begin{bmatrix} \alpha_{11} & \alpha_{12} \\ \alpha_{21} & \alpha_{22} \end{bmatrix} = [\alpha_{11}x + \alpha_{21}y, \alpha_{12}x + \alpha_{22}y],$$

so

$$Q(x, y) = [\alpha_{11}x + \alpha_{21}y, \alpha_{12}x + \alpha_{22}y]\begin{bmatrix} x \\ y \end{bmatrix} a_{11} = \left[\alpha_{11}x^2 + (\alpha_{12} + \alpha_{21})xy + \alpha_{22}y^2\right],$$

and as the last quantity on the right is a matrix containing only a single element, it can be written as a scalar quantity, so $Q(x, y)$ becomes

$$Q(x, y) = \alpha_{11}x^2 + (\alpha_{12} + \alpha_{21})xy + \alpha_{22}y^2.$$

It is always possible to express an arbitrary quadratic form in the n variables $x_1, x_2, \ldots, x_n$ in terms of a *symmetric* matrix $\mathbf{A}$, and an n element column vector $\mathbf{x} = [x_1, x_2, \ldots, x_n]^{\mathrm{T}}$. To achieve this, when the quadratic form is expressed as in (3.10), with a matrix $\tilde{\mathbf{A}}$ having the coefficients α_{ij}, the required symmetric matrix $\mathbf{A} = [a_{ij}]$ with elements a_{ij} defined in terms of the elements α_{ij} is given by

$$a_{ij} = \begin{cases} \alpha_{ij}, & i = j, \\ \frac{1}{2}(\alpha_{ij} + \alpha_{ji}), & i \neq j. \end{cases} \tag{3.14}$$

Once diagonalization has been discussed, it will be shown how any real quadratic form can be reduced to its *diagonal form*

$$Q(\mathbf{x}) = \lambda_1 x_1^2 + \lambda_2 x_2^2 + \cdots + \lambda_n x_n^2, \tag{3.15}$$

with $\lambda_1 \geq \lambda_2 \geq \cdots \geq \lambda_n$, where some of the n numbers λ_i may be negative, and some may be zero.

Example 3.4. Find the quadratic form defined by the matrix $\tilde{\mathbf{A}}$.

$$\tilde{\mathbf{A}} = \begin{bmatrix} 7 & 4 & 4 \\ -8 & 1 & 12 \\ -8 & -4 & 1 \end{bmatrix}.$$

Define the symmetric matrix $\mathbf{A}$ with coefficients a_{ij} determined by (3.13), and confirm that it generates the same quadratic form as matrix $\tilde{\mathbf{A}}$.

Solution. Setting $\mathbf{x} = [x_1, x_2, x_3]^{\mathrm{T}}$ we have

$$\mathbf{x}^{\mathrm{T}}\tilde{\mathbf{A}} = [7x_1 - 8x_2 - 8x_3, 4x_1 + x_2 - 4x_3, 4x_1 + 12x_2 + x_3],$$

so after evaluating the inner product of $\mathbf{x}^{\mathrm{T}}\tilde{\mathbf{A}}$ and $\mathbf{x}$, we find the required quadratic form is

$$Q(x) = \mathbf{x}^{\mathrm{T}}\tilde{\mathbf{A}}x = 7{x_1}^2 - 4x_1x_2 - 4x_1x_3 + {x_2}^2 + 8x_2x_3 + {x_3}^2.$$

From (3.13) the coefficients a_{ij} of a symmetric matrix $\mathbf{A}$ are $a_{11} = 7$, $a_{12} = a_{21} = \frac{1}{2}(4-8) = -2$, $a_{13} = a_{32} = \frac{1}{2}(4-8) = -2$, $a_{22} = 1$, $a_{23} = a_{32} = \frac{1}{2}(-4+12) = 4$, $a_{33} = 1$.
So the required symmetric matrix $\mathbf{A}$ becomes

$$\mathbf{A} = \begin{bmatrix} 7 & -2 & -2 \\ -2 & 1 & 4 \\ -2 & 4 & 1 \end{bmatrix}.$$

Repeating the previous calculation, but this time with $\mathbf{A}$ in place of $\tilde{\mathbf{A}}$, gives

$$\mathbf{x}^{\mathrm{T}}\mathbf{A} = [7x_1 - 2x_2 - 2x_3\,, -2x_1 + x_2 + 4x_3\,, -2x_1 + 4x_2 + x_3],$$

so that

$$Q(\mathbf{x}) = \mathbf{x}^{\mathrm{T}}\mathbf{A}\mathbf{x} = 7{x_1}^2 - 4x_1x_2 - 4x_1x_3 + {x_2}^2 + 8x_2x_3 + {x_3}^2,$$

confirming that $\mathbf{x}^{\mathrm{T}}\tilde{\mathbf{A}}x = \mathbf{x}^{\mathrm{T}}\mathbf{A}\mathbf{x}$.

Before leaving this example we mention that the linear change of variable

$$x_1 = \tfrac{1}{\sqrt{3}}y_2 - \tfrac{2}{\sqrt{6}}y_3,\ x_2 = -\tfrac{1}{\sqrt{2}}y_1 + \tfrac{1}{\sqrt{3}}y_2 + \tfrac{1}{\sqrt{6}}y_3,\ x_3 = \tfrac{1}{\sqrt{2}}y_1 + \tfrac{1}{\sqrt{3}}y_2 + \tfrac{1}{\sqrt{6}}y_3$$

reduces the quadratic form $Q(\mathbf{x})$ in terms of $\mathbf{x} = [x_1, x_2, x_3]^T$ to the much simpler form $Q(\mathbf{y}) = -3y_1^2 + 3y_2^2 + 9y_3^2$ involving only a sum of squares of the elements of $\mathbf{y} = [y_1, y_2, y_3]^T$. However, the way to find such a change of variable will be described later once the diagonalization of a matrix has been discussed.

◇

3.4 The Inverse Matrix

Previously, in connection with the system of Eq. (3.7), an $n \times n$ matrix $\mathbf{A}^{-1}$ with the property that $\mathbf{A}^{-1}\mathbf{A} = \mathbf{I}$ was introduced in a purely formal calculation to show how, when this matrix exists, the solution of the system of n nonhomogeneous algebraic equations

$$\mathbf{Ax} = \mathbf{b} \tag{3.16}$$

can be solved for $\mathbf{x}$ by pre-multiplying the equation by $\mathbf{A}^{-1}$, because $\mathbf{A}^{-1}\mathbf{Ax} = \mathbf{A}^{-1}\mathbf{b}$, but $\mathbf{A}^{-1}\mathbf{A} = \mathbf{I}$, and $\mathbf{Ix} = \mathbf{x}$, so

$$\mathbf{x} = \mathbf{A}^{-1}\mathbf{b}. \tag{3.17}$$

Consequently, given an $n \times n$ matrix $\mathbf{A}$, it is necessary to discover when and how an associated $n \times n$ matrix denoted by $\mathbf{A}^{-1}$ can be found with the property that

$$\mathbf{AA}^{-1} = \mathbf{A}^{-1}\mathbf{A} = \mathbf{I}. \tag{3.18}$$

As already stated, division by matrices is *not* defined, but when an $n \times n$ matrix $\mathbf{A}^{-1}$ associated with a matrix $\mathbf{A}$ can be found satisfying (3.18) it is called the matrix *inverse* of $\mathbf{A}$, or more simply the *inverse* of matrix $\mathbf{A}$. As the inverse matrix $\mathbf{A}^{-1}$ occurs in (3.18) both as a pre- and a post-multiplier of $\mathbf{A}$ it is, more properly described as the *multiplicative inverse* of $\mathbf{A}$, though for conciseness the term *multiplicative* is almost always omitted.

To see how, if $\mathbf{A}$ is an $n \times n$ matrix, a formal definition of the inverse matrix $\mathbf{A}^{-1}$ can be obtained, we consider the matrix product $\mathbf{M} = \mathbf{AC}^T$, where $\mathbf{C}^T$ is the transpose of the matrix of cofactors associated with $\mathbf{A}$. When written out formally this becomes

$$\mathbf{M} = \begin{bmatrix} a_{11} & a_{12} & \cdots & a_{1n} \\ a_{21} & a_{22} & \cdots & a_{2n} \\ \vdots & \vdots & \vdots & \vdots \\ a_{n1} & a_{n2} & \cdots & a_{nn} \end{bmatrix} \begin{bmatrix} C_{11} & C_{21} & \cdots & C_{n1} \\ C_{12} & C_{22} & \cdots & C_{n2} \\ \vdots & \vdots & \vdots & \vdots \\ C_{1n} & C_{2n} & \cdots & C_{nn} \end{bmatrix}. \tag{3.19}$$

Appeal to Theorem 2.2 (the Laplace expansion of a determinant) shows that each element on the diagonal of $\mathbf{M}$ is simply det $\mathbf{A}$, while every off-diagonal element is zero, because each off-diagonal element is obtained as the sum of the products of the elements of a row of $\mathbf{A}$ with the cofactors of a *different* row of $\mathbf{A}$. This allows us to write

$$\mathbf{M} = \begin{bmatrix} \det \mathbf{A} & 0 & \cdots & 0 \\ 0 & \det \mathbf{A} & \cdots & 0 \\ \vdots & \vdots & \vdots & \vdots \\ 0 & 0 & \cdots & \det \mathbf{A} \end{bmatrix}, \tag{3.20}$$

so from Property 1 of Theorem 2.1 with $k = \det \mathbf{A}$, because $\mathbf{M} = \mathbf{A}\mathbf{C}^{\mathrm{T}}$, it follows that

$$\mathbf{A}\mathbf{C}^{\mathrm{T}} = (\det \mathbf{A})\mathbf{I}. \tag{3.21}$$

A similar argument using the product $\mathbf{C}^{\mathrm{T}}\mathbf{A}$ leads to the result

$$\mathbf{C}^{\mathrm{T}}\mathbf{A} = (\det \mathbf{A})\mathbf{I}, \tag{3.22}$$

so

$$\mathbf{C}^{\mathrm{T}}\mathbf{A} = \mathbf{A}\mathbf{C}^{\mathrm{T}} = (\det \mathbf{A})\mathbf{I}. \tag{3.23}$$

When $\det \mathbf{A} \neq 0$, a comparison of (3.23) and (3.18) leads to the definition

$$\mathbf{A}^{-1} = \frac{1}{\det \mathbf{A}}\mathbf{C}^{\mathrm{T}}, \tag{3.24}$$

provided the scalar divisor $\det \mathbf{A} \neq 0$.

Because of its importance and frequent occurrence, the matrix $\mathbf{C}^{\mathrm{T}}$ defined as the transpose of the matrix of cofactors $\mathbf{C}$ of $\mathbf{A}$, is given a name and called the *adjoint* of $\mathbf{A}$, written adj $\mathbf{A}$, so that

$$\operatorname{adj} \mathbf{A} = \mathbf{C}^{\mathrm{T}}. \tag{3.25}$$

Thus the formal definition of the inverse matrix $\mathbf{A}^{-1}$ in terms of the adjoint of $\mathbf{A}$ is

$$\mathbf{A}^{-1} = \frac{1}{\det \mathbf{A}}\operatorname{adj} \mathbf{A}, \qquad \det \mathbf{A} \neq 0. \tag{3.26}$$

Matrix $\mathbf{A}$ is said to be *invertible*, meaning its inverse exists, when $\mathbf{A}^{-1}$ exists. This in turn shows that for $\mathbf{A}^{-1}$ to exist it is necessary for $\mathbf{A}$ to be *nonsingular*; so $\det \mathbf{A} \neq 0$.

We have proved that $\mathbf{AA}^{-1} = \mathbf{A}^{-1}\mathbf{A} = \mathbf{I}$. Notice that the exponent notation adopted in (3.9) applies equally well to (3.26) provided we use the definitions $\mathbf{A}^1 = \mathbf{A}$, and $\mathbf{A}^0 = \mathbf{I}$, because then $\mathbf{AA}^{-1} = \mathbf{A}^1\mathbf{A}^{-1} = \mathbf{A}^{(1-1)} = \mathbf{A}^0 = \mathbf{I}$.

When n is large the computation of det $\mathbf{A}$ and the elements of adj $\mathbf{A}$ is time consuming, so this definition of the inverse matrix is mainly of theoretical importance, though it can be useful when n is small. If systems of algebraic equations like (3.16) need to be solved, instead of computing $\mathbf{A}^{-1}$, a different and more efficient approach must be used. In this method the equations in the system are reduced to an upper triangular form, with suitable modifications to the nonhomogeneous terms, after which the system is solved using a process called *back substitution*. In back substitution, x_n is found first, and then this value is used to find x_{n-1}, after which the value of x_{n-2} is found from the values of x_n and x_{n-1}, and so on, until finally x_1 is found in terms of $x_n, x_{n-1}, \ldots, x_2$.

Example 3.5. Find $\mathbf{A}^{-1}$ given that

$$\mathbf{A} = \begin{bmatrix} 3 & 1 & 2 \\ 2 & -1 & 1 \\ 1 & 3 & -1 \end{bmatrix}.$$

Solution. A straightforward calculation shows the matrix $\mathbf{C}$ of cofactors is

$$\mathbf{C} = \begin{bmatrix} -2 & 3 & 7 \\ 7 & -5 & -8 \\ 3 & 1 & -5 \end{bmatrix}, \text{ so adj } \mathbf{A} = \mathbf{C}^{\mathrm{T}} = \begin{bmatrix} -2 & 7 & 3 \\ 3 & -5 & 1 \\ 7 & -8 & -5 \end{bmatrix}, \text{ and det } \mathbf{A} = 11,$$

so from (3.26)

$$\mathbf{A}^{-1} = \frac{1}{\det \mathbf{A}}\mathbf{C}^{\mathrm{T}} = \begin{bmatrix} -\frac{2}{11} & \frac{7}{11} & \frac{3}{11} \\ \frac{3}{11} & -\frac{5}{11} & \frac{1}{11} \\ \frac{7}{11} & -\frac{8}{11} & -\frac{5}{11} \end{bmatrix}.$$

A routine calculation confirms that $\mathbf{A}^{-1}$ has the required properties, because $\mathbf{AA}^{-1} = \mathbf{A}^{-1}\mathbf{A} = \mathbf{I}$.

◊

Example 3.6. Use the result of Example 3.4 to solve the system

$$\begin{aligned} 3x_1 + x_2 + 2x_3 &= 4, \\ 2x_1 - x_2 + x_3 &= -3, \\ x_1 + 3x_2 - x_3 &= 5. \end{aligned}$$

Solution. The coefficient matrix in this case is matrix **A** in Example 3.4, so setting

$$\mathbf{x} = \begin{bmatrix} x_1 \\ x_2 \\ x_3 \end{bmatrix}, \quad \mathbf{b} = \begin{bmatrix} 4 \\ -3 \\ 5 \end{bmatrix},$$

the system becomes $\mathbf{Ax} = \mathbf{b}$, so that $\mathbf{x} = \mathbf{A}^{-1}\mathbf{b}$. Using the result of Example 3.4 we find that $\mathbf{x} = \mathbf{A}^{-1}\mathbf{b}$ becomes

$$\begin{bmatrix} x_1 \\ x_2 \\ x_3 \end{bmatrix} = \begin{bmatrix} -\frac{2}{11} & \frac{7}{11} & \frac{3}{11} \\ \frac{3}{11} & -\frac{5}{11} & \frac{1}{11} \\ \frac{7}{11} & -\frac{8}{11} & -\frac{5}{11} \end{bmatrix} \begin{bmatrix} 4 \\ -3 \\ 5 \end{bmatrix} = \begin{bmatrix} -\frac{14}{11} \\ \frac{32}{11} \\ \frac{27}{11} \end{bmatrix}.$$

Equating corresponding elements in the column vectors on the left and right shows that the elements of the solution set $\{x_1, x_2, x_3\}$ are given by

$$x_1 = -\tfrac{14}{11}, \quad x_2 = \tfrac{32}{11}, \quad x_3 = \tfrac{17}{11}.$$

◊

The two fundamental properties of the inverse matrix that follow can be deduced very simply; the first being that

$$\left(\mathbf{A}^{-1}\right)^{-1} = \mathbf{A}, \tag{3.27}$$

while the second is that if the square matrices **A** and **B** are conformable for multiplication, then

$$(\mathbf{AB})^{-1} = \mathbf{B}^{-1}\mathbf{A}^{-1}. \tag{3.28}$$

When $\mathbf{A}^{-1}$ exists, (3.27) follows from (3.18), because $\mathbf{A}^{-1}\mathbf{A} = \mathbf{AA}^{-1} = \mathbf{I}$ shows that **A** is the inverse of $\mathbf{A}^{-1}$, so $(\mathbf{A}^{-1})^{-1} = \mathbf{A}$. Result (3.28) follows by considering the product $\mathbf{B}^{-1}\mathbf{A}^{-1}\mathbf{AB}$, because $\mathbf{A}^{-1}\mathbf{A} = \mathbf{I}$ and $\mathbf{B}^{-1}\mathbf{B} = \mathbf{I}$, so $\mathbf{B}^{-1}\mathbf{A}^{-1}\mathbf{AB} = \mathbf{B}^{-1}\mathbf{IB} = \mathbf{B}^{-1}\mathbf{B} = \mathbf{I}$. Thus the matrix product **AB** is the inverse of the matrix product $\mathbf{B}^{-1}\mathbf{A}^{-1}$, confirming that $(\mathbf{AB})^{-1} = \mathbf{B}^{-1}\mathbf{A}^{-1}$.

When $\mathbf{A}^{-1}$ exists, it is *always* true that $\mathbf{AB} = \mathbf{0}$ implies $\mathbf{B} = \mathbf{0}$, and that $\mathbf{AB} = \mathbf{AC}$ implies $\mathbf{B} = \mathbf{C}$. Although cancellation of matrices is not permitted, in this case pre-multiplication by $\mathbf{A}^{-1}$ has a similar effect. These statements do not contradict the results of the two examples following Eq. (3.3), because in those cases matrix **A** was singular, so $\mathbf{A}^{-1}$ did not exist.

Two useful results involving the *multiplication of determinants* can be deduced from matrix multiplication. These are that if **A** and **B** are an $n \times n$ matrices, then

$$\left.\begin{array}{l} \det(\mathbf{AB}) = \det\mathbf{A}\det\mathbf{B} \\ \det\mathbf{A}^{-1} = 1/\det\mathbf{A}, \quad \det\mathbf{A} \neq 0. \end{array}\right\} \tag{3.29}$$

The second result follows from the first one by setting $\mathbf{B} = \mathbf{A}^{-1}$, when $\mathbf{A}\mathbf{A}^{-1} = \mathbf{I}$, and $\det \mathbf{I} = 1$, so it is only necessary to prove the first result. For simplicity, the proof will only be given when **A** and **B** are 2×2 matrices, because although the proof generalizes in an obvious way to include $n \times n$ matrices, the calculations become tedious.

Let

$$D_1 = \begin{vmatrix} a_{11} & a_{12} \\ a_{21} & a_{22} \end{vmatrix}, \quad D_2 = \begin{vmatrix} b_{11} & b_{12} \\ b_{21} & b_{22} \end{vmatrix}$$

and define D as

$$D = \begin{vmatrix} a_{11} & a_{12} & 0 & 0 \\ a_{21} & a_{22} & 0 & 0 \\ -1 & 0 & b_{11} & b_{12} \\ 0 & -1 & b_{21} & b_{22} \end{vmatrix}.$$

Expanding D in terms of the elements of its last column, and then expanding the two third-order determinants that arise in terms of elements of their last columns, gives

$$D = -b_{12}b_{21}\begin{vmatrix} a_{11} & a_{12} \\ a_{21} & a_{22} \end{vmatrix} + b_{11}b_{22}\begin{vmatrix} a_{11} & a_{12} \\ a_{21} & a_{22} \end{vmatrix} = \begin{vmatrix} a_{11} & a_{12} \\ a_{21} & a_{22} \end{vmatrix}(b_{11}b_{22} - b_{12}b_{21}),$$

and so $D = \begin{vmatrix} a_{11} & a_{12} \\ a_{21} & a_{22} \end{vmatrix}\begin{vmatrix} b_{11} & b_{12} \\ b_{21} & b_{22} \end{vmatrix} = D_1 D_2$.

It remains for us to show that $D_1 D_2$ is the determinant formed from the matrix product **AB**.

Determinant D will be unchanged if its first row is replaced by Row $1 + a_{11}$ Row $3 + a_{12}$ Row 4, and its second row is replaced by Row $2 + a_{21}$ Row $3 + a_{22}$ Row 4, when it becomes

$$D = D_1 D_2 = \begin{vmatrix} 0 & 0 & a_{11}b_{11} + a_{12}b_{21} & a_{11}b_{12} + a_{12}b_{22} \\ 0 & 0 & a_{21}b_{11} + a_{22}b_{21} & a_{21}b_{12} + a_{22}b_{22} \\ -1 & 0 & b_{11} & b_{12} \\ 0 & -1 & b_{21} & b_{22} \end{vmatrix}.$$

Expanding this determinant by the elements in its first column to obtain two 3×3 determinants, and then expanding these by the elements in their first columns gives

$$\begin{vmatrix} a_{11} & a_{12} \\ a_{212} & a_{22} \end{vmatrix}\begin{vmatrix} b_{11} & b_{12} \\ b_{21} & b_{22} \end{vmatrix} = \begin{vmatrix} a_{11}b_{11} + a_{12}b_{21} & a_{11}b_{12} + a_{12}b_{22} \\ a_{21}b_{11} + a_{22}b_{21} & a_{21}b_{12} + a_{22}b_{22} \end{vmatrix}.$$

However, the determinant on the right is the determinant of the matrix product **BA**, so the result is proved for the product of second-order determinants. As already mentioned, the equivalent result for the product of determinants of order n follows in a similar fashion.

A useful result involving adjoint matrices follows from the first result of (3.29). Replacing $\mathbf{A}$ by $\mathbf{A}^{-1}$ in (3.26) and forming the matrix product $\mathbf{A}\mathbf{A}^{-1}$ gives

$$\mathbf{A}\mathbf{A}^{-1} = \frac{1}{\det \mathbf{A} \det (\mathbf{A}^{-1})} \operatorname{adj} (\mathbf{A}^{-1}) \operatorname{adj} \mathbf{A},$$

but $\mathbf{A}\mathbf{A}^{-1} = \mathbf{I}$ and $\det \mathbf{A} \det (\mathbf{A}^{-1}) = 1$, so it follows that

$$\operatorname{adj} (\mathbf{A}^{-1}) \operatorname{adj} \mathbf{A} = \mathbf{I},$$

and repeating the argument, but this time using the product $\mathbf{A}^{-1}\mathbf{A}$ shows that

$$\operatorname{adj} \mathbf{A} \operatorname{adj} (\mathbf{A}^{-1}) = \mathbf{I},$$

so we have proved that

$$\operatorname{adj} (\mathbf{A}^{-1}) \operatorname{adj} \mathbf{A} = \operatorname{adj} \mathbf{A} \operatorname{adj} (\mathbf{A}^{-1}) = \mathbf{I}. \tag{3.30}$$

3.5 Orthogonal Matrices

At this point it is convenient to introduce the concept of an *orthogonal* matrix, and to relate orthogonal matrices to their geometrical properties involving rotations in space. A real nonsingular square matrix $\mathbf{Q}$ is said to be *orthogonal* if it is such that

$$\mathbf{Q}^{-1} = \mathbf{Q}^{\mathrm{T}}, \tag{3.31}$$

so when $\mathbf{Q}$ is orthogonal, $\mathbf{Q}\mathbf{Q}^{\mathrm{T}} = \mathbf{I}$. A typical orthogonal matrix is

$$\mathbf{Q} = \begin{bmatrix} \frac{2}{3} & \frac{2}{3} & -\frac{1}{3} \\ \frac{1}{\sqrt{2}} & -\frac{1}{\sqrt{2}} & 0 \\ \frac{1}{3\sqrt{2}} & \frac{1}{3\sqrt{2}} & \frac{4}{3\sqrt{2}} \end{bmatrix},$$

and this result is easily checked by showing that $\mathbf{Q}\mathbf{Q}^{\mathrm{T}} = \mathbf{I}$.

Orthogonal matrices are so named because they possess an important geometrical property. This property is that they characterize coordinate transformations that rotate coordinate axes about an origin, while preserving orthogonality between perpendicular lines, and also preserving shapes and the lengths of vectors.

In rectangular Cartesian coordinates the transformation

$$
\begin{aligned}
x'_P &= x_P \cos\theta + y_P \sin\theta, \\
y'_P &= -x_P \sin\theta + y_P \cos\theta,
\end{aligned}
\tag{3.32}
$$

illustrated in Fig. 3.1, describes how the coordinates (x_p, y_p) of a fixed point P relative to the O(x, y)-axes, become the coordinates $(x'_P, y'_P)$of the same point P relative to the O(x', y')-axes, when the O(x', y')-axes are obtained by a *counter-clockwise rotation* of the O(x, y)-axes through an angle θ.

It is obvious geometrically that this rotation preserves lengths from the origin to arbitrary points P in the plane, but we will give an analytical proof of this fact. To show the transformation preserves length, let l_1 be the length of the straight line from the origin to the point P with the (x, y) coordinates (x_p, y_p), and l_2 be the length of the straight line from the origin to the point P with the (x', y') coordinates(x'_P, y'_P), then

$$
(x'_P)^2 + (y'_P)^2 = (x_P \cos\theta + y_P \sin\theta)^2 + (-x_P \sin\theta + y_P \cos\theta)^2 = x_P^2 + y_P^2. \tag{3.33}
$$

This shows that $l_1{}^2 = l_2{}^2$, so as lengths are essentially nonnegative, the length l_1 (the norm) of the vector drawn to the point (x_p, y_p) is the same as the length l_2 (the norm) of the vector drawn to the point (x'_P, y'_P).

As (3.32) applies to any point P, we will drop the suffix P and write (3.32) as

$$
\mathbf{x}' = \mathbf{Q}\mathbf{x}, \text{with } \mathbf{x} = \begin{bmatrix} x \\ y \end{bmatrix}, \mathbf{x}' = \begin{bmatrix} x' \\ y' \end{bmatrix}, \mathbf{Q} = \begin{bmatrix} \cos\theta & \sin\theta \\ -\sin\theta & \cos\theta \end{bmatrix}, \tag{3.34}
$$

it is easily seen that $\mathbf{Q}$ is an orthogonal matrix, because $\mathbf{Q}^{-1} = \mathbf{Q}^{\mathrm{T}}$. Consequently, $\mathbf{x}$ is given in terms of $\mathbf{x}'$ by

$$
\mathbf{x} = \mathbf{Q}^{\mathrm{T}}\mathbf{x}', \text{where } \mathbf{Q}^{\mathrm{T}} = \begin{bmatrix} \cos\theta & -\sin\theta \\ \sin\theta & \cos\theta \end{bmatrix}. \tag{3.35}
$$

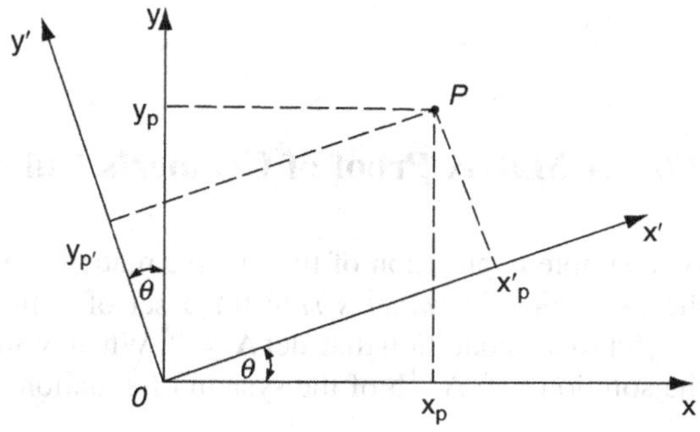

Fig. 3.1 The counterclockwise rotation of the O(x, y)-axes about the origin through an angle θ to become the O(x', y')-axes

Chapter 7 will show that Eq. (3.34) is a typical example of what is called a *linear transformation*. Two-dimensional linear transformations like (3.34) relate vectors in one plane, here the (x, y)-plane, to vectors in another plane, here considered to be the (x', y')-plane. Equation (3.34) is said to *map* a point (x, y) onto the point (x', y'), when the point (x', y') is then called the *image* of point (x, y).

Three important and useful properties of orthogonal matrices follow directly from definition (3.31).

Theorem 3.1 *Properties of Orthogonal Matrices*

(i) If $\mathbf{Q}$ is an orthogonal matrix, then $\det \mathbf{Q} = \pm 1$.
(ii) The columns of an orthogonal matrix $\mathbf{Q}$ are *orthonormal*, meaning that if $\mathbf{q}_i$ and $\mathbf{q}_j$ are any two columns of $\mathbf{Q}$, then

$$\mathbf{q}_i^{\mathrm{T}}\mathbf{q}_j = \begin{cases} 0, & i \neq j, \\ 1, & i = j. \end{cases}$$

(iii) If $\mathbf{Q}_1$ and $\mathbf{Q}_2$ are two $n \times n$ orthogonal matrices, then $\mathbf{Q}_1\mathbf{Q}_2$ is also an orthogonal matrix.

Proof. Property (i) follows from the results $\det \mathbf{Q}^{\mathrm{T}} = \det \mathbf{Q}$ and $\det \mathbf{Q}^{-1} = 1/\det \mathbf{Q}$, because when these results are combined they show that $(\det \mathbf{Q})^2 = 1$, so $\det \mathbf{Q} = \pm 1$. Property (ii) follows directly from the fact that $\mathbf{Q}^{\mathrm{T}}\mathbf{Q} = \mathbf{I}$, because if $\mathbf{q}_i$ and $\mathbf{q}_j$ are any two columns of $\mathbf{Q}$, then

$$\mathbf{q}_i^{\mathrm{T}}\mathbf{q}_j = \begin{cases} 0, & i \neq j, \\ 1, & i = j. \end{cases}$$

Finally, property (iii) follows from the fact that $\mathbf{Q}_1\mathbf{Q}_1^{\mathrm{T}} = 1$ and $\mathbf{Q}_2\mathbf{Q}_2^{\mathrm{T}} = 1$, because

$$(\mathbf{Q}_1\mathbf{Q}_2)^{\mathrm{T}}\mathbf{Q}_1\mathbf{Q}_2 = \mathbf{Q}_2^{\mathrm{T}}\mathbf{Q}_1^{\mathrm{T}}\mathbf{Q}_1\mathbf{Q}_2 = \mathbf{Q}_2^{\mathrm{T}}\mathbf{Q}_2 = \mathbf{I}.$$

◇

3.6 A Matrix Proof of Cramer's Rule

As a simple application of the inverse matrix, we now give the promised proof of the *generalized Cramer's rule* for a set of n linear nonhomogeneous equations, subject to the condition that $\det \mathbf{A} \neq 0$. When written in terms of the adjoint matrix, the solution $\mathbf{x} = \mathbf{A}^{-1}\mathbf{b}$ of the system of equations $\mathbf{A}\mathbf{x} = \mathbf{b}$ becomes

$$\begin{bmatrix} x_1 \\ x_2 \\ \vdots \\ x_n \end{bmatrix} = \frac{1}{\det \mathbf{A}} \begin{bmatrix} C_{11} & C_{21} & \cdots & C_{n1} \\ C_{12} & C_{22} & \cdots & C_{n2} \\ \vdots & \vdots & \vdots & \vdots \\ C_{1n} & C_{2n} & \cdots & C_{nn} \end{bmatrix} \begin{bmatrix} b_1 \\ b_2 \\ \vdots \\ b_n \end{bmatrix}, \tag{3.36}$$

where C_{ij} is the cofactor of the element a_{ij} in $\mathbf{A}$. The jth element on the left is x_j, and the jth element on the right is the sum of the products of the elements in the jth row of the matrix adj $\mathbf{A}$ on the right with the elements in column vector $\mathbf{b}$, so that

$$x_j = (C_{1j}b_1 + C_{2j}b_2 + \cdots + C_{nj})(1/\det \mathbf{A}). \tag{3.37}$$

However, if in matrix $\mathbf{A}$ the jth column is replaced by the elements in $\mathbf{b}$ to form a matrix $\mathbf{A}_j$, and if D_j is the determinant of this modified matrix, when the result is expanded in terms of elements of its jth column it becomes

$$D_j = b_1(\text{cofactor of } b_1) + b_2(\text{cofactor of } b_2) + \cdots + b_n(\text{cofactor of } b_n).$$

Because of the construction of matrix $\mathbf{A}_j$, the cofactor of b_i is simply the cofactor C_{ij} of the original element a_{ij}, so

$$D_j = C_{1j}b_1 + C_{2j}b_2 + \cdots + C_{nj} \text{ for } j = 1, 2, \cdots, n.$$

Thus D_j is the determinant Δ_j obtained from $\mathbf{A}$ by replacing its jth column by the elements of the nonhomogeneous vector $\mathbf{b}$, and we have established the extension of Cramer's rule to a system of n equations showing that

$$x_1 = \frac{\Delta_1}{\Delta},\ x_2 = \frac{\Delta_2}{\Delta}, \cdots,\ x_n = \frac{\Delta_n}{\Delta}, \text{ with } \Delta = \det \mathbf{A} \neq 0. \tag{3.38}$$

Example 3.7. Use Cramer's rule to solve the system of equations

$$x_1 + 2x_2 + x_3 = 8,\quad 2x_1 - x_2 + 2x_3 = 6, -x_1 + 3x_2 - 3x_3 = -4.$$

Solution. Here the matrix $\mathbf{A}$ is

$$\mathbf{A} = \begin{bmatrix} 1 & 2 & 1 \\ 2 & -1 & 2 \\ -1 & 3 & -3 \end{bmatrix} \text{ and } \Delta = \det \mathbf{A} = 10.$$

$$\Delta_1 = \det \begin{bmatrix} 8 & 2 & 1 \\ 6 & -1 & 2 \\ -4 & 3 & -3 \end{bmatrix} = 10,\ \Delta_2 = \det \begin{bmatrix} 1 & 8 & 1 \\ 2 & 6 & 2 \\ -1 & -4 & =3 \end{bmatrix} = 20,\ \Delta_3 = \begin{bmatrix} 1 & 2 & 8 \\ 2 & -1 & 6 \\ -1 & 3 & -4 \end{bmatrix} = 30,$$

so $x_1 = \Delta_1/\Delta = 1,\ x_2 = \Delta_2/\Delta = 2,\ x_3 = \Delta_3/\Delta = 3.$

◇

3.7 Partitioning of Matrices

In certain applications of matrices it is useful to divide a matrix into parts, by drawing dashed horizontal lines between some of its rows and some of its columns, as shown in the next example

Example 3.8. A typical partitioning of a 3×7 matrix $\mathbf{A}$ is

$$\mathbf{A} = \left[\begin{array}{ccc:cc:cc} 1 & 3 & -4 & 0 & 1 & 3 & 2 \\ 1 & -1 & 2 & 1 & 1 & -3 & \\ \hdashline 4 & 0 & 1 & 2 & -2 & 3 & -3 \end{array}\right].$$

◇

In Example 3.8 the dashed the lines divide the matrix into the six sub-matrices and if, for the moment, each sub-matrix is treated as a single entry, matrix $\mathbf{A}$ can be written

$$\mathbf{A} = \begin{bmatrix} \mathbf{A}_{11} & \mathbf{A}_{12} & \mathbf{A}_{13} \\ \mathbf{A}_{21} & \mathbf{A}_{22} & \mathbf{A}_{23} \end{bmatrix},$$

where

$$\mathbf{A}_{11} = \begin{bmatrix} 1 & 3 & -4 \\ 1 & -1 & 2 \end{bmatrix}, \quad \mathbf{A}_{12} = \begin{bmatrix} 0 & 1 \\ 1 & 1 \end{bmatrix}, \quad \mathbf{A}_{13} = \begin{bmatrix} 3 & 2 \\ -3 & 1 \end{bmatrix},$$

$$\mathbf{A}_{21} = [4 \quad 0 \quad 1], \quad \mathbf{A}_{22} = [2 \quad -2] \text{ and } \mathbf{A}_{23} = [3 \quad -3].$$

Each of these sub-matrices is called a *block matrix*, and the process of subdividing $\mathbf{A}$ into block matrices is called *partitioning* matrix $\mathbf{A}$. The numbering of the subscripts used to identify the block matrices is the same as that used to identify individual elements in a matrix, because $\mathbf{A}_{i\,j}$ identifies the block matrix in the ith row and the jth column of matrix $\mathbf{A}$ once it has been partitioned. Remember that, in effect, block matrices were used in (3.2) when defining a matrix product in (3.3).

A typical practical example of the use of partitioned matrices occurs in applications where each block matrix governs the behavior of a specific part of a complicated system described by linear first-order differential equations. In such cases partitioning often makes it easier to identify the contribution to the overall performance of the system that is made by a specific block matrix. A different use of partitioning happens when seeking the numerical solution of partial differential equations by finite difference or finite element methods, which usually produces very large matrices within which many blocks contain only zeros. The effect of partitioning then makes it possible to avoid performing unnecessary calculations on blocks that only contain zeros, since these contribute nothing to the final solution. Partitioning is also used when a matrix is extremely large, as may happen in some linear programming problems where an optimum solution is required involving very many variables with complicated constraint conditions. In such cases, if the calculation is properly organized, matrix operations can be performed more efficiently block by block, instead of at all times working with the entire matrix.

It is evident from the definitions of the linearity, scaling and summation of matrices, that if two $m \times n$ matrices $\mathbf{A}$ and $\mathbf{B}$ are partitioned in similar fashion, the scaling of matrix $\mathbf{A}$ corresponds to the scaling of its block matrices, while $\mathbf{A} \pm \mathbf{B}$, corresponds to the sum or difference of the corresponding block matrices. For example, if

$$\mathbf{A} = \left[\begin{array}{cc|c} 1 & 2 & 4 \\ 0 & 3 & 1 \\ \hline -2 & 2 & 1 \end{array}\right] \text{ then } k\mathbf{A} = \begin{bmatrix} k\mathbf{A}_{11} & k\mathbf{A}_{12} \\ k\mathbf{A}_{21} & k\mathbf{A}_{22} \end{bmatrix}, \text{ where } \mathbf{A}_{11} = \begin{bmatrix} 1 & 2 \\ 0 & 3 \end{bmatrix}, \quad \mathbf{A}_{12} = \begin{bmatrix} 4 \\ 1 \end{bmatrix},$$

$\mathbf{A}_{21} = [-2 \quad 2]$ and $\mathbf{A}_{22} = [1] = 1$.

Similarly, if

$$\mathbf{B} = \left[\begin{array}{cc|c} 1 & 2 & -2 \\ 0 & 3 & 4 \\ \hline 1 & 0 & 2 \end{array}\right] = \begin{bmatrix} \mathbf{B}_{11} & \mathbf{B}_{12} \\ \mathbf{B}_{21} & \mathbf{B}_{22} \end{bmatrix}, \text{ with } \mathbf{B}_{11} = \begin{bmatrix} 1 & 2 \\ 0 & 3 \end{bmatrix}, \ \mathbf{B}_{12} = \begin{bmatrix} -2 \\ 4 \end{bmatrix}, \ \mathbf{B}_{21} = [1 \quad 0]$$

and $\mathbf{B}_{22} = [2] = 2$, it follows that

$$\mathbf{A} \pm \mathbf{B} = \begin{bmatrix} \mathbf{A}_{11} \pm \mathbf{B}_{11} & \mathbf{A}_{12} \pm \mathbf{B}_{12} \\ \mathbf{A}_{21} \pm \mathbf{B}_{21} & \mathbf{A}_{22} \pm \mathbf{B}_{22} \end{bmatrix}.$$

Let the two matrices $\mathbf{A}$ and $\mathbf{B}$ by conformable for multiplication. Then, provided the matrices are partitioned in a suitable fashion, the product of two blocks involves ordinary matrix multiplication, and consequently the product $\mathbf{AB}$ in block matrix form obeys the usual rule for matrix multiplication. These can be described simply as the result of "the product of rows of block matrices with columns of block matrices", where now it is the block matrices that form the rows and columns of the partitioned matrices $\mathbf{A}$ and $\mathbf{B}$. The conditions to be satisfied if this result is to be true are that the partitioning of the matrices $\mathbf{A}$ and $\mathbf{B}$ must be such that all the resulting products of block matrices are defined, and the order in which the block matrices are multiplied is preserved. This last condition is obvious because, in general, matrix products are not commutative.

Example 3.9. Form the matrix product $\mathbf{AB}$, given that $\mathbf{A}$ is the partitioned matrix

$$\mathbf{A} = \left[\begin{array}{ccc|cc|cc} 1 & 3 & -4 & 0 & 1 & 3 & 2 \\ 1 & -1 & 2 & 1 & 1 & -3 & 1 \\ \hline 4 & 0 & 1 & 2 & -2 & 3 & -3 \end{array}\right]$$

used in Example 3.8, and **B** is the partitioned matrix

$$\mathbf{B} = \left[\begin{array}{c|c} 1 & 2 \\ 0 & 1 \\ 2 & 2 \\ \hline 1 & 1 \\ 2 & 1 \\ \hline -1 & 2 \\ 3 & 1 \end{array}\right].$$

Solution. Notice first that **A** is a 3×7 matrix and **B** is a 7×2 matrix, so the matrix product **AB** will be a 3×2 matrix. Let **A** be partitioned as $\mathbf{A} = \begin{bmatrix} \mathbf{A}_{11} & \mathbf{A}_{12} & \mathbf{A}_{13} \\ \mathbf{A}_{21} & \mathbf{A}_{22} & \mathbf{A}_{23} \end{bmatrix}$, where $\mathbf{A}_{11} = \begin{bmatrix} 1 & 3 & -4 \\ 1 & -1 & 2 \end{bmatrix}$, $\mathbf{A}_{12} = \begin{bmatrix} 0 & 1 \\ 1 & 1 \end{bmatrix}$, $\mathbf{A}_{13} = \begin{bmatrix} 3 & 2 \\ -3 & 1 \end{bmatrix}$, and $\mathbf{A}_{21} = [4 \;\; 0 \;\; 1]$, $\mathbf{A}_{22} = [2 \;\; -2]$ and $\mathbf{A}_{23} = [3 \;\; -3]$, and let **B** be partitioned as

$$\mathbf{B} = \left[\begin{array}{c} \mathbf{B}_{11} \\ \hline \mathbf{B}_{21} \\ \hline \mathbf{B}_{31} \end{array}\right], \text{ with } \mathbf{B}_{11} = \begin{bmatrix} 1 & 2 \\ 0 & 1 \\ 2 & 2 \end{bmatrix}, \; \mathbf{B}_{21} = \begin{bmatrix} 1 & 1 \\ 2 & 1 \end{bmatrix} \text{ and } \mathbf{B}_{31} = \begin{bmatrix} -1 & 2 \\ 3 & 1 \end{bmatrix}.$$

This partitioning permits the "product of rows with columns" to proceed in the usual way, because the blocks are compatible for multiplication. The result is the block matrix product

$$\mathbf{AB} = \begin{bmatrix} \mathbf{A}_{11}\mathbf{B}_{11} + \mathbf{A}_{12}\mathbf{B}_{21} + \mathbf{A}_{13}\mathbf{B}_{31} \\ \mathbf{A}_{21}\mathbf{B}_{11} + \mathbf{A}_{22}\mathbf{B}_{21} + \mathbf{A}_{23}\mathbf{B}_{31} \end{bmatrix} = \left[\begin{array}{cc} -2 & 6 \\ 14 & 2 \\ \hline -8 & 13 \end{array}\right].$$

The final result has been partitioned because it shows how the partitioning of **A** and **B** leads to the partitioning of the final matrix product. To see this, notice that the sum of the products in the top row produces a 2×2 matrix, while the sum of the products in the bottom row produces a 1×2 matrix.

◇

A special case of the next example will be needed in Chapter 5 when partitioned matrices are used to reduce a special type of 2×2 matrix with real elements to what is called its *Jordan normal form*.

Example 3.10. Let **R** and **S** be $2n \times 2n$ matrices, each partitioned into the four $n \times n$ block matrices

$$\mathbf{R} = \left[\begin{array}{c|c} \mathbf{P} & -\mathbf{Q} \\ \hline \mathbf{Q} & \mathbf{P} \end{array}\right] \text{ and } \mathbf{S} = \left[\begin{array}{c|c} \mathbf{P} & \mathbf{Q} \\ \hline -\mathbf{Q} & \mathbf{P} \end{array}\right].$$

Find the form of the matrix product **RS** when $\mathbf{P} = \mathbf{A} - \alpha\mathbf{I}_n$, and $\mathbf{Q} = -\beta\mathbf{I}_n$, with α and β real numbers and $\beta > 0$, where **A** is an $n \times n$ block matrices, and $\mathbf{I}_n$ is the $n \times n$ unit matrix. Comment on the relationship between the result of the product **RS** and det **M**, when $\mathbf{M} = \begin{bmatrix} \alpha - \lambda & -\beta \\ \beta & \alpha - \lambda \end{bmatrix}$, with λ a scalar parameter.

Solution.

$$\mathbf{RS} = \left[\begin{array}{c|c} \mathbf{P} & -\mathbf{Q} \\ \hline \mathbf{Q} & \mathbf{P} \end{array}\right]\left[\begin{array}{c|c} \mathbf{P} & \mathbf{Q} \\ \hline -\mathbf{Q} & \mathbf{P} \end{array}\right] = \left[\begin{array}{c|c} \mathbf{P}^2 + \mathbf{Q}^2 & \mathbf{0} \\ \hline \mathbf{0} & \mathbf{P}^2 + \mathbf{Q}^2 \end{array}\right].$$

Setting $\mathbf{P} = \mathbf{A} - \alpha\mathbf{I}_n$ and $\mathbf{Q} = \beta\mathbf{I}_n$ we find that

$$\mathbf{P}^2 + \mathbf{Q}^2 = (\mathbf{A} - \alpha\mathbf{I})^2 + \beta^2\mathbf{I}_n = \mathbf{A}^2 - 2\alpha\mathbf{A} + \left(\alpha^2 + \beta^2\right)\mathbf{I}_n,$$

so

$$\mathbf{RS} = \left[\begin{array}{c|c} \mathbf{A}^2 - 2\alpha\mathbf{A} + \left(\alpha^2 + \beta^2\right)\mathbf{I}_n & \mathbf{0} \\ \hline \mathbf{0} & \mathbf{A}^2 - 2\alpha\mathbf{A} + \left(\alpha^2 + \beta^2\right)\mathbf{I}_n \end{array}\right].$$

Expanding det **M** gives the quadratic expression

$$p(\lambda) = \det\begin{bmatrix} \alpha - \lambda & -\beta \\ \beta & \alpha - \lambda \end{bmatrix} = \lambda^2 - 2\alpha\lambda + \alpha^2 + \beta^2.$$

So if λ is replaced by **A** and$\alpha^2 + \beta^2$ by $\left(\alpha^2 + \beta^2\right)\mathbf{I}_n$, the ordinary quadratic polynomial $p(\lambda)$ becomes the **matrix polynomial** $p(\mathbf{A})$ given by $p(\mathbf{A}) = \mathbf{A}^2$ $-2\alpha\mathbf{A} + \left(\alpha^2 + \beta^2\right)\mathbf{I}_n$, which is seen to occur in each of the nonzero block matrices in the product **RS**. As $p(\lambda) = \det \mathbf{M}$ is the *characteristic polynomial* associated with **M**, the above result shows that matrix **A** satisfies the same polynomial expression $p(\lambda)$ as the scalar parameter λ in **M**.

◇

Finding the inverse of a nonsingular partitioned $n \times n$ matrix **A** in terms of block matrices is lengthy, but a simplification occurs when **A** can be partitioned such that the nonzero blocks themselves form an upper triangular block matrix. Let us consider the case when $\mathbf{A}$ can be partitioned into nine blocks, as follows

$$\mathbf{A} = \left[\begin{array}{c|c|c} \mathbf{A}_{11}^{(p\times p)} & \mathbf{A}_{12}^{(p\times q)} & \mathbf{A}_{13}^{(p\times r)} \\ \hline \mathbf{0} & \mathbf{A}_{22}^{(q\times q)} & \mathbf{A}_{23}^{(q\times r)} \\ \hline \mathbf{0} & \mathbf{0} & \mathbf{A}_{33}^{(r\times r)} \end{array}\right], \tag{3.39}$$

where $\mathbf{A}_{11}^{(p\times p)}$ is a $p \times p$ matrix, $\mathbf{A}_{22}^{(q\times q)}$ is a $q \times q$ matrix, $\mathbf{A}_{33}^{(r\times r)}$ is an $r \times r$ matrix, and $p + q + r = n$. The superscript on each off-diagonal block matrix shows the shape of the block so, for example $(q \times r)$ signifies a block matrix with q rows and r columns.

From this point onward, it will be assumed that matrix $\mathbf{A}$, its inverse $\mathbf{B} = \mathbf{A}^{-1}$, and the $n \times n$ unit matrix $\mathbf{I}_n$ are all partitioned in this manner. To simplify what follows, the superscripts will be omitted, and we will seek a partitioned matrix $\mathbf{B}$ such that $\mathbf{AB} = \mathbf{I}_n$, which is equivalent to the block matrix equation

$$\left[\begin{array}{c|c|c} \mathbf{A}_{11} & \mathbf{A}_{12} & \mathbf{A}_{13} \\ \hline \mathbf{0} & \mathbf{A}_{22} & \mathbf{A}_{23} \\ \hline \mathbf{0} & \mathbf{0} & \mathbf{A}_{33} \end{array}\right] \left[\begin{array}{c|c|c} \mathbf{B}_{11} & \mathbf{B}_{12} & \mathbf{B}_{13} \\ \hline \mathbf{B}_{21} & \mathbf{B}_{22} & \mathbf{B}_{23} \\ \hline \mathbf{B}_{31} & \mathbf{B}_{32} & \mathbf{B}_{33} \end{array}\right] = \left[\begin{array}{c|c|c} \mathbf{I}_p & \mathbf{0} & \mathbf{0} \\ \hline \mathbf{0} & \mathbf{I}_q & \mathbf{0} \\ \hline \mathbf{0} & \mathbf{0} & \mathbf{I}_r \end{array}\right]. \tag{3.40}$$

Here, $\mathbf{I}_p$ is a $p \times p$ element unit matrix, while the unit matrices $\mathbf{I}_q$ and $\mathbf{I}_r$ are, respectively, $q \times q$ and $r \times r$ unit matrices. This product is equivalent to the following nine block matrix equations from which the sub-matrices $\mathbf{B}_{ij}$ must be determined:

$$\begin{aligned} &\mathbf{A}_{11}\mathbf{B}_{11} + \mathbf{A}_{12}\mathbf{B}_{21} + \mathbf{A}_{13}\mathbf{B}_{31} = \mathbf{I}_p, \\ &\mathbf{A}_{11}\mathbf{B}_{12} + \mathbf{A}_{12}\mathbf{B}_{22} + \mathbf{A}_{13}\mathbf{B}_{32} = \mathbf{0}, \\ &\mathbf{A}_{11}\mathbf{B}_{13} + \mathbf{A}_{12}\mathbf{B}_{23} + \mathbf{A}_{13}\mathbf{B}_{33} = \mathbf{0}, \\ &\mathbf{A}_{22}\mathbf{B}_{21} + \mathbf{A}_{23}\mathbf{B}_{31} = \mathbf{0}, \\ &\mathbf{A}_{22}\mathbf{B}_{22} + \mathbf{A}_{23}\mathbf{B}_{32} = \mathbf{I}_p, \\ &\mathbf{A}_{22}\mathbf{B}_{23} + \mathbf{A}_{23}\mathbf{B}_{33} = \mathbf{0}, \\ &\mathbf{A}_{33}\mathbf{B}_{31} = \mathbf{0}, \\ &\mathbf{A}_{33}\mathbf{B}_{32} = \mathbf{0}, \\ &\mathbf{A}_{33}\mathbf{B}_{33} = \mathbf{I}_r. \end{aligned} \tag{3.41}$$

Notice first that matrix $\mathbf{A}$ is assumed to be nonsingular, so the sub-matrices $\mathbf{A}_{11}$, $\mathbf{A}_{22}$ and $\mathbf{A}_{33}$ must all have inverses. These equations can be solved recursively by *back substitution*, starting with the last equation that shows $\mathbf{B}_{33} = \mathbf{A}_{33}^{-1}$, so because $\mathbf{A}_{33}^{-1}$ exists, the next two homogeneous equations show that $\mathbf{B}_{32} = \mathbf{B}_{31} = \mathbf{0}$. Proceeding in this manner all of the sub-matrices $\mathbf{B}_{ij}$ can be found, though as the

$\mathbf{B}_{ij}$ are matrices, it is necessary to preserve the order in which matrix products occur. Determining all of these sub-matrices, and incorporating them into the matrix $\mathbf{B} = \mathbf{A}^{-1}$ leads to the result

$$\mathbf{B} = \mathbf{A}^{-1} = \left[\begin{array}{c|c|c} \mathbf{A}_{11}^{-1} & -\mathbf{A}_{11}^{-1}\mathbf{A}_{12}\mathbf{A}_{22}^{-1} & \mathbf{A}_{11}^{-1}\left[\mathbf{A}_{12}\mathbf{A}_{22}^{-1}\mathbf{A}_{23}\mathbf{A}_{33}^{-1} - \mathbf{A}_{13}\mathbf{A}_{33}^{-1}\right] \\ \hline \mathbf{0} & \mathbf{A}_{22}^{-1} & -\mathbf{A}_{22}^{-1}\mathbf{A}_{23}\mathbf{A}_{33}^{-1} \\ \hline \mathbf{0} & \mathbf{0} & \mathbf{A}_{33}^{-1} \end{array}\right]. \tag{3.42}$$

Example 3.11. Find $\mathbf{A}^{-1}$ given that $\mathbf{A}$ is the partitioned matrix

$$\mathbf{A} = \left[\begin{array}{cc|c|c} 2 & 2 & 0 & -1 \\ 0 & 3 & -1 & -2 \\ \hline 0 & 0 & 1 & -2 \\ \hline 0 & 0 & 0 & 1 \end{array}\right].$$

Solution. The sub-matrices are

$$\mathbf{A}_{11} = \begin{bmatrix} 2 & 2 \\ 0 & 3 \end{bmatrix}, \ \mathbf{A}_{12} = \begin{bmatrix} 0 \\ -1 \end{bmatrix}, \ \mathbf{A}_{13} = \begin{bmatrix} -1 \\ -2 \end{bmatrix}, \ \mathbf{A}_{21} = [0 \ \ 0], \ \mathbf{A}_{22} = [1],$$
$$\mathbf{A}_{23} = [-2], \ \ \mathbf{A}_{31} = [0 \ \ 0], \ \mathbf{A}_{32} = [0] \ \text{and} \ \mathbf{A}_{33} = [1].$$

A routine calculation shows that $\mathbf{A}_{11}^{-1} = \begin{bmatrix} \frac{1}{2} & -\frac{1}{3} \\ 0 & \frac{1}{3} \end{bmatrix}$, and after substituting into result (3.42) it is found that

$$\mathbf{A}^{-1} = \begin{bmatrix} \frac{1}{2} & -\frac{1}{3} & -\frac{1}{3} & -\frac{5}{6} \\ 0 & \frac{1}{3} & \frac{1}{3} & \frac{4}{3} \\ 0 & 0 & 1 & 2 \\ 0 & 0 & 0 & 1 \end{bmatrix}.$$

Routine matrix multiplication confirms this result, because $\mathbf{AA}^{-1} = \mathbf{I}$. ◇

3.8 Matrices and Least-Squares Curve Fitting

A record of experimental or statistical data it is usually in the form of n discrete pairs of measurements $[x_1, y_1], [x_2, y_2], \ldots, [x_n, y_n]$ that show how a quantity y of interest depends on an argument x, where often both the x_i and y_i are subject to experimental error. We will call these pairs of measurements *data points*.

When it is necessary to infer values of y for values of the argument x that lie intermediate between the discrete values $x_1, x_2, \ldots, x_n$, or when the set of data points is to be approximated by a smooth curve, this is most easily accomplished by approximating the discrete observations by a continuous curve $y = f(x)$.

When representing experimental data points by a curve, it is usual to choose a curve in the form of a polynomial of low degree, and to fit it by using the method of *least squares*. If the plot of data points can reasonably be represented by a straight line, the equation $y = a_0 + a_1 x$ can be fitted, but if a plot of the data points appears to be parabolic in shape a quadratic equation of the form $Y = a_0 + a_1 x + a_2 x^2$ can be used. Polynomials of still higher degree can also be fitted, though a cubic is usually the highest degree equation that is used. This is because when a higher-degree polynomial is fitted, the coefficients of the polynomial become very sensitive to the errors in the data points which can lead to a poor approximation.

Because the measurements contain errors of observation, a curve cannot be expected to pass through each data point, so some compromise becomes necessary. The idea underlying the *least-squares approximation* involves choosing the coefficients in the equation to be fitted, like a_0, a_1 and a_2 in a quadratic (parabolic) approximation, in such a way that the sum of the squares S of the differences between the points Y_i on the curve $Y_i = a_0 + a_1 x_i + a_2 x_i^2$ at the points x_i, and the actual measurements y_i at the points x_i is minimized. So the expression S that is to be minimized is given by

$$S = \sum_{i=1}^{n} (Y_i - y_i)^2 = \sum_{i=1}^{n} \left(a_0 + a_1 x_i + a_2 x_i^2 - y_i\right)^2. \tag{3.43}$$

The quantity S is simply the sum of the squares of the vertical distances between Y_i and the actual measurement y_i at each of the n values x_i is minimized. Here S is defined as the sum of the *squares* of these distances, because the quantities $(Y_i - y_i)^2$ take account of the magnitude of the differences between the Y_i and the y_i, without regard to the signs of the differences.

If the equation $Y = a_0 + a_1 x + a_2 x^2$ is to be fitted, the sum S of the squares will be minimized when a_0, a_1 and a_2 are chosen such that $\partial S/\partial a_0 = 0$, $\partial S/\partial a_1 = 0$ and $\partial S/\partial a_2 = 0$. After differentiation with respect to a_0 we find that

$$\begin{aligned}
\frac{\partial S}{\partial a_0} &= 2\sum_{i=1}^{n} \left(a_0 + a_1 x_i + a_2 x_i^2 - y_i\right) \\
&= 2\left(a_0 \sum_{i=1}^{n} 1 + a_1 \sum_{i=1}^{n} x_i + a_2 \sum_{i=1}^{n} x_i^2 - \sum_{i=1}^{n} y_i\right) \\
&= 2\left\{ n a_0 + a_1 \sum_{i=1}^{n} x_i + a_2 \sum_{i=1}^{n} x_i^2 - \sum_{i=1}^{n} y_i \right\}.
\end{aligned}$$

Setting $\partial S/\partial a_0 = 0$, the first equation from which a_0, a_1 and a_2 are to be found becomes

$$na_0 + a_1 \sum_{i=1}^{n} x_i + a_2 \sum_{i=1}^{n} x_i^2 = \sum_{i=1}^{n} y_i.$$

Similar reasoning involving $\partial S/\partial a_1$ and $\partial S/\partial a_2$ yields two further equations, and the system of equations from which a_0, a_1 and a_2 are to be found by least squares becomes

$$\begin{aligned} na_0 + a_1 \sum_{i=1}^{n} x_i + a_2 \sum_{i=1}^{n} x_i^2 &= \sum_{i=1}^{n} y_i, \\ a_0 \sum_{i=1}^{n} x_i + a_1 \sum_{i=1}^{n} x_i^2 + a_2 \sum_{i=1}^{n} x_i^3 &= \sum_{i=1}^{n} x_i y_i, \\ a_0 \sum_{i=1}^{n} x_i^2 + a_1 \sum_{i=1}^{n} x_i^3 + a_2 \sum_{i=1}^{n} x_i^4 &= \sum_{i=1}^{n} x_i^2 y_i. \end{aligned} \tag{3.44}$$

Instead of finding a_0, a_1 and a_2 from these equations, we now show how a matrix argument can generalize these results. This approach has the advantage that the same form of matrix computation will enable a polynomial of *any* degree to be fitted to a set of data points.

Let a quadratic be fitted to the n sets of data points $[x_1, y_1], [x_2, y_2], \ldots, [x_n, y_n]$ using the quadratic approximation

$$Y = a_0 + a_1 x + a_2 x^2. \tag{3.45}$$

Consider for the moment the over-determined system of equations

$$\begin{aligned} a_0 + a_1 x_1 + a_2 x_1^2 &= y_1, \\ a_0 + a_1 x_2 + a_2 x_2^2 &= y_2, \\ a_0 + a_1 x_3 + a_2 x_3^2 &= y_3, \\ \vdots \quad \vdots \quad \vdots \quad &\vdots \\ a_0 + a_1 x_n + a_2 x_n^2 &= y_n, \end{aligned}$$

which can be written in the matrix form

$$\mathbf{Xa} = \mathbf{y}, \text{ where } \mathbf{X} = \begin{bmatrix} 1 & x_1 & x_1^2 \\ 1 & x_2 & x_2^2 \\ \vdots & \vdots & \vdots \\ 1 & x_n & x_n^2 \end{bmatrix}, \ \mathbf{a} = \begin{bmatrix} a_0 \\ a_1 \\ a_2 \end{bmatrix}, \ \mathbf{y} = \begin{bmatrix} y_1 \\ y_2 \\ \vdots \\ y_n \end{bmatrix}. \tag{3.46}$$

Clearly the equation $\mathbf{Xa} = \mathbf{y}$ cannot be solved for $\mathbf{a}$ as it stands, but it can be solved if it is pre-multiplied by a $3 \times n$ matrix $\mathbf{M}$, because then both $\mathbf{MX}$ and $\mathbf{My}$ become 3×3 matrices, and for a suitable matrix $\mathbf{M}$ the matrix $(\mathbf{MX})^{-1}$ will exist, leading to the result $\mathbf{a} = (\mathbf{MX})^{-1}\mathbf{My}$, though vector $\mathbf{a}$ will then depend on the choice of $\mathbf{M}$. To avoid introducing an arbitrary matrix $\mathbf{M}$, let us try setting $\mathbf{M} = \mathbf{X}^{\mathrm{T}}$, when after pre-multiplication by $\mathbf{X}^{\mathrm{T}}$ Eq. (3.46) becomes

$$\mathbf{X}^{\mathrm{T}}\mathbf{Xa} = \mathbf{X}^{\mathrm{T}}\mathbf{y}\,. \tag{3.47}$$

We must now see if this result is in any way relevant to the least-squares curve fitting of a quadratic, and to do this we need to consider the matrix product $\mathbf{X}^{\mathrm{T}}\mathbf{X}$, which becomes

$$\mathbf{X}^{\mathrm{T}}\mathbf{X} = \begin{bmatrix} n & \sum_{i=1}^{n} x_i & \sum_{i=1}^{n} x_i^2 \\ \sum_{i=1}^{n} x_i & \sum_{i=1}^{n} x_i^2 & \sum_{i=1}^{n} x_i^3 \\ \sum_{i=1}^{n} x_i^2 & \sum_{i=1}^{n} x_i^3 & \sum_{i=1}^{n} x_i^4 \end{bmatrix}. \tag{3.48}$$

The matrix product $\mathbf{X}^{\mathrm{T}}\mathbf{Xa}$ is now seen to be the left side of Eq. (3.44), while $\mathbf{X}^{\mathrm{T}}\mathbf{y}$ becomes the right side of the equations. Thus the matrix equation

$$\mathbf{X}^{\mathrm{T}}\mathbf{Xa} = \mathbf{X}^{\mathrm{T}}\mathbf{y} \tag{3.49}$$

is precisely the matrix form of the system of Eq. (3.44) that determine the least-squares values of a_0, a_1, and a_2. So, in terms of matrices, the coefficients a_0, a_1 and a_2 are the elements of a vector $\mathbf{a} = [a_0,\ a_1,\ a_2]^{\mathrm{T}}$ where

$$\mathbf{a} = \left(\mathbf{X}^{\mathrm{T}}\mathbf{X}\right)^{-1}\mathbf{X}^{\mathrm{T}}\mathbf{y}\,. \tag{3.50}$$

If, instead of a parabola a straight line $Y = a_0 + a_1 x$ is to be fitted to the data points by least squares, $\mathbf{X}$ simplifies to the $n \times 2$ matrix

$$\mathbf{X} = \begin{bmatrix} 1 & x_1 \\ 1 & x_2 \\ \vdots & \vdots \\ 1 & x_n \end{bmatrix}, \text{ with } \mathbf{a} = \begin{bmatrix} a_0 \\ a_1 \end{bmatrix}. \tag{3.51}$$

In statistics the fitting of a straight line to a data set by least squares is called *regression*, and the straight line itself is called the *regression line*, and the coefficient a_1 that measures the slope of the regression line is called the *regression coefficient*. In general, when a set of data points $[x_1,\ y_1]$, $[x_2,\ y_2], \ldots,\ [x_n,\ y_n]$ is involved, the regression line is described by saying it is the regression of y on x.

If a cubic $Y = a_0 + a_1x + a_2x^2 + a_3x^3$ is to be fitted to data points by least squares, the previous argument is easily generalized to show that $\mathbf{X}$ becomes the $n \times 4$ matrix

$$\mathbf{X} = \begin{bmatrix} 1 & x_1 & x_1^2 & x_1^3 \\ 1 & x_2 & x_2^2 & x_2^3 \\ \vdots & \vdots & \vdots & \vdots \\ 1 & x_n & x_n^2 & x_n^3 \end{bmatrix}, \text{ with } \mathbf{a} = \begin{bmatrix} a_0 \\ a_1 \\ a_2 \\ a_3 \end{bmatrix}. \tag{3.52}$$

In all cases, the vector $\mathbf{a}$ is given by the matrix expression in (3.50)

$$\mathbf{a} = \left(\mathbf{X}^{\mathrm{T}}\mathbf{X}\right)^{-1}\mathbf{X}^{\mathrm{T}}\mathbf{y},$$

and if a polynomial of degree $m > 3$ is to be fitted, it is only necessary to generalize matrix $\mathbf{X}$ and vector $\mathbf{a}$ in an obvious manner.

Example 3.12. Use the method of least squares to fit the quadratic $Y = a_0 + a_1x + a_2x^2$ to the set of data points $[-2, 3.45]$, $[-1, 1.71]$, $[0, 0.03]$, $[1, -0.29]$, $[2, -0.55]$, $[3, 0.62]$.

Solution.

$$\mathbf{X} = \begin{bmatrix} 1 & -2 & 4 \\ 1 & -1 & 1 \\ 1 & 0 & 0 \\ 1 & 1 & 1 \\ 1 & 2 & 4 \\ 1 & 3 & 9 \end{bmatrix}, \ \mathbf{X}^{\mathrm{T}} = \begin{bmatrix} 1 & 1 & 1 & 1 & 1 & 1 \\ -2 & -1 & 0 & 1 & 2 & 3 \\ 4 & 1 & 0 & 1 & 4 & 9 \end{bmatrix}, \ \mathbf{y} = \begin{bmatrix} 3.45 \\ 1.71 \\ 0.03 \\ -0.29 \\ -0.55 \\ 0.62 \end{bmatrix}, \ \mathbf{a} = \begin{bmatrix} a_0 \\ a_1 \\ a_2 \end{bmatrix}.$$

$$\mathbf{X}^{\mathrm{T}}\mathbf{X} = \begin{bmatrix} 6 & 3 & 19 \\ 3 & 19 & 27 \\ 19 & 27 & 115 \end{bmatrix}, \ \left(\mathbf{X}^{\mathrm{T}}\mathbf{X}\right)^{-1} = \begin{bmatrix} 0.371 & 0.043 & -0.071 \\ 0.043 & 0.084 & -0.027 \\ -0.071 & -0.027 & 0.027 \end{bmatrix},$$

so

$$\mathbf{a} = \left(\mathbf{X}^{\mathrm{T}}\mathbf{X}\right)^{-1}\mathbf{X}^{\mathrm{T}}\mathbf{y} = \begin{bmatrix} 0.169 \\ -0.968 \\ 0.361 \end{bmatrix}.$$

Thus the least-squares quadratic approximation becomes

$$Y = 0.169 - 0.968x + 0.361x^2.$$

A plot of the least-squares quadratic approximation is shown in Fig. 3.2, to which have been added the data points shown as large dots.

◇

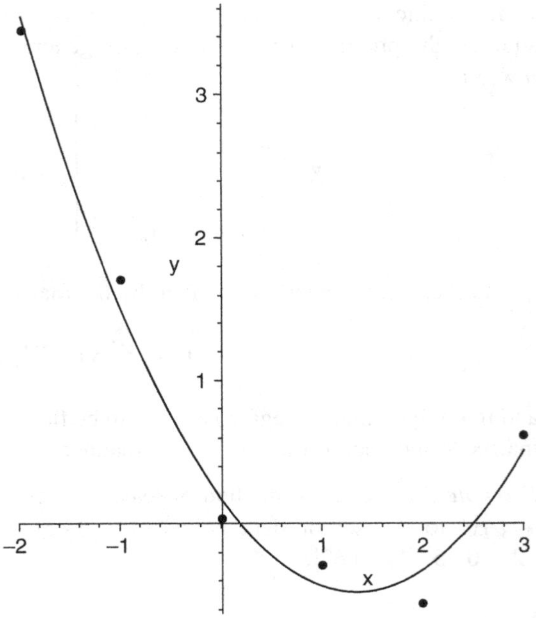

Fig. 3.2 The solid line is the least-squares quadratic approximation, and the data points are shown as dots

3.9 Matrices and the Laplace Equation

Many different types of problem can lead to the generation of very large augmented matrices, and a typical example will be considered here. It will demonstrate how such a matrix can be generated when seeking a numerical solution of a boundary-value problem for the Laplace equation. The augmented matrix produced in this example represents a set of nonhomogeneous simultaneous algebraic equations, whose solution will give numerical approximations for the solution of the Laplace equation at a network of discrete points throughout the region where the Laplace equation is to be solved. However, the augmented matrix produced in this example has been kept sufficiently small for the solution of the equations to be found by elementary means, though the example will nevertheless make perfectly clear how such a problem can give rise to very large augmented matrices that will need sophisticated numerical techniques when seeking a solution.

The *two-dimensional Laplace equation* for a function $u(x, y)$ is the linear second-order partial differential equation

$$\frac{\partial^2 u}{\partial x^2} + \frac{\partial^2 u}{\partial y^2} = 0, \tag{3.53}$$

and a *boundary-value* problem for this equation involves finding its solution in a region of the (x, y)-plane when the value of $u(x, y)$ is specified on the boundary of the region. This is called a *Dirichlet boundary-value problem*, and in physical examples the solution could represent the steady-state temperature distribution in a solid heat conducting material when the temperature is prescribed on its surface, or the *electric potential* in a cavity when the potential is prescribed on the walls of the cavity, though there are many other physical situations that give rise to this equation. The equation is called an *elliptic equation*, though the term *elliptic* is simply a means of classifying the type of partial differential equation to which the Laplace equation belongs, and the name has no geometrical implication for the actual solution.

Before proceeding further, the relationship between this two-dimensional problem and a solution in a three-dimensional world must be made clear. The region in the (x, y)-plane plane where $u(x, y)$ is to be determined should be thought of as a cross-section of a long volume in space with its z-axis perpendicular to the (x, y)-plane, where the cross-section of the volume is the same for all planes $z = \text{constant}$. For convenience, $u(x, y)$ is usually considered to be the solution of the Laplace equation in the plane $z = 0$. The solution of a boundary-value problem for the Laplace equation can be found analytically when the shape of the region and the boundary conditions are simple, though in all other cases it must be found by numerical methods. The numerical method to be outlined here, which is only one of the ways of finding a numerical solution, is called a *finite difference* method, and it determines the approximate solution at the points where two sets of parallel lines intersect, that will be called a *grid points*. To construct the grid of points, one set of lines will be drawn parallel to the x-axis, and the other parallel to the y-axis. In general the separation of the $x = \text{constant}$ lines is h and the separation of the $y = \text{constant}$ lines is k, but for the purpose of this example both separations will be taken equal to h. A typical part of a grid of points is shown in Fig. 3.3.

Let the coordinates of P be (x_i, y_j), where $x_i = ih$ and $y_j = jh$, then the coordinates of Q are (x_{i+1}, y_j) with $x_{i+1} = (i + 1)h$ and $y_j = jh$ while the coordinates of S are (x_{i-1}, y_j),

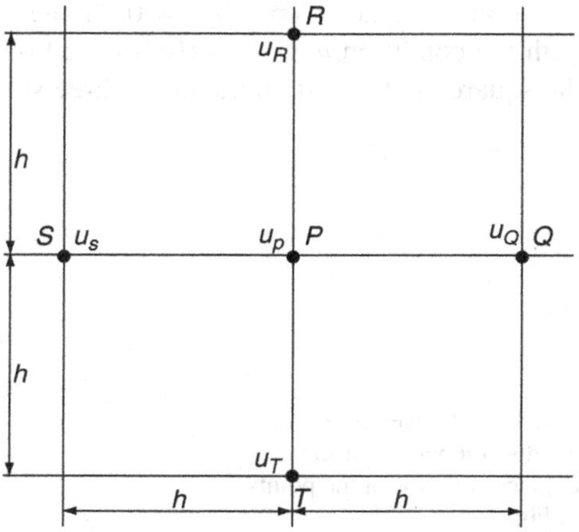

Fig. 3.3 A typical grid of five points with a central point at P and four immediate neighboring points at Q, R, S and T

with $x_{i-1} = (i-1)h$ and $y_j = jh$. The truncated two variable Taylor series expansion of $u(x, y)$ about (x_i, y_j) in the x-direction can be written

$$u_{i+1,j} = u_{i,j} + h(\partial u/\partial x)_{(i,j)} + \left(h^2/2\right)\left(\partial^2 u/\partial x^2\right)_{(i,j)} + \text{ a remainder term.} \quad (3.54)$$

Similarly,

$$u_{i-1,j} = u_{i,j} - h(\partial u/\partial x)_{(i,j)} + \left(h^2/2\right)\left(\partial^2 u/\partial x^2\right)_{(i,j)} + \text{ a remainder term.} \quad (3.55)$$

Referring to the letters in Fig. 3.3, the result of adding (3.54) and (3.55) and ignoring the remainder term enables the result to be given in the abbreviated form

$$2u_P - u_Q - u_S = \left(h^2/2\right)\left(\partial^2 u/\partial x^2\right)_P. \quad (3.56)$$

An application of the Taylor series expansion between the points R and T gives the corresponding result

$$2u_P - u_R - u_s = \left(h^2/2\right)\left(\partial^2 u/\partial y^2\right)_P. \quad (3.57)$$

The addition of (3.56) and (3.57), coupled with the fact that because of the Laplace equation $\left(\partial^2 u/\partial x^2 + \partial^2 u/\partial y^2\right)_P = 0$, gives the *finite difference* approximation for the Laplace equation at point P

$$4u_P - u_Q - u_R - u_s - u_T = 0. \quad (3.58)$$

Thus the sum of discrete solutions of the Laplace equation at the points Q, R, S and T is seen to be four times the solution at P.

The weight to be attributed by result (3.58) to each point in Fig. 3.3 is shown diagrammatically in Fig. 3.4.

Now consider the boundary-value problem illustrated in Fig. 3.5 for the Laplace equation $\partial^2 u/\partial x^2 + \partial^2 u/\partial y^2 = 0$ in the unit square $0 \le x \le 1$ and $0 \le y \le 1$, with the condition $u(x, 1) = 10x^3(1 - x)$ on the top boundary $y = 1$, $0 \le x \le 1$ of the square, and $u = 0$ on the other three sides.

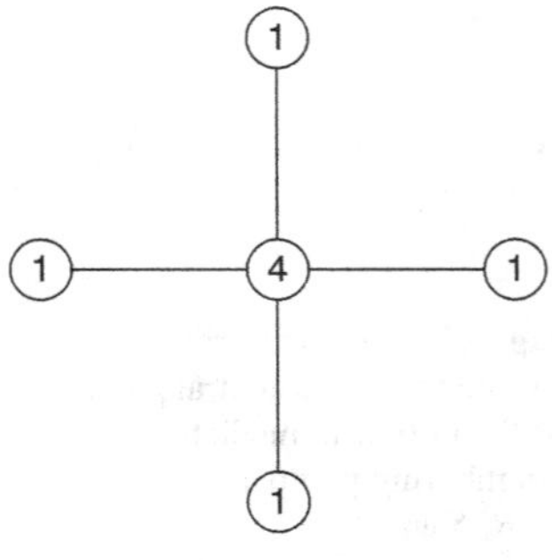

Fig. 3.4 The weighting for the discrete values of the Laplace equation at the points in Fig. 3.3

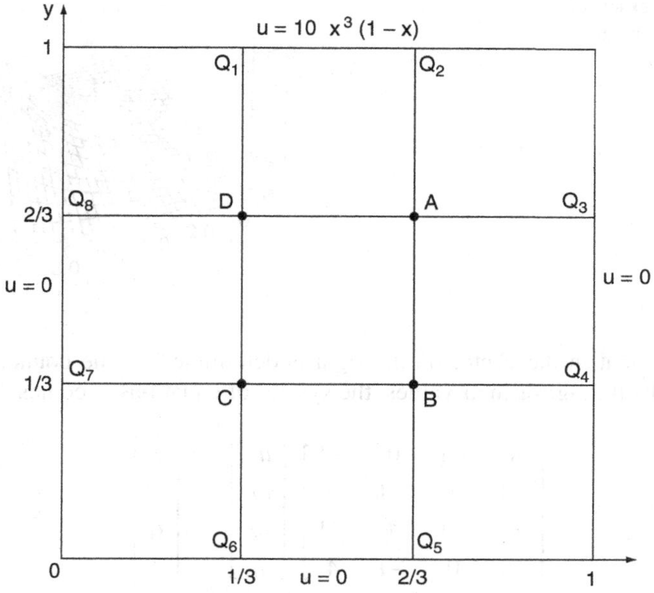

Fig. 3.5 The Dirichlet problem for the Laplace equation in a unit square

The grid points are equally spaced throughout the unit square with $h = \frac{1}{3}$, so there are four internal grid points and twelve grid points on the boundary of the unit square at each of which the value of $u(x, y)$ is determined by the boundary conditions.

Apply the difference equation to each of the internal grid points leads to the four equations

$$\begin{aligned} 4u_A &= u_{Q_2} + u_{Q_3} + u_B + u_D, \\ 4u_B &= u_A + u_{Q_4} + u_{Q_5} + u_C, \\ 4u_C &= u_D + u_B + u_{Q_6} + u_{Q_7}, \\ 4u_D &= u_{Q_1} + u_A + u_C + u_{Q_8}. \end{aligned} \tag{3.59}$$

Notice that due to the weighting shown in Fig. 3.4, the values of the boundary conditions at the corners of the square do not occur in the calculations. We are now in a position to show how a symmetric matrix enters into the calculations, because (3.59) can be written in the matrix form

$$\begin{bmatrix} 4 & -1 & 0 & -1 \\ -1 & 4 & -1 & 0 \\ 0 & -1 & 4 & -1 \\ -1 & 0 & -1 & 4 \end{bmatrix} \begin{bmatrix} u_A \\ u_B \\ u_6 \\ u_D \end{bmatrix} = \begin{bmatrix} u_{Q2} + u_{Q3} \\ u_{Q4} + u_{Q5} \\ u_{Q6} + u_{Q7} \\ u_{Q1} + u_{Q8} \end{bmatrix}. \tag{5.60}$$

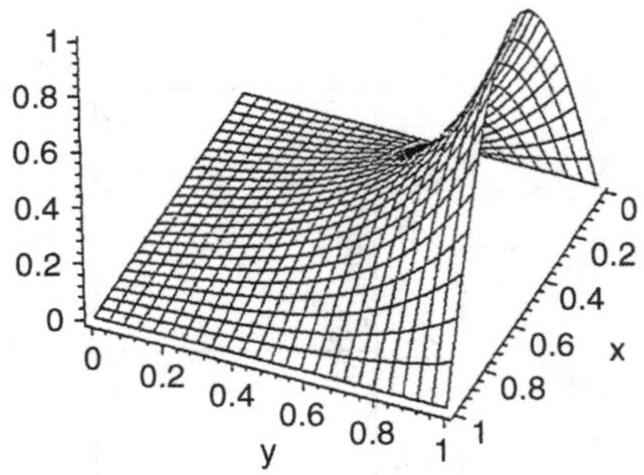

Fig. 3.6 The exact solution of the Laplace boundary-value problem

Each element in the vector on the right is determined by the boundary conditions, so substituting for their values, the system of equations becomes

$$\begin{bmatrix} 4 & -1 & 0 & -1 \\ -1 & 4 & -1 & 0 \\ 0 & -1 & 4 & -1 \\ -1 & 0 & -1 & 4 \end{bmatrix} \begin{bmatrix} u_A \\ u_B \\ u_6 \\ u_D \end{bmatrix} = \begin{bmatrix} \frac{20}{81} \\ \frac{80}{81} \\ 0 \\ 0 \end{bmatrix}.$$

This system is simple enough to solve by elimination and the result is $u_A = 0.1543$, $u_B = 0.3086$, $u_C = 0.0926$ and $u_D = 0.0617$.

These approximate values of the solution should be compared with the exact analytical values

$$u_{A\,(\text{exact})} = 0.1689, u_{B\,(\text{exact})} = 0.2705, u_{C\,(\text{exact})} = 0.0749 \text{ and } u_{D\,(\text{exact})} = 0.0624.$$

For the reference, a plot of the exact solution is shown in Fig. 3.6.

Considering the large value of h that was used, the agreement between the approximate and exact solutions is surprisingly good. To obtain more accurate approximations it will be necessary to use a much smaller value of h, with the result that the number of equations will increase dramatically. If, for example, the value $h = 0.05$ were to be used, the number of internal points would increase from 4 to 324, increasing the number of equations to be solved by a factor 81. A more complicated boundary shape with boundary conditions that change rapidly along each boundary would require an even smaller value of h, leading to an even larger number of equations. When large numbers of equations are involved special numerical techniques becomes necessary when solving them, like an optimized computer form of Gaussian elimination, or an iterative method.

Exercises

1. Find $\mathbf{xy}$ and $\mathbf{yx}$ if $\mathbf{x} = [1, -2, 4, 3]$ and $\mathbf{y} = [2, 4, -3, 1]^{\mathrm{T}}$.
2. Find $\mathbf{xA}$ if $\mathbf{x} = [2, 3, -2, 4]$ and $\mathbf{A} = \begin{bmatrix} 1 & -1 & 2 & 3 \\ 1 & 2 & 3 & 0 \\ 3 & 0 & 1 & -1 \\ 4 & -2 & 3 & 2 \end{bmatrix}$.

3. Find **AB** if

$$\mathbf{A}=\begin{bmatrix}1 & -1 & 2 & 3\\ 1 & 2 & 3 & 0\\ 3 & 0 & 1 & -1\\ 4 & -2 & 3 & 2\end{bmatrix},\quad \mathbf{B}=\begin{bmatrix}1 & -1 & 1\\ 2 & 3 & -1\\ 5 & -1 & 4\\ 1 & 0 & -1\end{bmatrix}.$$

4. Find **AB** if

$$\mathbf{A}=\begin{bmatrix}6 & -2 & 3\\ 4 & -2 & 5\end{bmatrix}\ \text{and}\ \mathbf{B}=\begin{bmatrix}7 & -1\\ 2 & 3\\ 6 & -2\end{bmatrix},$$

and verify that $(\mathbf{AB})^{\mathrm{T}}=\mathbf{B}^{\mathrm{T}}\mathbf{A}^{\mathrm{T}}$.

5. A quadratic form $Q(\mathbf{x})$ in the variables x_1, x_2, x_3, x_4 is defined as $Q(\mathbf{x})=\mathbf{x}^{\mathrm{T}}\tilde{\mathbf{A}}x$, where

$$\mathbf{x}=[x_1,x_2,x_3,x_4]^{\mathrm{T}}\ \text{and}\ \tilde{\mathbf{A}}=\begin{bmatrix}2 & 4 & 3 & 0\\ -6 & 1 & 3 & 7\\ 0 & 2 & 4 & 1\\ 1 & 2 & 1 & -1\end{bmatrix}.$$

Write down $Q(\mathbf{x})$, and express it in the form $Q(\mathbf{x})=\mathbf{x}^{\mathrm{T}}\mathbf{A}\mathbf{x}$, where **A** is a symmetric matrix.

6. Use Definition (3.24) to find $\mathbf{A}^{-1}$ if $\mathbf{A}=\begin{bmatrix}a & b\\ c & d\end{bmatrix}$, stating any condition necessary for $\mathbf{A}^{-1}$ to exist.

7. Find $\mathbf{A}^{-1}$ if

$$\mathbf{A}=\begin{bmatrix}2 & -3 & 1\\ 4 & 3 & -2\\ 1 & 2 & -1\end{bmatrix}.$$

8. Find $\mathbf{A}^{-1}$ if

$$\mathbf{A}=\begin{bmatrix}1 & 4 & 2\\ 2 & 3 & 2\\ 1 & 0 & -1\end{bmatrix}.$$

9. Find $\mathbf{A}^{-1}$ if

$$\mathbf{A}=\begin{bmatrix}1 & 4 & -2\\ 4 & -1 & 3\\ -2 & 3 & 4\end{bmatrix}.$$

10. Verify result (3.30) given that

$$\mathbf{A} = \begin{bmatrix} 1 & 3 & -6 \\ 4 & 1 & 2 \\ 3 & -2 & 1 \end{bmatrix}.$$

11. Use the result $(\mathbf{AB})^{-1} = \mathbf{B}^{-1}\mathbf{A}^{-1}$ to prove that $(\mathbf{A}^{-1})^n = (\mathbf{A}^n)^{-1}$.
12. If $\mathbf{A}$ is a nonsingular matrix, show that $(\mathbf{A}^{-1})^{\mathrm{T}} = (\mathbf{A}^{\mathrm{T}})^{-1}$, and verify the result using a 3×3 nonsingular matrix of your own choice.
13. Use the generalization of Cramer's rule to solve

$$\begin{aligned} &x_1 + 2x_2 - 2x_3 + x_4 = 1, \\ &3x_1 - 3x_2 + x_3 + 2x_4 = 3, \\ &x_1 - x_2 + x_3 + x_4 = 4, \\ &x_1 - 3x_2 + 2x_3 + 4x_4 = 6\ . \end{aligned}$$

14. Use the generalization of Cramer's rule to solve

$$\begin{aligned} &4x_1 + 2x_2 - x_3 + x_4 = 5, \\ &2x_1 + 3x_2 - 2x_3 + 4x_4 = 1, \\ &-4x_1 + x_2 - 5x_3 + 2x_4 = 7, \\ &3x_1 + 2x_2 - x_4 = 6\ . \end{aligned}$$

15. Any matrix derived from an identity (unit) matrix by interchanging two or more of its rows or columns is called a *permutation matrix*. Describe the effect on the matrix $\mathbf{A}$ of forming the matrix products $\mathbf{PA}$ and $\mathbf{AP}$ if

$$\mathbf{P} = \begin{bmatrix} 1 & 0 & 0 & 0 \\ 0 & 0 & 1 & 0 \\ 0 & 1 & 0 & 0 \\ 0 & 0 & 0 & 1 \end{bmatrix} \text{ and } \mathbf{A} = \begin{bmatrix} a & b & c & d \\ e & f & g & h \\ i & j & k & l \\ m & n & o & p \end{bmatrix}.$$

Find matrix $\mathbf{P}$ if the first and last rows of $\mathbf{A}$ are to be interchanged by the product $\mathbf{PA}$, and find matrix $\mathbf{P}$ if the second and fourth columns of $\mathbf{A}$ are to be interchanged by the product $\mathbf{AP}$.
16. If $\mathbf{P}$ is any permutation matrix prove that $\mathbf{PP}^{\mathrm{T}} = \mathbf{P}^{\mathrm{T}}\mathbf{P} = \mathbf{I}$.
17. Write the system

$$\begin{aligned} &2x_1 + x_2 - x_3 + 4x_4 = 9, \\ &x_1 + 6x_2 + 2x_3 - 2x_4 = -3, \\ &7x_1 - 3x_2 + 4x_3 + x_4 = 2, \\ &5x_1 + 2x_2 + 3x_3 - 5x_4 = 6 \end{aligned}$$

in the matrix form $\mathbf{Ax} = \mathbf{b}$. Find the form of the permutation matrix $\mathbf{P}$ such that in the equivalent system $\mathbf{PAx} = \mathbf{Pb}$ the coefficients of x_1 in the first column of $\mathbf{A}$ are arranged in the decreasing order of magnitude 1, 2, 5, 7.

18. If $\mathbf{A}$ is an $n \times n$ matrix and $\mathbf{P}$ is an $n \times n$ permutation matrix, how is det $\mathbf{A}$ related to det ($\mathbf{PA}$) and to det ($\mathbf{AP}$)? If $\mathbf{P}$ is a permutation matrix, give a simple explanation why $\mathbf{P}^{-1} = \mathbf{P}$, and confirm this result by applying it to a permutation matrix of your own construction.
19. Find which of the following matrices is orthogonal:

$$\text{(a)} \begin{bmatrix} \cos\theta & -\sin\theta & 0 \\ \sin\theta & \cos\theta & 0 \\ 0 & 0 & 1 \end{bmatrix} \quad \text{(b)} \begin{bmatrix} \cos\theta & 0 & -\sin\theta \\ 1 & 1 & 0 \\ \sin\theta & 0 & \cos\theta \end{bmatrix} \quad \text{(c)} \begin{bmatrix} \cos\theta & 0 & -\sin\theta \\ 0 & 1 & 0 \\ \sin\theta & 0 & \cos\theta \end{bmatrix}.$$

20. Confirm that

$$\mathbf{Q} = \begin{bmatrix} 0 & \frac{1}{\sqrt{3}} & -\frac{2}{\sqrt{6}} \\ -\frac{1}{\sqrt{2}} & \frac{1}{\sqrt{3}} & \frac{1}{\sqrt{6}} \\ \frac{1}{\sqrt{2}} & \frac{1}{\sqrt{3}} & \frac{1}{\sqrt{6}} \end{bmatrix}$$

is an orthogonal matrix. Permute any two rows or columns of $\mathbf{Q}$ to obtain another matrix $\mathbf{Q}_1$. Show that $\mathbf{Q}_1$ is also orthogonal, and verify Property 3 of orthogonal matrices stated above that $\mathbf{QQ}_1$ is also orthogonal. Explain why permuting rows or columns of an orthogonal matrix yields another orthogonal matrix.

21. Solve the equations (a) using the inverse matrix, and (b) by Cramer's rule.

$$\begin{aligned} x_1 + 3x_2 - x_3 &= -5, \\ 2x_1 - x_2 + x_3 &= 9, \\ -x_1 + x_2 + 2x_3 &= 5. \end{aligned}$$

22. Given that

$$\mathbf{A} = \begin{bmatrix} 1 & 2 \\ 2 & 1 \end{bmatrix},$$

find the eigenvalues λ of $\mathbf{A}$ by solving det$[\mathbf{A} - \lambda\mathbf{I}] = 0$ (see the end of Chapter 2). For each eigenvalue λ_i of $\mathbf{A}$, with $i = 1, 2$, find the column vector $\mathbf{x}^{(i)} = \left[x_1^{(i)}, x_2^{(i)}\right]$ that satisfies the matrix equation $[\mathbf{A} - \lambda_i\mathbf{I}]\mathbf{x}^{(i)} = \mathbf{0}$. The vectors $\mathbf{x}^{(i)}$ for $i = 1, 2$ are called, respectively, the *eigenvectors* of matrix $\mathbf{A}$ associated with the eigenvalues λ_i. (Hint: To find an eigenvector write out in full the system of equations involved, and then solve them by elimination. Don't be surprised to discover that the scaling of the eigenvectors is arbitrary.)

23. Given that

$$\mathbf{A} = \begin{bmatrix} 1 & 3 & 0 \\ 3 & -1 & 0 \\ -2 & -2 & 1 \end{bmatrix},$$

find the eigenvalues λ of $\mathbf{A}$ by solving $\det[\mathbf{A} - \lambda\mathbf{I}] = 0$. For each eigenvalue λ_i of $\mathbf{A}$, with $i = 1, 2, 3$, find the column vector $\mathbf{x}^{(i)} = \left[x_1^{(i)}, x_2^{(i)}, x_3^{(i)}\right]^{\mathrm{T}}$ that satisfies the matrix equation $[\mathbf{A} - \lambda_i\mathbf{I}]\mathbf{x}^{(i)} = \mathbf{0}$. As in Exercise 22, the vectors $\mathbf{x}^{(i)}$ for $i = 1, 2, 3$ are called, respectively, the *eigenvectors* of matrix $\mathbf{A}$ associated with the eigenvalues λ_i. (Hint: Proceed as suggested in Exercise 22.)

24. In Example 3.7, make a different partitioning of the matrices $\mathbf{A}$ and $\mathbf{B}$, and use the result to verify that the product $\mathbf{AB}$ is still given by

$$\mathbf{AB} = \begin{bmatrix} -2 & 6 \\ 14 & 2 \\ -8 & 13 \end{bmatrix}.$$

25. Let the row block matrix $\mathbf{A} = \left[\mathbf{A}_1 \,\vdots\, \quad \vdots\, \mathbf{A}_n\right]$ and the column block matrix $\mathbf{B} = \begin{bmatrix} \mathbf{B}_1 \\ \hdashline \\ \hdashline \mathbf{B}_n \end{bmatrix}$ each be partitioned into n blocks such that the product $\mathbf{AB}$ is defined. What is the form of the block matrix products $\mathbf{AB}$ and $\mathbf{BA}$?

26. If both $\mathbf{A}$ and $\mathbf{B}$ are 2×2 block matrices for which the product $\mathbf{AB}$ is defined, show that $(\mathbf{AB})^{\mathrm{T}} = \mathbf{B}^{\mathrm{T}}\mathbf{A}^{\mathrm{T}}$, where the superscript ^{T} denotes the block matrix transpose operation.

27. Partition a nonsingular $n \times n$ matrix $\mathbf{A}$ of the form

$$\mathbf{A} = \left[\begin{array}{c:c} \mathbf{A}_{11} & \mathbf{A}_{12} \\ \hdashline \mathbf{0} & \mathbf{A}_{22} \end{array}\right]$$

into four blocks, where $\mathbf{A}_{11}$ is a $p \times p$ matrix and $\mathbf{A}_{22}$ is a $q \times q$ matrix, with $p + q = n$. By following the reasoning in the text, show from first principles that the block matrix form of $\mathbf{A}^{-1}$ is

$$\mathbf{A}^{-1} = \left[\begin{array}{c:c} \mathbf{A}_{11}^{-1} & -\mathbf{A}_{11}^{-1}\mathbf{A}_{12}\mathbf{A}_{22}^{-1} \\ \hdashline \mathbf{0} & \mathbf{A}_{22}^{-1} \end{array}\right].$$

How could this result have been deduced from Eq. (3.41)?

Partition matrix $\mathbf{A}$ in Example 3.9 into four 2×2 block matrices, and use the above expression for $\mathbf{A}^{-1}$ to confirm the expression for $\mathbf{A}^{-1}$ found in the example.

28. Let $\mathbf{A}$ and $\mathbf{B}$ be two nonsingular $n \times n$ matrices of the form

$$\mathbf{A} = \left[\begin{array}{c|c} \mathbf{A}_{11} & \mathbf{A}_{12} \\ \hline \mathbf{0} & \mathbf{A}_{22} \end{array}\right] \text{ and } \mathbf{B} = \left[\begin{array}{c|c} \mathbf{B}_{11} & \mathbf{B}_{12} \\ \hline \mathbf{0} & \mathbf{B}_{22} \end{array}\right]$$

that are partitioned in such a way that the product $\mathbf{AB}$ is defined. Show that

$$(\mathbf{AB})^{-1} = \mathbf{B}^{-1}\mathbf{A}^{-1}.$$

29. Given a nonsingular $n \times n$ block matrix $\mathbf{A}$ of the form $\mathbf{A} = \left[\begin{array}{c|c} \mathbf{A} & \mathbf{I} \\ \hline \mathbf{0} & \mathbf{I} \end{array}\right]$, find a block matrix expression for $\mathbf{A}^{-1}$. Apply the result to find $\mathbf{A}^{-1}$ given that

$$\mathbf{A} = \left[\begin{array}{cc|cc} 1 & 0 & 1 & 2 \\ 0 & 1 & -3 & 1 \\ \hline 0 & 0 & 1 & 0 \\ 0 & 0 & 0 & 1 \end{array}\right].$$

Check your result by using ordinary matrix multiplication to confirm that $\mathbf{AA}^{-1} = \mathbf{I}$.

30. An $n \times n$ matrix $\mathbf{A}$ is said to be *idempotent* if $\mathbf{A}^2 = \mathbf{A}$. Obvious examples of $n \times n$ idempotent matrices are the unit matrix $\mathbf{I}$ and the zero matrix $\mathbf{0}$, while a nontrivial example of a 2×2 idempotent matrix is $\mathbf{A} = \begin{bmatrix} 4 & -6 \\ 2 & -3 \end{bmatrix}$.

(a) If $\mathbf{A}$ is idempotent, prove that $\mathbf{A}^k = \mathbf{A}$ for all positive integers $k \geq 2$.
(b) If $\mathbf{A}$ is idempotent, what are the possible values of $\det \mathbf{A}$?
(c) What are the conditions on the elements of an $n \times n$ diagonal matrix $\mathbf{D} = \operatorname{diag}\{\lambda_1, \lambda_2, \ldots, \lambda_n\}$ in order that it is idempotent?
(d) If $\mathbf{A}$ and $\mathbf{B}$ are idempotent, and $\mathbf{AB} = \mathbf{BA} = \mathbf{0}$ show $\mathbf{A} + \mathbf{B}$ is idempotent.
(e) If $\mathbf{A}$ is idempotent, show that $\mathbf{A} - \mathbf{I}$ is idempotent.
(f) If $\mathbf{A}$ is idempotent, show that either $\det \mathbf{A} = 0$ or $\det(\mathbf{A} - \mathbf{I}) = 0$.

31. Let matrix $\mathbf{A} = \begin{bmatrix} a & b \\ c & d \end{bmatrix}$. Find conditions on the elements a and d in terms of the elements b and c, in order that $\mathbf{A}$ is idempotent. Use your result to construct a numerical example and verify that it is idempotent. Does your result determine all possible 2×2 idempotent matrices $\mathbf{A}$, with the exception of the matrices $\mathbf{I}$ and $\mathbf{0}$.

32. The following inequality provides a useful overestimate of the magnitude of a determinant in terms of the inner products of its columns.

The Hadamard overestimate for $|\det \mathbf{A}|$

Let $\mathbf{A} = [\mathbf{a}_1, \mathbf{a}_2, \ldots, \mathbf{a}_n]$ be an arbitrary $n \times n$ matrix with columns $\mathbf{a}_1, \mathbf{a}_2, \ldots, \mathbf{a}_n$.

Then

$$|\det \mathbf{A}|^2 \le (\mathbf{a}_1^{\mathrm{T}}\mathbf{a}_1)(\mathbf{a}_2^{\mathrm{T}}\mathbf{a}_2)\cdots(\mathbf{a}_n^{\mathrm{T}}\mathbf{a}_n).$$

The equality sign holds only if **A** has an inverse (it is *invertible*) and the columns of **A** are orthogonal.

$$\text{Given } \mathbf{A} = \begin{bmatrix} 1 & 2 & -1 & 4 \\ 2 & -1 & 0 & 2 \\ 1 & 3 & 1 & -1 \\ 2 & 3 & -2 & -1 \end{bmatrix},$$

find det **A** and use the result to verify the Hadamard overestimate of det **A**.

Notice that the expression $\Pi_{j=1}^{n}\, \mathbf{a}_j^{\mathrm{T}}\mathbf{a}_j$ is also used for the continued product $(\mathbf{a}_1^{\mathrm{T}}\mathbf{a}_1)(\mathbf{a}_2^{\mathrm{T}}\mathbf{a}_2)\cdots(\mathbf{a}_n^{\mathrm{T}}\mathbf{a}_n)$, so the Hadamard inequality can be written more concisely as $|\det \mathbf{A}|^2 \le \Pi_{j=1}^{n}\, \mathbf{a}_j^{\mathrm{T}}\mathbf{a}_j$.

In Exercises 33 and 34, use the method of least squares to fit a straight line to the given data sets. In each case graph the straight line approximation and superimpose the data points to show how the straight line has approximated the spread of data points.

33. The data set is [0, −0.8], [1, 0.3], [2, 0.3], [3, 1.3], [4, 1.7].
34. The data set is [−2, 1.93], [−1, 1.63], [0, 0.75], [1, 0.71], [2, 0.47], [3, −0.27].

Chapter 4
Systems of Linear Algebraic Equations

4.1 The Augmented Matrix and Elementary Row Operations

The solution of a system of n first-order linear algebraic equations with constant coefficients requires knowledge of certain properties of an $n \times n$ coefficient matrix and the nonhomogeneous matrix vector **b** belonging to the system. So the main purpose of this chapter is to provide an introduction to the solution of systems of m nonhomogeneous linear algebraic equations in the n unknown real variables $x_1, x_2, \ldots, x_n$. Associated with this is the solution of a special type of homogeneous algebraic problem involving n homogeneous linear algebraic equations in n unknowns and a parameter λ, that leads to the study of the *eigenvalues* and *eigenvectors* of an $n \times n$ matrix. It will be recalled that an eigenvalue was introduced briefly at the end of Chapter 2, and encountered again in Exercises 22 and 23 at the end of Chapter 3. The formal definition of the eigenvalues and the associated eigenvectors of square matrices will be given in this chapter, though the properties and use of eigenvectors will be studied in greater detail in Chapter 5.

Consider the system of linear algebraic equations

$$\begin{aligned}
a_{11}x_1 + a_{12}x_2 + \cdots + a_{1n}x_n &= b_1, \\
a_{21}x_1 + a_{22}x_2 + \cdots + a_{2n}x_n &= b_2, \\
\cdots\cdots\cdots \\
a_{m1}x_1 + a_{m2}x_2 + \cdots + a_{mn}x_m &= b_m,
\end{aligned} \tag{4.1}$$

where the a_{ij} and b_i are real constants. It will be recalled from Chapter 1 that system (4.1) is said to be *underdetermined* when $m < n$, *properly determined* when $m = n$, and *overdetermined* when $m > n$. The method that will be used to find a solution set for system (4.1) is called *Gaussian elimination*. As well as showing when a solution exists, and enabling it to be found in a computationally efficient manner, the method also shows when the equations are inconsistent, and so have no solution set.

A. Jeffrey, *Matrix Operations for Engineers and Scientists*,
DOI 10.1007/978-90-481-9274-8_4, © Springer Science+Business Media B.V. 2010

To develop the Gaussian elimination method, it is convenient to represent system (4.1) in terms of what is called the *augmented* matrix **A|b**, comprising the coefficient matrix $\mathbf{A} = [a_{ij}]$, to which is adjoined on the right the nonhomogeneous vector $\mathbf{b} = [b_1, b_2, \ldots, b_n]^{\mathrm{T}}$, so that

$$\mathbf{A}|\mathbf{b} = \begin{bmatrix} a_{11} & a_{12} & a_{13} & \cdots & a_{1n} & b_1 \\ a_{21} & a_{22} & a_{23} & \cdots & a_{2n} & b_2 \\ a_{31} & a_{32} & a_{33} & \cdots & a_{3n} & b_3 \\ \vdots & \vdots & \vdots & \vdots & \vdots & \vdots \\ a_{m1} & a_{m2} & a_{m3} & \cdots & a_{mn} & b_m \end{bmatrix}. \tag{4.2}$$

This matrix contains all of the information in (4.1), because in the *i*th row of **A|b**, for $i = 1, 2, \ldots, m$, the element a_{ij} is associated with the variable x_j, while b_i is the corresponding nonhomogeneous term on the right of (4.1). When **A|b** is interpreted as the system of equations in (4.1), it *implies* the presence of the unknowns x_1, $x_2, \ldots, x_n$, and an equality sign between the terms on the left represented by **A**, and the nonhomogeneous terms on the right represented by **b**. So the augmented matrix is a representation of $\mathbf{Ax} = \mathbf{b}$, without explicitly showing the variables $x_1, x_2, \ldots, x_n$.

The idea underlying Gaussian elimination is simple, and it depends for its success on the following obvious facts.

1. The order in which the equations appear in (4.1) can be changed without altering the solution set.
2. Individual equations can be multiplied throughout by a constant without altering the solution set.
3. Multiples of equations in (4.1) can be added to or subtracted from other equations in (4.1) without altering the solution set.

When working with the augmented matrix **A|b**, which is equivalent to the original set of Eq. (4.1), performing these operations on the original system of equations in (4.1) corresponds to performing what are called *elementary row operations* on the augmented matrix to produce a modified, but equivalent, augmented matrix. The elementary row operations that can be performed on an augmented matrix derived from Eq. (4.1) and operations 1–3 above are as follows.

Elementary row operations on a matrix

1. Interchanging rows.
2. Multiplying each element in a row by a constant k.
3. Adding a multiple of a row to another row, or subtracting a multiple of a row from another row.

The effect of performing these elementary row operations on an augmented matrix **A|b** is to produce a modified augmented matrix that is equivalent in all respects to the original system of equations in (4.1).

The approach starts by assuming that in (4.2) the coefficient $a_{11} \neq 0$. This is no limitation, because if this is not the case the order of the equations can be changed to bring into the first row of (4.1) an equation for which this condition is true. The method then proceeds by subtracting multiples of row 1 of (4.2) from each of the $m-1$ rows below it in such a way that the coefficient of the variable x_1 is made to vanish from each of the subsequent $m-1$ equations. Thus a_{21}/a_{11} times row 1 is subtracted from row 2, a_{31}/a_{11} times row 1 is subtracted from row 3 and so on, until finally a_{m1}/a_{11} times row 1 is subtracted from row m, leading to a modified augmented matrix $\mathbf{A}|\mathbf{b}^{(1)}$ of the form

$$\mathbf{A}|\mathbf{b}^{(1)} = \begin{bmatrix} a_{11} & a_{12} & a_{13} & \cdots & a_{1n} & b_1 \\ 0 & a_{22}^{(1)} & a_{23}^{(1)} & \cdots & a_{2n}^{(1)} & b_2^{(1)} \\ 0 & a_{32}^{(1)} & a_{33}^{(1)} & \cdots & a_{3n}^{(1)} & b_3^{(1)} \\ \vdots & \vdots & \vdots & \vdots & \vdots & \vdots \\ 0 & a_{m2}^{(1)} & a_{m3}^{(1)} & \cdots & a_{mn}^{(1)} & b_m^{(1)} \end{bmatrix}, \tag{4.3}$$

where the superscript (1) indicates an element that has been modified.

This same process is now repeated starting with row 2 of $\mathbf{A}|\mathbf{b}^{(1)}$. Now, row 2, with its first nonzero element $a_{22}^{(1)}$, is used to reduced to zero all elements in the column below it, leading to a modification of $\mathbf{A}|\mathbf{b}^{(1)}$denoted by $\mathbf{A}|\mathbf{b}^{(2)}$, that typically is of the form

$$\mathbf{A}|\mathbf{b}^{(2)} = \begin{bmatrix} a_{11} & a_{12} & a_{13} & \cdots & a_{1n} & b_1 \\ 0 & a_{22}^{(1)} & a_{23}^{(1)} & \cdots & a_{2n}^{(1)} & b_2^{(1)} \\ 0 & 0 & a_{33}^{(2)} & \cdots & a_{3n}^{(2)} & b_3^{(2)} \\ \vdots & \vdots & \vdots & \vdots & \vdots & \vdots \\ 0 & 0 & 0 & \cdots & a_{mn}^{(2)} & b_m^{(2)} \end{bmatrix}, \tag{4.4}$$

where the superscript (2) indicates a modification of an entry with a superscript (1).

This process will lead to a simplification of the original system of equations, though the pattern of zeros will depend on the values of m and n. This method is illustrated below using examples involving different values of m and n. The numbers $a_{11}, a_{22}^{(1)}, a_{33}^{(2)}, a_{44}^{(3)}, \ldots$ used to reduce to zero the entries in the columns below them are called the *pivots* for the Gaussian elimination process. If it happens that at some intermediate stage a pivot becomes zero, and so cannot be used to reduce to zero all entries in the column below it, the difficulty is overcome by interchanging the row with the zero pivot with a row below it in which the corresponding entry is nonzero, after which the process continues as before. This amounts to changing the order of the equations in system (4.1), and so does not influence the solution set. The reduction terminates if at some stage a complete row of zeros is produced, indicating that the corresponding equation is a linear combination of the ones above it.

The pattern of entries attained by Gaussian elimination in the final modification of a matrix is said to be the *echelon form* of the matrix. The formal definition of an *echelon form* is given below.

4.2 The Echelon and Reduced Echelon Forms of a Matrix

A matrix **A** is said to be in *echelon form*, denoted by $\mathbf{A_E}$, if

1. All rows of **A** containing nonzero elements lie above any rows that contain only zero elements.
2. The first nonzero entry in a row of **A**, called the *leading entry* in the row, lies in a column to the right of the leading entry in the row above.

Notice that condition 2 implies that all entries in the column below a leading entry are zero.

A typical pattern of entries in the echelon form of a matrix **A** generated by the application of Gaussian elimination to a 6×8 matrix is shown below, where the symbol $\bullet$ represents a leading entry that is always nonzero, while and the symbol $\square$ represents an entry that may, or may not, be nonzero.

$$\mathbf{A_E} = \begin{bmatrix} \bullet & \square & \square & \square & \square & \square & \square & \square \\ 0 & \bullet & \square & \square & \square & \square & \square & \square \\ 0 & 0 & \bullet & \square & \square & \square & \square & \square \\ 0 & 0 & 0 & \bullet & \square & \square & \square & \square \\ 0 & 0 & 0 & 0 & \bullet & \square & \square & \square \\ 0 & 0 & 0 & 0 & 0 & 0 & 0 & 0 \end{bmatrix}.$$

If this matrix represents the transformation of a nonhomogeneous system of six equations in the seven variables $x_1, x_2, \ldots, x_7$ to its echelon form (remember that the eighth column represents the transformed nonhomogeneous terms), then the row of zeros tells us that the sixth equation is linearly dependent on the five previous equations, and so can be discarded (ignored). Furthermore, the fifth row represents an equation relating x_5, x_6, x_7 and the modification of the nonhomogeneous term b_5, so that x_5 can only be found if x_6 and x_7 are assigned arbitrary values.

A matrix **A** is said to be in *reduced echelon form*, denoted by $\mathbf{A_{ER}}$, if the value of every pivot in $\mathbf{A_{ER}}$ is 1. This reduction is obtained if, after the *echelon form* $\mathbf{A_E}$ has been obtained, each element in a row of $\mathbf{A_E}$ is divided by the value of the pivot that belongs to the row. Clearly, when a nonsingular square matrix **A** is involved, its *reduced echelon form* $\mathbf{A_{ER}}$ will have 1s on its leading diagonal.

For example, if the echelon form $\mathbf{A_E}$ of a 3×4 matrix **A** is

$$\mathbf{A_E} = \begin{bmatrix} 3 & 1 & 0 & 2 \\ 0 & 2 & 3 & 4 \\ 0 & 0 & -1 & 5 \end{bmatrix} \text{ then its } \textit{reduced echelon form } \mathbf{A_{ER}} = \begin{bmatrix} 1 & \frac{1}{3} & 0 & \frac{2}{3} \\ 0 & 1 & \frac{3}{2} & 2 \\ 0 & 0 & 1 & -5 \end{bmatrix}.$$

Specific examples of the echelon forms generated by Gaussian elimination applied to systems of equations now follow, together with their associated solution sets.

Example 4.1. Use Gaussian elimination, in the form of elementary row operations applied to the augmented matrix, to solve the system of equations

$$\begin{aligned} x_1 - 2x_2 + 4x_3 + x_4 &= 4, \\ 2x_1 + x_2 + 2x_3 + 4x_4 &= 0, \\ -x_1 + 4x_2 + 2x_3 + 2x_4 &= 1. \end{aligned}$$

Solution. In this case $m = 3$ and $n = 4$, so the system is *underdetermined*. The application of elementary row operations transforms the augmented matrix as follows:

$$\mathbf{A}|\mathbf{b} = \begin{bmatrix} 1 & -2 & 4 & 1 & 4 \\ 2 & 1 & 2 & 4 & 0 \\ -1 & 4 & 2 & 2 & 1 \end{bmatrix} \quad \begin{array}{c}\text{subtracting } 2 \times \text{row 1 from row2} \\ \text{and adding row 1 to row 3}\end{array}$$

$$\rightarrow \mathbf{A}|\mathbf{b}^{(1)} = \begin{bmatrix} 1 & -2 & 4 & 1 & 4 \\ 0 & 5 & -6 & 2 & -8 \\ 0 & 2 & 6 & 3 & 5 \end{bmatrix},$$

$$\text{subtracting } 2/5 \times \text{row 2 from row 3} \rightarrow \mathbf{A}|\mathbf{b}^{(2)} = \begin{bmatrix} 1 & -2 & 5 & 1 & 4 \\ 0 & 5 & -6 & 2 & -8 \\ 0 & 0 & \frac{42}{5} & \frac{11}{5} & \frac{41}{5} \end{bmatrix},$$

$$5 \times \text{row } 3 \rightarrow \mathbf{A}|\mathbf{b}^{(3)} = \begin{bmatrix} 1 & -2 & 4 & 1 & 4 \\ 0 & 5 & -6 & 2 & -8 \\ 0 & 0 & 42 & 11 & 41 \end{bmatrix}.$$

The last operation involving the multiplication of row 3 by the factor 5 was not strictly necessary, but it was included because the determination of x_1, x_2, x_3 and x_4 is simplified if fractions are cleared after performing an elementary row operation on the augmented matrix.

The reduction can proceed no further, so $\mathbf{A}|\mathbf{b}^{(3)}$ is the echelon form of $\mathbf{A}|\mathbf{b}$. Setting $x_4 = k$, an arbitrary parameter, the third row of $\mathbf{A}|\mathbf{b}^{(3)}$ is seen to be equivalent to $42x_3 + 11\,k = 41$, so $x_3 = \frac{41}{42} - \frac{11}{42}\,k$. The second row is equivalent to $5x_2 - 6x_3 + 2k = -8$, so substituting for x_3 gives $x_2 = -\frac{3}{7} - \frac{5}{7}k$. Finally, the first row is equivalent to $x_1 - 2x_2 + 4x_3 + k = 4$, so substituting for x_3 and x_2 gives $x_1 = -\frac{16}{21} - \frac{29}{21}k$. Thus we have found a one parameter solution set $\{x_1, x_2, x_3, x_4\}$ for the original set of equations with its elements given by $x_1 = -\frac{16}{21} - \frac{29}{21}\,k$, $x_2 = -\frac{3}{7} - \frac{5}{7}\,k$, $x_3 = \frac{41}{42} - \frac{11}{42}\,k$, $x_4 = k$, with k an arbitrary parameter.

The process of first finding x_3, then using it to find x_2, and finally using x_3 and x_2 to find x_1 is the *back substitution* procedure mentioned previously. Modifications of this method designed to maintain the highest accuracy are made in computer routines that use, the Gaussian elimination process to solve systems of linear algebraic equations. ◇

A typical modification of the Gaussian elimination process used in computer routines involves changing the order of the equations at each stage of the process, so the absolute value of the pivot to be used has the largest of the absolute values of the coefficients in the column that contains it. This has the effect that at no stage is a pivot with a small absolute value used to reduce to zero a coefficient below it with a much larger absolute value, thereby reducing the buildup of round-off errors that would otherwise accumulate as the computation proceeds.

4.3 The Row Rank of a Matrix

It is now necessary to introduce a new definition that describes an important property of a matrix. The *row rank* of a matrix **M** is defined as the number of linearly independent rows in the matrix, denoted by row rank(**M**). Thus, if matrix **A** is the coefficient matrix of a homogeneous set of linear algebraic equations, row rank(**A**) represents the number of linearly independent equations in the system. The augmented matrix **A**|**b** represents a combination of two matrices, namely the matrix of coefficients **A** and the matrix **A**|**b** which also describes the nonhomogeneous system with vector **b**, and it is not necessarily the case that the row ranks of **A** and **A**|**b** are equal. The implications of the row ranks of **A** and **A**|**b** will become clear from the following examples.

In Example 4.1 it can be seen from the reduction to the echelon form $\mathbf{A}|\mathbf{b}^{(3)}$ that row rank(**A**) = row rank(**A**|**b**) = 3, because both the matrix **A** represented by its first three columns, and the matrix $\mathbf{A}|\mathbf{b}^{(3)}$ itself, each have three nonzero rows. We have seen that a solution set could be found for this example, but as there were only three linearly independent equations and four unknowns, it was only possible for three of the unknowns to be found in terms of the fourth unknown, the value of which was assigned as an arbitrary parameter.

Example 4.2. Use Gaussian elimination, in the form of elementary row operations applied to the augmented matrix, to solve the system of equations

$$\begin{aligned}
2x_1 + x_3 + 2x_4 &= 1,\\
x_1 + x_3 &= 2,\\
-2x_1 + x_2 - x_3 + 2x_4 &= 1,\\
x_1 + 2x_2 - 2x_3 - x_4 &= 1\,.
\end{aligned}$$

Solution. In this nonhomogeneous system $m = n = 4$, so the system is *properly determined*, and provided there is no linear dependence between equations a unique solution can be expected. The augmented matrix is

$$\mathbf{A}|\mathbf{b} = \begin{bmatrix} 2 & 0 & 1 & 2 & 1 \\ 1 & 0 & 1 & 0 & 2 \\ -2 & 1 & -1 & 2 & 1 \\ 1 & 2 & -2 & -1 & 1 \end{bmatrix}.$$

After performing elementary row operations on the augmented matrix, where now we use the symbol $\sim$ in place of $\rightarrow$ to denote "is equivalent to", the matrix is reduced to the echelon form

$$\begin{bmatrix} 2 & 0 & 1 & 2 & 1 \\ 0 & 2 & 0 & 8 & 4 \\ 0 & 0 & 1 & -2 & 3 \\ 0 & 0 & 0 & -15 & 4 \end{bmatrix}.$$

Inspection shows that row rank($\mathbf{A}$) = row rank($\mathbf{A}|\mathbf{b}$) = 4, so the equations are consistent and a unique solution exists. The last row of the echelon form corresponds to the equation $-15x_4 = 4$, so $x_4 = -\frac{4}{15}$. Proceeding with back substitution we arrive at the unique solution set $\{x_1, x_2, x_3, x_4\}$ where the elements are $x_1 = -\frac{7}{15}$, $x_2 = \frac{46}{15}$, $x_3 = \frac{37}{15}$, $x_4 = -\frac{4}{15}$.

◇

Example 4.3. Use Gaussian elimination, in the form of elementary row operations applied to the augmented matrix, to solve the system of equations

$$\begin{aligned} &x_1 - x_2 + x_3 + 2x_4 = 1, \\ &-x_1 + 2x_2 + x_3 - x_4 = 0, \\ &2x_1 - 2x_2 - x_3 + 2x_4 = 1, \\ &-2x_1 + 4x_2 + 2x_3 - 2x_4 = 0, \\ &4x_1 - 4x_2 + x_3 + 6x_4 = 3. \end{aligned}$$

Solution. In this nonhomogeneous system $m = 5$ and $n = 4$, so the system is *overdetermined*. Consequently, as there are more equations (constraints on the unknowns) than there are unknowns, no solution can exist unless there is linear dependence between the equations. The augmented matrix is

$$\mathbf{A}|\mathbf{b} = \begin{bmatrix} 1 & -1 & 1 & 2 & 1 \\ -1 & 2 & 1 & -1 & 0 \\ 2 & -2 & -1 & 2 & 1 \\ -2 & 4 & 2 & -2 & 0 \\ 4 & -4 & 1 & 6 & 3 \end{bmatrix}.$$

After the use of elementary row operations this reduces to the echelon form

$$\mathbf{A}|\mathbf{b} \sim \begin{bmatrix} 1 & -1 & 1 & 2 & 1 \\ 0 & 1 & 2 & 1 & 1 \\ 0 & 0 & -3 & -2 & -1 \\ 0 & 0 & 0 & 0 & 0 \\ 0 & 0 & 0 & 0 & 0 \end{bmatrix}.$$

Inspection shows that row rank(**A**) = row rank(**A**|**b**) = 3, so here also the equations are consistent so a solution is possible. However, as in Example 4.1, there are only three linearly independent equations imposing constraints on the four unknowns x_1, x_2, x_3 and x_4. So if we allow x_4, say, to be arbitrary and set $x_4 = k$, we can solve for x_1, x_2 and x_3 in terms of $x_4 = k$. Using back substitution the solution set $\{x_1,\ x_2,\ x_3,\ x_4\}$ is found have the elements $x_1 = 1 - k$, $x_2 = \frac{1}{3} + \frac{1}{3}k$, $x_3 = \frac{1}{3} - \frac{2}{3}k$, $x_4 = k$, with k an arbitrary parameter. So, in this case, only three of the five equations were linearly independent, with the solution set being determined in terms of the arbitrarily assigned parameter $x_4 = k$.

◇

Example 4.4. Use Gaussian elimination, in the form of elementary row operations applied to the augmented matrix, to solve the system of equations

$$\begin{aligned} &x_1 - x_2 + x_3 + 2x_4 = 1, \\ &-x_1 + 2x_2 + x_3 - x_4 = 0, \\ &2x_1 - 2x_2 - x_3 + 2x_4 = 1, \\ &x_1 + x_2 + x_3 - x_4 = 2, \\ &4x_1 - 4x_2 + x_3 + 6x_4 = 3\ . \end{aligned}$$

Solution. In this nonhomogeneous system $m = 5$ and $n = 4$, so the system is *overdetermined.* Unless there is linear dependence between the equations, the constraints imposed by the five equations on the four unknowns will make a solution impossible. The augmented matrix is

$$\mathbf{A}|\mathbf{b} = \begin{bmatrix} 1 & -1 & 1 & 2 & 1 \\ -1 & 2 & 1 & -1 & 0 \\ 2 & -2 & -1 & 2 & 1 \\ 1 & 1 & 1 & -1 & 2 \\ 4 & -4 & 1 & 6 & 3 \end{bmatrix}.$$

After the use of elementary row operations this reduces to the echelon form

$$\mathbf{A}|\mathbf{b} \sim \begin{bmatrix} 1 & -1 & 1 & 2 & 1 \\ 0 & 1 & 2 & 1 & 1 \\ 0 & 0 & -3 & -2 & -1 \\ 0 & 0 & 0 & 7 & -1 \\ 0 & 0 & 0 & 0 & 0 \end{bmatrix}.$$

Inspection shows that row rank(**A**) = row rank(**A**|**b**) = 4, so once again the equations are consistent, and the final row of zeros indicates that the fifth equation is expressible as a linear combination of the other four equations, and so may be disregarded since it is redundant, though the nature of the linear dependence is immaterial. Back substitution shows the system has the unique solution set $\{x_1, x_2, x_3, x_4\}$ with its elements given by $x_1 = \frac{4}{7}$, $x_2 = \frac{2}{7}$, $x_3 = \frac{3}{7}$, $x_4 = -\frac{1}{7}$. ◇

Example 4.5. Use Gaussian elimination, in the form of elementary row operations applied to the augmented matrix, to solve the system of equations

$$\begin{aligned} x_1 - x_2 + x_3 + 2x_4 &= 1, \\ -x_1 + 2x_2 + x_3 - x_4 &= 0, \\ 2x_1 - x_2 - x_3 + 2x_4 &= 1, \\ 11x_1 + x_2 + x_3 - x_4 &= 2, \\ 3x_1 + x_2 + 4x_3 + 5x_4 &= 2\,. \end{aligned}$$

Solution. In this nonhomogeneous system again $m = 5$ and $n = 4$, so the system is *overdetermined*. So, unless there is linear dependence between the equations, the constraints imposed by the five equations on four unknowns will make a solution impossible. The augmented matrix is

$$\mathbf{A}|\mathbf{b} = \begin{bmatrix} 1 & -1 & 1 & 2 & 1 \\ -1 & 2 & 1 & -1 & 0 \\ 2 & -1 & -1 & 2 & 1 \\ 11 & 1 & 1 & -1 & 2 \\ 3 & 1 & 4 & 5 & 2 \end{bmatrix}.$$

After the use of elementary row operations, this reduces to the echelon form

$$\mathbf{A}|\mathbf{b} \sim \begin{bmatrix} 1 & -1 & 1 & 2 & 1 \\ 0 & 1 & 2 & 1 & 1 \\ 0 & 0 & -5 & -3 & -2 \\ 0 & 0 & 0 & 4 & 1 \\ 0 & 0 & 0 & 0 & 131 \end{bmatrix}.$$

In this case we see that row rank($\mathbf{A}$) = 4 while row rank($\mathbf{A}|\mathbf{b}$) = 5, so row rank ($\mathbf{A}$) $\neq$ row rank($\mathbf{A}|\mathbf{b}$) showing that the equations are inconsistent. This is easily seen to be so, because the fourth row implies $4x_4 = 1$, while the fifth row implies that $0 \times x_4 = 131$, which is impossible.

◇

The implications of the row ranks of **A** and **A|b** illustrated by the previous examples can be summarized as follows.

4.3.1 Row Rank of an Augmented Matrix and the Nature of a Solution Set

---

Let the coefficient matrix **A** of equations in (4.1) be an $m \times n$ matrix, and let **b** be an m element column vector.

1. A solution set exists if row rank($\mathbf{A}$) = row rank($\mathbf{A}|\mathbf{b}$). The solution will be unique if row rank($\mathbf{A}$) = row rank($\mathbf{A}|\mathbf{b}$) = n, but if row rank($\mathbf{A}$) = row rank($\mathbf{A}|\mathbf{b}$) = $r < n$, then r of the unknowns x_1, $x_2, \ldots, x_n$ can be expressed in terms of the remaining $n - r$ unknowns when specified as arbitrary parameters.
2. No solution set exists if row rank($\mathbf{A}$) < row rank($\mathbf{A}|\mathbf{b}$).

The number of linearly independent *rows* in a matrix, called its *row rank*, has been shown to be of fundamental importance when solving linear systems of equations. Similarly, the number of linearly independent *columns* of a matrix, is called its *column rank.* A key result to be proved in Chapter 7 is that row rank($\mathbf{A}$) = column rank($\mathbf{A}$). So in future, and without ambiguity, we need only to refer to the rank of a matrix.

Example 4.6. Verify the equivalence of the row and column ranks of

$$\mathbf{A} = \begin{bmatrix} 1 & 2 & 3 & 6 \\ 2 & 1 & 0 & 4 \\ 0 & -3 & -6 & -8 \end{bmatrix}.$$

Solution. The echelon form of **A** is $\mathbf{A}_E = \begin{bmatrix} 1 & 0 & -1 & \frac{2}{3} \\ 0 & 1 & 2 & \frac{8}{3} \\ 0 & 0 & 0 & 0 \end{bmatrix}$, so row rank($\mathbf{A}$) = 2.

Transposing **A**, finding the echelon form of $\mathbf{A}^{\mathrm{T}}$, and then transposing again to display the linearly independent columns of **A**, gives

$$\begin{bmatrix} 1 & 0 & 0 & 0 \\ 0 & 1 & 0 & 0 \\ -2 & 1 & 0 & 0 \end{bmatrix},$$ showing, as expected, that column rank(**A**) = 2.

◇

The definition of rank leads directly to the following test for linear independence.

4.3.2 *Testing the Linear Independence of the Rows (Columns) of an* n × n *Matrix A*

The rows (columns) of an $n \times n$ matrix **A** will be linearly independent if, and only if, $\det \mathbf{A} \neq 0$.

4.4 Elementary Row Operations and the Inverse Matrix

Before considering an important general problem in the study of matrices, it is useful to show how, when n is small, elementary row operations provide a way of finding the inverse of an $n \times n$ matrix. Once again the idea is simple, and it starts by writing side by side the square matrix **A**, and an identity matrix **I** of the same size, where the juxtaposition of the matrices does *not* imply their multiplication. Operations are performed row by row on matrix **A** on the left to reduce it to an identity matrix while, simultaneously, and in the same order, the same row operations are performed row by row on the identity matrix on the right. When **A** has been reduced to the identity matrix, the original identity matrix **I** on the right will have been transformed into the inverse matrix $\mathbf{A}^{-1}$. If during this procedure a row of zeros is produced during the modification of matrix **A**, the reduction process will terminate, indicating that $\mathbf{A}^{-1}$ does *not* exist. This will occur if one or more rows of **A** are linearly dependent on its other rows, causing matrix **A** to be singular, in which case $\det \mathbf{A} = 0$.

Example 4.7. Apply elementary row operations on matrix **A** to find $\mathbf{A}^{-1}$, given that

$$\mathbf{A} = \begin{bmatrix} -1 & 1 & 3 \\ 1 & -1 & -2 \\ 1 & 0 & -3 \end{bmatrix}.$$

Solution. We start with **A** and **I** side by side, and perform elementary row operations on **A** to reduce it to the unit matrix **I**, while at the same time performing the *same* elementary row operations on the unit matrix **I** on the right, leading to the results

$$\begin{bmatrix} -1 & 1 & 3 \\ 1 & -1 & -2 \\ 1 & 0 & -3 \end{bmatrix} \begin{bmatrix} 1 & 0 & 0 \\ 0 & 1 & 0 \\ 0 & 0 & 1 \end{bmatrix} \begin{matrix} \text{add row 1 to row 2} \\ \text{and row 1 to row 3} \end{matrix} \rightarrow \begin{bmatrix} -1 & 1 & 3 \\ 0 & 0 & 1 \\ 0 & 1 & 0 \end{bmatrix} \begin{bmatrix} 1 & 0 & 0 \\ 1 & 1 & 0 \\ 1 & 0 & 1 \end{bmatrix}$$

$$\text{interchange rows 2 and 3} \rightarrow \begin{bmatrix} -1 & 1 & 3 \\ 0 & 1 & 0 \\ 0 & 0 & 1 \end{bmatrix} \begin{bmatrix} 1 & 0 & 0 \\ 1 & 0 & 1 \\ 1 & 1 & 0 \end{bmatrix}$$

$$\text{subtract row } 2 + 3 \times \text{row 3 from row 1} \rightarrow \begin{bmatrix} -1 & 0 & 0 \\ 0 & 1 & 0 \\ 0 & 0 & 1 \end{bmatrix} \begin{bmatrix} -3 & -3 & -1 \\ 1 & 0 & 1 \\ 1 & 1 & 0 \end{bmatrix}$$

$$\text{change the sign of row 1} \rightarrow \begin{bmatrix} 1 & 0 & 0 \\ 0 & 1 & 0 \\ 0 & 0 & 1 \end{bmatrix} \begin{bmatrix} 3 & 3 & 1 \\ 1 & 0 & 1 \\ 1 & 1 & 0 \end{bmatrix}.$$

The required inverse matrix exists because the reduction of **A** to the identity matrix has been successful, and $\mathbf{A}^{-1}$ is given by the matrix on the right so that as

$$\mathbf{A} = \begin{bmatrix} -1 & 1 & 3 \\ 1 & -1 & -2 \\ 1 & 0 & -3 \end{bmatrix}, \text{ then } \mathbf{A}^{-1} = \begin{bmatrix} 3 & 3 & 1 \\ 1 & 0 & 1 \\ 1 & 1 & 0 \end{bmatrix}.$$

The result is easily checked by confirming that $\mathbf{AA}^{-1} = \mathbf{I}$

◊

4.5 LU Factorization of a Matrix and Its Use When Solving Linear Systems of Algebraic Equations

This section examines the possibility of expressing a nonsingular $n \times n$ matrix **A** as the product $\mathbf{A} = \mathbf{LU}$ of an $n \times n$ lower triangular matrix **L** with 1s along its leading diagonal, and an $n \times n$ upper triangular matrix **U**. This factorization is particularly useful when a system of equations $\mathbf{Ax} = \mathbf{b}$ has to be solved repeatedly with the same matrix **A**, but with different column vectors **b**. This is because for a given matrix **A**, the matrices **L** and **U** are unique, so they can be used repeatedly to solve the system of equations $\mathbf{Ax} = \mathbf{b}$ for different vectors **b**. To see how this factorization works when solving systems of algebraic equations, let the column vector **y** be defined as the solution of the system of equations $\mathbf{Ly} = \mathbf{b}$, from which the required solution vector **x** then follows by solving the system of equations $\mathbf{y} = \mathbf{Ux}$. Although at first sight this method of solution may appear to be unnecessarily complicated in fact, the method which is based on Gaussian elimination actually offers several advantages over ordinary Gaussian elimination.

The first advantage offered by this method is because triangular matrices are involved. The elements $y_1, y_2, \dots, y_n$ of the column vector $\mathbf{y}$ are obtained very simply by *forward substitution* in the system $\mathbf{Ay} = \mathbf{b}$, after which the elements $x_1, x_2, \dots, x_n$ of the solution vector $\mathbf{x}$ follow immediately by *backward substitution* in the system $\mathbf{Ux} = \mathbf{y}$.

The second advantage offered by this method is that the *LU* factorization of a matrix $\mathbf{A}$ need only be performed *once*, after which the matrices $\mathbf{L}$ and $\mathbf{U}$ can be used repeatedly to find solution vectors $\mathbf{x}$ that correspond to various *different* vectors $\mathbf{b}$.

The determination of the upper triangular matrix $\mathbf{U}$ follows directly from the Gaussian elimination process, after which the lower triangular matrix $\mathbf{L}$, which is in *reduced echelon form* (see Section 4.2), then follows from the elementary row operations used to find $\mathbf{U}$.

The method is best illustrated by applying it to a 4×4 matrix $\mathbf{A}$ for which *no* row interchanges are necessary during the Gaussian elimination process when finding the matrix $\mathbf{U}$. The modification that is necessary if row interchanges are needed during the Gaussian elimination process used to find $\mathbf{U}$ will be explained later. We will presuppose that the first element on the leading diagonal of $\mathbf{A}$ does not vanish. This is no restriction, because if it is not so, the order of the equations can be changed by interchanging the first equation with one that satisfies this condition. Let us take for our example the matrix

$$\mathbf{A} = \begin{bmatrix} 2 & 1 & 2 & 3 \\ 1 & 0 & 1 & 1 \\ -2 & 1 & -1 & 1 \\ 2 & 1 & -1 & 0 \end{bmatrix},$$

which is nonsingular because $\det \mathbf{A} = -3$.

The first stage in the Gaussian elimination process applied to column 1 subtracts $\frac{1}{2}$ of row 1 from row 2 to produce a zero as the first element of the modified row 2. The second step adds row 1 to row 3 to produce a zero as the first element in row 3, while subtracting row 1 from row 4 produces a zero as the first element of the modified row 4. The result is

$$\mathbf{A}_1 = \begin{bmatrix} 2 & 1 & 2 & 3 \\ 0 & -\frac{1}{2} & 0 & -\frac{1}{2} \\ 0 & 2 & 1 & 4 \\ 0 & 0 & -3 & -3 \end{bmatrix}.$$

These elementary row operations can be represented in the matrix form

$$\mathbf{M}_1 = \begin{bmatrix} 1 & 0 & 0 & 0 \\ -\frac{1}{2} & 1 & 0 & 0 \\ 1 & 0 & 1 & 0 \\ -1 & 0 & 0 & 1 \end{bmatrix},$$

because pre-multiplication of $\mathbf{A}$ by $\mathbf{M}_1$ gives $\mathbf{A}_1 = \mathbf{M}_1\mathbf{A}$, where the suffix 1 shows that $\mathbf{M}_1$ is the first matrix multiplier of $\mathbf{A}$ used to modify the first column of $\mathbf{A}$ to arrive at $\mathbf{A}_1$.

The second stage in the Gaussian elimination process is applied to $\mathbf{A}_1$ when four times row 2 is added to row 3 to produce a zero as the second element in the modified row 3. There is already a zero as the second element of row 4, so no further modifications are necessary in this second stage of the Gaussian elimination process. The result is

$$\mathbf{A}_2 = \begin{bmatrix} 2 & 1 & 2 & 3 \\ - & -\frac{1}{2} & 0 & -\frac{1}{2} \\ 0 & 0 & 1 & 2 \\ 0 & 0 & -3 & -3 \end{bmatrix}.$$

The elementary row operations that produce this result are described by the matrix

$$\mathbf{M}_2 = \begin{bmatrix} 1 & 0 & 0 & 0 \\ 0 & 1 & 0 & 0 \\ 0 & 4 & 1 & 0 \\ 0 & 0 & 0 & 1 \end{bmatrix},$$

because $\mathbf{A}_2 = \mathbf{M}_2\mathbf{A}_1 = \mathbf{M}_2\mathbf{M}_1\mathbf{A}$.

The third and final stage of the Gaussian elimination process involves adding three times row 3 to row 4 to produce a zero in the third element of the modified row 4. The result is

$$\mathbf{A}_3 = \begin{bmatrix} 2 & 1 & 2 & 3 \\ 0 & -\frac{1}{2} & 0 & -\frac{1}{2} \\ 0 & 0 & 1 & 2 \\ 0 & 0 & 0 & -3 \end{bmatrix}.$$

The elementary row operations that produced this result are described by the matrix

$$\mathbf{M}_3 = \begin{bmatrix} 1 & 0 & 0 & 0 \\ 0 & 1 & 0 & 0 \\ 0 & 0 & 1 & 0 \\ 0 & 0 & -3 & 1 \end{bmatrix},$$

because $\mathbf{A}_3 = \mathbf{M}_3\mathbf{A}_2 = \mathbf{M}_3\mathbf{M}_2\mathbf{M}_1\mathbf{A}$. Consequently the upper triangular matrix $\mathbf{U} = \mathbf{M}_3$, so

$$\mathbf{U} = \begin{bmatrix} 2 & 1 & 2 & 3 \\ 0 & -\frac{1}{2} & 0 & -\frac{1}{2} \\ 0 & 0 & 1 & 2 \\ 0 & 0 & 0 & -3 \end{bmatrix}.$$

Next, as $(\mathbf{M}_3\mathbf{M}_2\mathbf{M}_1)^{-1} = \mathbf{M}_1^{-1}\mathbf{M}_2^{-1}\mathbf{M}_3^{-1}$, it follows that $\mathbf{A} = \mathbf{M}_1^{-1}\mathbf{M}_2^{-1}\mathbf{M}_3^{-1}\mathbf{U}$, so the factorization will be completed if we can show that $\mathbf{M}_1^{-1}\mathbf{M}_2^{-1}\mathbf{M}_3^{-1} = \mathbf{U}$. This follows from the special structure of the matrix row operations $\mathbf{M}_i$, and from the definition of inverse matrices $\mathbf{M}_i^{-1}$ in terms of cofactors. Because the inverse of $\mathbf{M}_i$ follows immediately by reversing the signs of the elements in its ith column that lie below the element 1.

Applying this result to the factors $\mathbf{M}_1$, $\mathbf{M}_2$ and $\mathbf{M}_3$ gives

$$\mathbf{L} = \begin{bmatrix} 1 & 0 & 0 & 0 \\ \frac{1}{2} & 1 & 0 & 0 \\ -1 & 0 & 1 & 0 \\ 1 & 0 & 0 & 1 \end{bmatrix} \begin{bmatrix} 1 & 0 & 0 & 0 \\ 0 & 1 & 0 & 0 \\ 0 & -4 & 1 & 0 \\ 0 & 0 & 0 & 1 \end{bmatrix} \begin{bmatrix} 1 & 0 & 0 & 0 \\ 0 & 1 & 0 & 0 \\ 0 & -4 & 1 & 0 \\ 0 & 0 & 0 & 1 \end{bmatrix}$$

$$= \begin{bmatrix} 1 & 0 & 0 & 0 \\ \frac{1}{2} & 1 & 0 & 0 \\ -1 & -4 & 1 & 0 \\ 1 & 0 & -3 & 1 \end{bmatrix}.$$

Thus the factorization has been achieved, and we have $\mathbf{A} = \mathbf{LU}$ in the form

$$\underbrace{\begin{bmatrix} 2 & 1 & 2 & 3 \\ 1 & 0 & 1 & 1 \\ -2 & 1 & -1 & 1 \\ 2 & 1 & -1 & 0 \end{bmatrix}}_{\mathbf{A}} = \underbrace{\begin{bmatrix} 1 & 0 & 0 & 0 \\ \frac{1}{2} & 1 & 0 & 0 \\ -1 & -4 & 1 & 0 \\ 1 & 0 & -3 & 1 \end{bmatrix}}_{\mathbf{L}} \underbrace{\begin{bmatrix} 2 & 1 & 2 & 3 \\ 0 & -\frac{1}{2} & 0 & -\frac{1}{2} \\ 0 & 0 & 1 & 2 \\ 0 & 0 & 0 & 3 \end{bmatrix}}_{\mathbf{U}}.$$

Example 4.8. Use *LU* factorization to solve the system $\mathbf{Ax} = \mathbf{b}$, given that $\mathbf{A}$ is the matrix that has just been factorized and $\mathbf{x} = [x_1, x_2, x_3, x_4]^{\mathrm{T}}$ with (a) $\mathbf{b} = [1, 2, -1, 1]^{\mathrm{T}}$ and (b) $\mathbf{b} = [2, 0, 1, 1]^{\mathrm{T}}$.

Solution. (a) Setting $\mathbf{y} = [y_1, y_2, y_3, y_4]^{\mathrm{T}}$, the equation $\mathbf{Ly} = \mathbf{b}$ with $\mathbf{b} = [1, 2, -1, 1]^{\mathrm{T}}$ becomes

$$y_1 = 1, \tfrac{1}{2}y_1 + y_2 = 2, -y_1 - 4y_2 + y_3 = -1, \ y_1 - 3y_3 + y_4 = 1,$$

with the solution obtained by *forward substitution* $y_1 = 1, \ y_2 = \frac{3}{2}, \ y_3 = 6, y_4 = 18.$ The equation $\mathbf{Ux} = \mathbf{y}$ then gives the set of equations

$$2x_1 + x_2 + 2x_3 + 3x_4 = 1, -\tfrac{1}{2}x_2 - \tfrac{1}{2}x_4 = \tfrac{3}{2}, \ x_3 + 2x_4 = 6, \ 3x_4 = 18,$$

with the solution obtained by *backward substitution* $x_1 = 2, \ x_2 = -9, \ x_3 = -6, \ x_4 = 6,$ so the solution set has been found.
(b) Using the same $\mathbf{L}$ and $\mathbf{U}$, but this time with $\mathbf{b} = [2, 0, 1, 1]^{\mathrm{T}}$, the equation $\mathbf{Ly} = \mathbf{b}$ becomes

$$y_1 = 2, \tfrac{1}{2}y_1 + y_2 = 0, -y_1 - 4y_2 + y_3 = 1, \ y_1 - 3y_3 + y_4 = 1,$$

with the solution obtained by *forward substitution* $y_1 = 2, \ y_2 = -1, \ y_3 = -1, \ y_4 = -4$. The equation $\mathbf{Ux} = \mathbf{y}$ then gives the set of equations

$$2x_1 + x_2 + 2x_3 + 3x_4 = 2, -\tfrac{1}{2}x_2 - \tfrac{1}{2}x_4 = -1, \ x_3 + 2x_4 = -1, \ 3x_4 = -4,$$

with the solution obtained by *backward substitution* $x_1 = -\frac{1}{3}, \ x_2 = \frac{10}{3}, x_3 = \frac{5}{3}, \ x_4 = -\frac{4}{3},$ so the new solution set has been found by using the *same* matrices $\mathbf{L}$ and $\mathbf{U}$.

◇

It may happen during the Gaussian elimination process leading to the derivation of the matrix $\mathbf{U}$ that a zero occurs on the leading diagonal at, say, the ith position, where a nonzero pivot is required. If this happens, it is necessary to interchange the row concerned with one *below* it which has a nonzero element in its ith position to allow the reduction process to continue. This is always possible, because matrix $\mathbf{A}$ is nonsingular. In this case, when the reduction process is completed, the previous result $\mathbf{A} = \mathbf{LU}$ must be modified to $\mathbf{A} = \mathbf{PLU}$, where $\mathbf{P}$ is a permutation matrix (like a matrix $\mathbf{M}$) that describes the row interchanges that have been made (see Chapter 3, Exercises 15 through 18).

As a simple example, consider a set of four equations that is to be solved by *LU* factorization where the first element on the leading diagonal of $\mathbf{A}$ is zero, but the element immediately below it is nonzero. Instead of interchanging the first two equations by hand a permutation matrix $\mathbf{P}$ can be used. In this case the permutation matrix $\mathbf{P}_1$ can be used where

$$\mathbf{P}_1 = \begin{bmatrix} 0 & 1 & 0 & 0 \\ 1 & 0 & 0 & 0 \\ 0 & 0 & 1 & 0 \\ 0 & 0 & 0 & 1 \end{bmatrix},$$

because pre-multiplication by $\mathbf{P}_1$ interchanges rows 1 and 2 to bring a nonzero pivot into the position of the first element in the first row of $\mathbf{A}$. If this happens when solving a system of equations $\mathbf{Ax} = \mathbf{b}$, the calculation proceeds as before, except for the fact that it then becomes necessary to set $\mathbf{Ly} = \mathbf{P}_1\mathbf{b}$ instead of $\mathbf{Ly} = \mathbf{b}$, with the introduction of other permutation matrices in the appropriate order if further equation interchanges become necessary.

If the solution by *LU* factorization is programmed for a computer, provision must be made for an interchange of equations at any stage of the calculations, including an initial equation interchange like the one represented by the permutation matrix $\mathbf{P}_1$.

4.6 Eigenvalues and Eigenvectors

A problem of fundamental importance that occurs in many applications of matrices can be formulated as follows. When system (4.1) is properly determined ($m = n$), how can a solution be found in which the nonhomogeneous vector $\mathbf{b} = [b_1, b_2, \ldots, b_n]^{\mathrm{T}}$ is proportional to the unknown vector $\mathbf{x} = [x_1, x_2, \ldots, x_n]^{\mathrm{T}}$? One reason for this seemingly odd question will become clear in Chapter 6. Denoting the constant of proportionality by λ, the problem involves finding column vector $\mathbf{x}$ such that $\mathbf{b} = \lambda\mathbf{x}$, in which case system (4.1) becomes the matrix equation $\mathbf{Ax} = \lambda\mathbf{x}$. When written out in full, the system $\mathbf{Ax} = \lambda\mathbf{x}$ is seen to be

$$\begin{aligned} a_{11}x_1 + a_{12}x_2 + \cdots + a_{1n}x_n &= \lambda x_1, \\ a_{21}x_1 + a_{22}x_2 + \cdots + a_{2n}x_n &= \lambda x_2, \\ \cdots\cdots\cdots\cdots \\ a_{n1}x_1 + a_{n2}x_2 + \cdots + a_{nn}x_n &= \lambda x_n. \end{aligned} \tag{4.5}$$

At first sight this appears to be a nonhomogeneous system. However, in each equation the term on the right of the equality sign can be combined with a corresponding term in the expression on the left, leading to the following homogeneous system of algebraic equations, in which λ appears as a parameter

$$\begin{aligned} (a_{11} - \lambda)x_1 + a_{12}x_2 + \cdots + a_{1n}x_n &= 0, \\ a_{21}x_1 + (a_{22} - \lambda)x_2 + \cdots + a_{2n}x_n &= 0, \\ \cdots\cdots\cdots\cdots \\ a_{n1}x_1 + a_{n2}x_2 + \cdots + (a_{nn} - \lambda)x_n &= 0. \end{aligned} \tag{4.6}$$

In matrix notation, after introducing the identity matrix **I**, system (4.5) becomes

$$[\mathbf{A} - \lambda\mathbf{I}]\mathbf{x} = \mathbf{0}. \tag{4.7}$$

This system is homogeneous, so there are two possible types of solution. The first is the obvious *trivial solution* $\mathbf{x} = \mathbf{0}$. The second type of solution is nontrivial (one in which $\mathbf{x} \neq \mathbf{0}$), though it can only be found if the determinant of the coefficient matrix $\mathbf{A} - \lambda\mathbf{I}$ in (4.7) vanishes, in which case there is linear dependence between the rows. So we see that the condition for the existence of nontrivial solution vectors **x** is

$$\det[\mathbf{A} - \lambda\mathbf{I}] = 0. \tag{4.8}$$

In general the determinant of a coefficient matrix will not vanish. However, in this case the parameter λ occurs in each element of the leading diagonal of the matrix

$\mathbf{A} - \lambda\mathbf{I}$, so when det $[\mathbf{A} - \lambda\mathbf{I}]$ is expanded it will give rise to a polynomial $p(\lambda)$ in λ of degree n. This polynomial in λ is called the *characteristic polynomial* associated with matrix **A**, and the characteristic polynomial will vanish when λ is any one of its n zeros. When expanded, (4.8) is called the *characteristic equation* associated with **A**, and it is a polynomial equation $p(\lambda) = 0$ in λ of degree n, with n roots $\lambda_1, \lambda_2, \ldots, \lambda_n$. The roots λ_i are called the *eigenvalues* of **A**, and from (4.7) it follows that to each eigenvalue λ_i of **A** there corresponds a column vector $\mathbf{x}^{(i)}$ such that

$$[\mathbf{A} - \lambda_i\mathbf{I}]\mathbf{x}^{(i)} = \mathbf{0}. \tag{4.9}$$

Vector $\mathbf{x}^{(i)}$ is called the *eigenvector* of **A** corresponding to the eigenvalue λ_i, and in general an $n \times n$ matrix **A** will have n different eigenvectors $\mathbf{x}^{(1)}, \mathbf{x}^{(2)}, \ldots, \mathbf{x}^{(n)}$.

We mention that older terms for *eigenvalues* and *eigenvectors* that are still in use are *characteristic values* and *characteristic vectors*.

It can happen that a matrix has an eigenvalue λ_j that is repeated r times, in which case $(\lambda - \lambda_j)^r$ is a factor of the characteristic equation. Such a repeated root of the characteristic equation is said to be an eigenvalue with **algebraic multiplicity** r, often abbreviated to **multiplicity** r. Our main concern will be with the case when **A** has n linearly independent eigenvectors (there is *no* proportionality between them), even though some of the eigenvalues may be repeated. The more complicated situation that arises when **A** has fewer than n distinct eigenvectors will be examined later.

Expanding $p(\lambda) = \det[\mathbf{A} - \lambda\mathbf{I}]$, the eigenvalues λ_i are seen to be the roots of the polynomial of degree n in λ given by

$$p(\lambda) = \det[\mathbf{A} - \lambda\mathbf{I}] = \begin{vmatrix} a_{11} - \lambda & a_{12} & \cdots & a_{1n} \\ a_{21} & a_{22} - \lambda & \cdots & a_{2n} \\ \vdots & \vdots & \vdots & \vdots \\ a_{n1} & a_{n2} & \vdots & a_{nn} - \lambda \end{vmatrix} = 0\,, \tag{4.10}$$

so $p(\lambda) = \det[\mathbf{A} - \lambda\mathbf{I}]$ can be factored and written as

$$p(\lambda) = \det(\mathbf{A} - \lambda\mathbf{I}) = (\lambda_1 - \lambda)(\lambda_2 - \lambda)\cdots(\lambda_n - \lambda)\,, \tag{4.11}$$

where the λ_i with $i = 1, 2, \ldots, n$ are the n eigenvalues of $\mathbf{A}$. Setting $\lambda = 0$ in identity (4.11) gives the useful result that the product of the eigenvalues is equal to det $\mathbf{A}$, so

$$\lambda_1\lambda_2\cdots\lambda_n = \det\mathbf{A}\,. \tag{4.12}$$

The coefficient of λ^{n-1} on the left of (4.11) can be seen to be $(-1)^n(\lambda_1 + \lambda_2 + \cdots + \lambda_n)$, and a little thought shows the coefficient of λ^{n-1} on the right is given by $(-1)^n(a_{11} + a_{12} + \cdots + a_{nn})$, so equating these two expressions we arrive at another useful result

$$\lambda_1 + \lambda_2 + \cdots + \lambda_n = a_{11} + a_{22} + \cdots + a_{nn}. \tag{4.13}$$

Thus the sum of the eigenvalues of $\mathbf{A}$ is seen to be equal to the sum of the elements on the leading diagonal of $\mathbf{A}$. Because of its importance in the study of eigenvalues, and elsewhere, the sum of the elements on the leading diagonal of a square matrix $\mathbf{A}$ is given a name and called the *trace* of $\mathbf{A}$, written tr$\mathbf{A}$, so we have the definition

$$\mathrm{tr}\mathbf{A} = a_{11} + a_{22} + \cdots + a_{nn}. \tag{4.14}$$

Apart from various other uses, result (4.12), and result (4.13) in the form

$$\lambda_1 + \lambda_2 + \cdots + \lambda_n = \mathrm{tr}\ \mathbf{A}, \tag{4.15}$$

are useful when checking the values of eigenvalues that have been computed, with (4.15) being particularly simple to apply.

A result that is also useful when considering 2×2 matrices $\mathbf{A} = [a_{ij}]$ is that the eigenvalues of $\mathbf{A}$ are given by

$$\lambda_\pm = \tfrac{1}{2}\left[\mathrm{tr}\mathbf{A} \pm \sqrt{(\mathrm{tr}\mathbf{A})^2 - 4\det\mathbf{A}}\right]. \tag{4.16}$$

This result follows directly by expanding the characteristic determinant

$$\begin{vmatrix} a_{11}-\lambda & a_{12} \\ a_{21} & a_{22}-\lambda \end{vmatrix} = 0,$$

solving the resulting quadratic equation for λ, and using the definitions $\mathrm{tr}\mathbf{A} = a_{11} + a_{22}$ and $\det \mathbf{A} = a_{11}a_{22} - a_{12}a_{21}$.

The quantity

$$\Delta = (\mathrm{tr}\mathbf{A})^2 - 4\det \mathbf{A}, \tag{4.17}$$

in terms of which (4.16) can be written

$$\lambda_{\pm} = \tfrac{1}{2}\left[\mathrm{tr}\mathbf{A} \pm \sqrt{\Delta}\right], \tag{4.18}$$

is called the *discriminant*, because when the elements a_{ij} are all real it shows that the two eigenvalues $\lambda_{\pm}$ will be *real* if $\Delta \geq 0$, but they will be complex conjugates if $\Delta < 0$.

Let us now return to consider matrix (4.9) that defines the eigenvectors of $\mathbf{A}$, and in terms of its elements this can be written $\mathbf{x}^{(i)} = [x_1^{(i)}, x_2^{(i)}, \cdots, x_n^{(i)}]^{\mathrm{T}}$, for $i = 1, 2, \ldots, n$. When written out in full, (4.9) shows the $\mathbf{x}^{(i)}$ are the solutions of the homogeneous system of equations

$$\begin{bmatrix} a_{11}-\lambda_i & a_{12} & \cdots & a_{1n} \\ a_{21} & a_{22}-\lambda_i & \cdots & a_{2n} \\ \vdots & \vdots & \vdots & \vdots \\ a_{n1} & a_{n2} & \cdots & a_{nn}-\lambda_i \end{bmatrix} \begin{bmatrix} x_1^{(i)} \\ x_2^{(i)} \\ \vdots \\ x_n^{(i)} \end{bmatrix} = \begin{bmatrix} 0 \\ 0 \\ \vdots \\ 0 \end{bmatrix}, \quad i = 1, 2, \ldots, n. \tag{4.19}$$

The homogeneity of (4.19) means that the absolute values of the n quantities $x_1^{(i)}, x_2^{(i)}, \ldots, x_n^{(i)}$ cannot be determined, so instead, $n - 1$ of the elements must be expressed in terms of the remaining element, say $x_r^{(i)}$, the value of which may be assigned arbitrarily. So (4.19) only determines the *ratios* of the elements of $\mathbf{x}^{(i)}$ with respect to $x_r^{(i)}$ as a parameter. This means that once an eigenvector has been found, it can be multiplied by an arbitrary constant $k \neq 0$ (scaled by k) and still remain an eigenvector. This fact can be seen directly from (4.9), because replacing $\mathbf{x}^{(i)}$ by $k\mathbf{x}^{(i)}$ with $k \neq 0$ an arbitrary number, cancellation of the multiplicative factor k leads directly to (4.19).

Finding the characteristic polynomial $p(\lambda)$ of an $n \times n$ matrix is straightforward, but unless the characteristic polynomial can be factored, finding its roots when $n > 2$ usually requires the use of numerical methods. To simplify the calculations, in the examples and exercises that follow, the 3×3 matrices $\mathbf{A}$ have been constructed so that once the characteristic equation has been determined, at least one of its roots (eigenvalues), say $\tilde{\lambda}$, can be found by inspection. Then, removing the factor $(\lambda - \tilde{\lambda})$ from the characteristic equation by long division, the remaining two roots can be found by using the quadratic formula.

Example 4.9. Find the characteristic polynomial of $\mathbf{A}$ and its eigenvalues, given that

$$\mathbf{A} = \begin{bmatrix} 1 & 0 & -1 \\ -2 & -1 & 2 \\ -1 & 2 & 1 \end{bmatrix}.$$

Solution. The characteristic polynomial is found by expanding the determinant

$$p(\lambda) = \det[\mathbf{A} - \lambda\mathbf{I}] = \begin{vmatrix} 1-\lambda & 0 & -1 \\ -2 & -1-\lambda & 2 \\ 1 & 2 & 1-\lambda \end{vmatrix} = 6\lambda + \lambda^2 - \lambda^3 .$$

The eigenvalues of $\mathbf{A}$ are the roots of the characteristic equation, that is the roots of $p(\lambda) = \det[\mathbf{A} - \lambda\mathbf{I}] = 0$, which is equivalent to finding the roots of $6\lambda + \lambda^2 - \lambda^3 = 0$. The expression on the left has the obvious factor $\lambda = 0$, so the characteristic equation can be factored and written as $\lambda(\lambda + 2)(3 - \lambda) = 0$. Its roots are 0, -2 and 3, so when for convenience they are arranged in numerical order, the eigenvalues of $\mathbf{A}$ are seen to be $\lambda_1 = -2$, $\lambda_2 = 0$, $\lambda_3 = 3$.

These results can be checked using results (4.12) and (4.14). A simple calculation shows that $\det \mathbf{A} = 0$, and from (4.12) we have

$$\lambda_1\lambda_2\lambda_3 = (-2)(0)(3) = 0 = \det \mathbf{A}.$$

Simpler still, from (4.14) we have

$$\lambda_1 + \lambda_2 + \lambda_3 = -2 + 0 + 3 = 1 = \operatorname{tr}(\mathbf{A}) = 1 - 1 + 1 = 1.$$

◇

4.7 The Companion Matrix and the Characteristic Polynomial

Given an $n \times n$ matrix $\mathbf{A}$, the characteristic polynomial $p(\lambda)$ associated with $\mathbf{A}$ is determined by $p(\lambda) = \det|\mathbf{A} - \lambda\mathbf{I}|$. In this section we now reverse this process and ask how, given a polynomial

$$p(\lambda) = \lambda^n + a_1\lambda^{n-1} + a_2\lambda^{n-2} + \cdots + a_{n-1}\lambda + a_n, \tag{4.20}$$

can a matrix $\mathbf{A}$ be constructed with $p(\lambda)$ as its characteristic polynomial. There is no unique answer to this question, but a standard approach to this problem, which is useful in certain circumstances, is based on the matrix

$$\mathbf{A} = \begin{bmatrix} 0 & 0 & 0 & \cdots & 0 & -a_n \\ 1 & 0 & 0 & \cdots & 0 & -a_{n-1} \\ 0 & 1 & 0 & \cdots & 0 & -a_{n-2} \\ 0 & 0 & 1 & \cdots & 0 & -a_{n-3} \\ \vdots & \vdots & \vdots & \vdots & \vdots & \vdots \\ 0 & 0 & 0 & \cdots & 1 & -a_1 \end{bmatrix}, \tag{4.21}$$

called the *companion matrix* for $p(\lambda) = \lambda^n + a_1\lambda^{n-1} + a_2\lambda^{n-2} + \cdots + a_{n-1}\lambda + a_n$, where the elements a_i for $i = 1, 2, \ldots, n$ in the last column are the coefficients of $p(\lambda)$. The characteristic polynomial for matrix **A** is, by definition,

$$p(\lambda) = \det|\mathbf{A} - \lambda\mathbf{I}| = \begin{vmatrix} -\lambda & 0 & 0 & \cdots & 0 & -a_n \\ 1 & -\lambda & 0 & \cdots & 0 & -a_{n-1} \\ 0 & 1 & -\lambda & \cdots & 0 & -a_{n-2} \\ 0 & 0 & 1 & \cdots & 0 & -a_{n-3} \\ \vdots & \vdots & \vdots & \vdots & \vdots & \vdots \\ 0 & 0 & 0 & \cdots & 1 & -\lambda - a_1 \end{vmatrix}. \tag{4.22}$$

To show that a polynomial $p(\lambda)$ of the required form follows by expanding this determinant it is necessary to use the property of determinants that allows multiples of a row to be added to another row without changing the value of the determinant. The expansion of this determinant starts by multiplying the row n by λ, and adding the result to row $(n - 1)$. Next, the modified row $(n - 1)$ is multiplied by λ and added to row $(n - 2)$, and thereafter this process is continued until the first row is reached. The final result is

$$p(\lambda) = \begin{vmatrix} 0 & 0 & 0 & \cdots & 0 & -\lambda^n - a_1\lambda^{n-1} - a_2\lambda^{n-2} - \cdots - a_{n-1}\lambda - a_n \\ 1 & 0 & 0 & \cdots & 0 & -\lambda^{n-1} - a_1\lambda^{n-2} - \cdots - a_{n-1} \\ 0 & 1 & 0 & \cdots & 0 & \lambda^{n-2} - a_1\lambda^{n-3} - \cdots - a_{n-2} \\ \vdots & \vdots & \vdots & \vdots & \vdots & \vdots \\ 0 & 0 & 0 & \cdots & 0 & -\lambda^2 - a_1\lambda - a_2 \\ 0 & 0 & 0 & \cdots & 1 & -\lambda - a_1 \end{vmatrix}.$$

The result follows by expanding this determinant in terms of element in the first row, when it becomes

$$p(\lambda) = \lambda^n + a_1\lambda^{n-1} + a_2\lambda^{n-2} + \cdots + a_{n-1}\lambda + a_n.$$

◇

Exercises

In Exercises 1 through 6 use Gaussian elimination, in the form of elementary row operations applied to the augmented matrix, to find a solution set when it exists, and comment on the values of rank(**A**) and rank(**A**|**b**).

1.

$$\begin{aligned}
&x_1 + x_2 - x_3 + 2x_4 = 1,\\
&2x_1 + 2x_2 - x_3 + 2x_4 = 3,\\
&-x_1 + 2x_2 + x_3 - x_4 = -2.
\end{aligned}$$

2.

$$\begin{aligned}
&x_1 + x_2 + 3x_3 + 2x_4 = -2,\\
&3x_1 + x_2 + 2x_4 = 3,\\
&2x_1 - x_2 + 2x_3 + 4x_4 = 1,\\
&x_1 + 2x_2 - x_3 - 2x_4 = 1.
\end{aligned}$$

3.

$$\begin{aligned}
&x_1 + 4x_2 + 2x_3 = 3,\\
&2x_1 + 3x_2 + x_3 = 1,\\
&x_1 + 3x_2 + 2x_3 = 4,\\
&3x_1 + x_2 - x_3 = 2.
\end{aligned}$$

4.

$$\begin{aligned}
&x_1 - x_2 = 2,\\
&3x_1 + x_2 - x_3 = 4,\\
&2x_1 + x_2 = 2,\\
&4x_1 - x_3 = 6,\\
&5x_1 - x_2 - x_3 = 8.
\end{aligned}$$

5.

$$\begin{aligned}
&2x_1 + x_3 = 4,\\
&2x_1 + 4x_2 + 3x_3 = 2,\\
&x_1 + 3x_3 = 1,\\
&5x_1 + 4x_2 + 7x_3 = 7.
\end{aligned}$$

6.

$$\begin{aligned}
&x_1 + 2x_2 + x_3 = 5,\\
&2x_1 + x_2 = 2,\\
&-x_1 + x_2 + 3x_3 = 1,\\
&2x_1 + x_3 = 2,\\
&x_1 + x_2 + 3x_3 = 0.
\end{aligned}$$

In Exercises 7 and 8 use elementary row operations applied to A and **I** to find $\mathbf{A}^{-1}$, and check that $\mathbf{AA}^{-1} = \mathbf{I}$.

7.

$$\mathbf{A} = \begin{bmatrix} 1 & 1 & -1 \\ 1 & 2 & 0 \\ 2 & 2 & 1 \end{bmatrix}.$$

8.

$$\mathbf{A} = \begin{bmatrix} 3 & 1 & 2 \\ 1 & -1 & 2 \\ 1 & 1 & -1 \end{bmatrix}.$$

In Exercises 9 and 10 find the characteristic polynomial of matrix **A**, but do not attempt to find the eigenvalues of **A**.

9.

$$\mathbf{A} = \begin{bmatrix} 1 & -1 & 3 \\ 2 & 1 & 4 \\ 3 & -1 & 1 \end{bmatrix}.$$

10.

$$\mathbf{A} = \begin{bmatrix} 1 & 1 & 3 & 1 \\ 0 & 2 & 1 & -1 \\ 1 & 1 & -1 & 2 \\ 1 & -1 & 3 & 1 \end{bmatrix}.$$

11. Find the condition on the real number α such that

$$\mathbf{A} = \begin{bmatrix} 1 & 2 \\ \alpha & 3 \end{bmatrix}$$

has (a) two real and distinct eigenvalues (b) two equal eigenvalues and (c) complex conjugate eigenvalues.

12. Verify that row rank(**A**) = column rank(**A**) = 3, given that

$$\mathbf{A} = \begin{bmatrix} 2 & 1 & 4 & 3 & 0 \\ 1 & 2 & 1 & 1 & 1 \\ 2 & 0 & 2 & 4 & 1 \\ 7 & 4 & 11 & 11 & 2 \end{bmatrix}.$$

In Exercises 13 through 16 find the *LU* factorization of the given matrix **A**, and use it to solve the system of equations $\mathbf{Ax} = \mathbf{b}$ for the given column vectors **b**.

13. $\mathbf{A} = \begin{bmatrix} 1 & 2 & 3 \\ 2 & -1 & 1 \\ 1 & 2 & -1 \end{bmatrix}$, (a) with $\mathbf{b} = \begin{bmatrix} -2 \\ 1 \\ -1 \end{bmatrix}$ and (b) with $\mathbf{b} = \begin{bmatrix} 3 \\ 0 \\ 2 \end{bmatrix}$.

14. $\mathbf{A} = \begin{bmatrix} 1 & -1 & 2 \\ 1 & 2 & -1 \\ -2 & -3 & -1 \end{bmatrix}$, $\mathbf{b} = \begin{bmatrix} 3 \\ 1 \\ -1 \end{bmatrix}$.

15. $\mathbf{A} = \begin{bmatrix} 2 & 1 & 1 & 2 \\ 1 & 1 & 1 & 0 \\ 1 & 0 & -1 & 1 \\ -1 & 1 & 1 & 1 \end{bmatrix}$, (a) with $\mathbf{b} = \begin{bmatrix} 1 \\ 2 \\ -2 \\ 0 \end{bmatrix}$ and (b) with $\mathbf{b} = \begin{bmatrix} 1 \\ -1 \\ 3 \\ -2 \end{bmatrix}$.

16. $\mathbf{A} = \begin{bmatrix} 0 & 1 & -1 & 1 \\ 2 & -1 & -1 & -1 \\ 1 & 2 & 0 & 1 \\ 1 & 1 & -1 & -1 \end{bmatrix}$, $\mathbf{b} = \begin{bmatrix} 2 \\ 1 \\ -1 \\ 3 \end{bmatrix}$.

17. Construct a polynomial of your choice of degree 5. By constructing the corresponding determinant in (4.22), use the row modifications described in Section 4.7 to simplify the determinant to the point where it is clear that it reproduces your polynomial.

Chapter 5
Eigenvalues, Eigenvectors, Diagonalization, Similarity, Jordan Normal Forms, and Estimating Regions Containing Eigenvalues

5.1 Finding Eigenvectors

It was shown in Chapter 4 that the *eigenvalues* $\lambda_1, \lambda_2, \ldots, \lambda_n$ of an $n \times n$ matrix $\mathbf{A}$ are the roots of the nth degree polynomial equation $p(\lambda) = 0$ in λ, called the *characteristic equation* of $\mathbf{A}$, and given by

$$p(\lambda) = \det[\mathbf{A} - \lambda\mathbf{I}] = 0. \tag{5.1}$$

The polynomial expression $p(\lambda) = \det[\mathbf{A} - \lambda\mathbf{I}]$ is called the *characteristic polynomial* of matrix $\mathbf{A}$. It may happen that some eigenvalues are repeated, though there will always be n eigenvalues provided an eigenvalue repeated r times it is counted as r eigenvalues. It is important to know that even when the elements of $\mathbf{A}$ are all real, so the coefficients of the characteristic equation are all real, the eigenvalues (the roots of $p(\lambda) = 0$) may be real or complex, though when they are complex they must occur in complex conjugate pairs.

The column vector $\mathbf{x}_i$ associated with the eigenvalue λ_i is called the *eigenvector* of $\mathbf{A}$ belonging to the eigenvalue λ_i, and it is a solution of the matrix equation

$$[\mathbf{A} - \lambda_i\mathbf{I}]\mathbf{x}_i = \mathbf{0}. \tag{5.2}$$

When the n eigenvalues are all distinct, there will always be n linearly independent eigenvectors $\mathbf{x}_1, \mathbf{x}_2, \ldots, \mathbf{x}_n$ that are solutions of Eq. (5.2) for $i = 1, 2, \ldots, n$.

However, when an eigenvalue λ_m, say, is repeated r times, it may or may not have associated with it r linearly independent eigenvectors.

The following typical example shows how the eigenvectors of a matrix are found once its eigenvalues are known.

Example 5.1. Find the eigenvalues and eigenvectors of the matrix

$$\mathbf{A} = \begin{bmatrix} 1 & 0 & -1 \\ -2 & -1 & 2 \\ -1 & 2 & 1 \end{bmatrix}.$$

A. Jeffrey, *Matrix Operations for Engineers and Scientists*,
DOI 10.1007/978-90-481-9274-8_5, © Springer Science+Business Media B.V. 2010

Solution. This is the matrix considered in Example 4.7 where the characteristic equation $p(\lambda) = 0$ was found to be

$$p(\lambda) = \det[\mathbf{A} - \lambda\mathbf{I}] = \begin{vmatrix} 1-\lambda & 0 & -1 \\ -2 & -1-\lambda & 2 \\ -1 & 2 & 1-\lambda \end{vmatrix} = 6\lambda + \lambda^2 - \lambda^3 = \lambda(\lambda+2)(3-\lambda).$$

So the eigenvalues of **A**, that is the roots of $p(\lambda) = 0$, are

$$\lambda_1 = -2, \lambda_2 = 0 \text{ and } \lambda_3 = 3.$$

To find the eigenvector $\mathbf{x}_1$ corresponding to λ_1 we must solve (5.2) with $\lambda_i = \lambda_1 = -2$. The matrix Eq. (5.2) then becomes

$$\begin{bmatrix} 1-(-2) & 0 & -1 \\ -2 & -1-(-2) & 2 \\ -1 & 2 & 1-(-2) \end{bmatrix} \begin{bmatrix} x_1^{(1)} \\ x_2^{(1)} \\ x_3^{(1)} \end{bmatrix} = \begin{bmatrix} 3 & 0 & -1 \\ -2 & 1 & 2 \\ -1 & 2 & 3 \end{bmatrix} \begin{bmatrix} x_1^{(1)} \\ x_2^{(1)} \\ x_3^{(1)} \end{bmatrix} = \begin{bmatrix} 0 \\ 0 \\ 0 \end{bmatrix}.$$

The superscript (1) attached to the unknowns $x_1^{(1)}$, $x_2^{(1)}$ and $x_3^{(1)}$ is used to show these are the elements of the eigenvector $\mathbf{x}_1$ corresponding to $\lambda_1 = -2$. When this matrix equation is written out in full it leads to the three equations

$$3x_1^{(1)} - x_3^{(1)} = 0, -2x_1^{(1)} + x_2^{(1)} + 2x_3^{(1)} = 0, -x_1^{(1)} + 2x_2^{(1)} + 3x_3^{(1)} = 0.$$

As these three equations are homogeneous, and the determinant of their coefficient matrix is zero, any one of the equations must be linearly dependent on the other two. Discarding one of these equations as redundant, and using the two that remain, it is only possible for two of the three unknowns $x_1^{(1)}$, $x_2^{(1)}$ and $x_3^{(1)}$ to be found in terms of a third unknown, say $x_1^{(1)}$, which can be assigned an arbitrary value. If in this case we take the third equation to be redundant, we are left with the first two equations from which to determine the three unknowns ${x_1}^{(1)}, {x_2}^{(1)}$ and ${x_3}^{(1)}$. It is easy to confirm that the first two equations are linearly independent, because they are not proportional. To proceed further, let us find ${x_2}^{(1)}$ and ${x_3}^{(1)}$ in terms ${x_1}^{(1)}$ by setting ${x_1}^{(1)} = k_1$, where $k_1 \neq 0$ is arbitrary (it can be regarded as a parameter). The first equation gives $x_3^{(1)} = 3k_1$, so using $x_1^{(1)} = k_1$ and $x_3^{(1)} = 3k_1$ in the second equation we find that $x_2^{(1)} = -4k_1$.

Of course, using $x_1^{(1)} = k_1$, and $x_3^{(1)} = 3k_1$ in the third equation will again give $x_2^{(1)} = -4k_1$, confirming it redundancy, since it is automatically compatible with the first two equations.

We have shown the eigenvector $\mathbf{x}_1$ can be taken to be $\mathbf{x}_1 = [k_1, -4k_1, \ 3k_1]^{\mathrm{T}}$, where as usual, to save space on a printed page, the column eigenvector $\mathbf{x}_1$ has been written as the transpose of a row vector. As the scaling of an eigenvector is arbitrary, it is usual to set the scale factor k_1 equal to a convenient numerical value, typically $k_1 = 1$, when the above eigenvector becomes $\mathbf{x}_1 = [1, -4, 3]^{\mathrm{T}}$.

To find the eigenvector $\mathbf{x}_2$, Eq. (5.2) must be solved with $\lambda_i = \lambda_2 = 0$, when the elements $x_1^{(2)}$, $x_2^{(2)}$ and $x_3^{(2)}$ of $\mathbf{x}_2$ are seen to satisfy the matrix equation

$$\begin{bmatrix} 1-(0) & 0 & -1 \\ -2 & -1-(0) & 2 \\ 1 & 2 & 1-(0) \end{bmatrix} \begin{bmatrix} x_1^{(2)} \\ x_2^{(2)} \\ x_3^{(2)} \end{bmatrix} = \begin{bmatrix} 1 & 0 & -1 \\ -2 & -1 & 2 \\ 1 & 2 & 1 \end{bmatrix} \begin{bmatrix} x_1^{(2)} \\ x_2^{(2)} \\ x_3^{(2)} \end{bmatrix} = \begin{bmatrix} 0 \\ 0 \\ 0 \end{bmatrix}.$$

When written out in full, the components of $\boldsymbol{x}_2$ are determined by the equations

$$x_1^{(2)} - x_3^{(2)} = 0\,, -2x_1^{(2)} - x_2^{(2)} + 2x_3^{(2)} = 0, -x_1^{(2)} + 2x_2^{(2)} + x_3^{(2)} = 0\,.$$

For convenience we will again take the third equation to be redundant, and set $x_1^{(2)} = k_2$ (an arbitrary parameter). Then, proceeding as before, we find that $x_1^{(2)} = k_2$, $x_2^{(2)} = 0$, $x_3^{(2)} = k_2$, so the eigenvector $\boldsymbol{x}_2$ becomes $\boldsymbol{x}_2 = [k_2, 0, k_2]^{\mathrm{T}}$. Assigning k_2 the arbitrary value $k_2 = 1$, the eigenvector $\boldsymbol{x}_2$ corresponding to $\lambda_2 = 0$ can be taken to be $\boldsymbol{x}_2 = [1, 0, 1]^{\mathrm{T}}$.

Finally, to find $\mathbf{x}_3$ we set $\lambda_i = \lambda_3 = 3$ in (5.2), when the matrix equation becomes

$$\begin{bmatrix} 1-(3) & 0 & -1 \\ -2 & -1-(3) & 2 \\ 1 & 2 & 1-(3) \end{bmatrix} \begin{bmatrix} x_1^{(3)} \\ x_2^{(3)} \\ x_3^{(3)} \end{bmatrix} = \begin{bmatrix} 2 & 0 & -1 \\ -2 & -4 & 2 \\ 1 & 2 & -2 \end{bmatrix} \begin{bmatrix} x_1^{(3)} \\ x_2^{(3)} \\ x_3^{(3)} \end{bmatrix} = \begin{bmatrix} 0 \\ 0 \\ 0 \end{bmatrix},$$

leading to the three scalar equations

$$2x_1^{(3)} - x_3^{(3)} = 0\,, -2x_1^{(3)} - 4x_2^{(3)} + 2x_3^{(3)} = 0,\ x_1^{(3)} + 2x_2^{(3)} - 2x_3^{(3)} = 0.$$

As before, one of these equations must be redundant, so once again taking this to be the third equation and setting $x_1^{(3)} = k_3$ (an arbitrary parameter), we find that $x_1^{(3)} = k_3$, $x_2^{(3)} = -\frac{3}{2}k_3$, and $x_3^{(3)} = -2k_3$. So the eigenvector $\mathbf{x}_3$ becomes $\mathbf{x}_3 = [k_3, -\frac{3}{2}k_3, -2k_3]^{\mathrm{T}}$. Again setting $k_3 = 1$, the eigenvector corresponding to $\lambda_3 = 3$ becomes $\mathbf{x}_3 = [1, -\frac{3}{2}, -2]^{\mathrm{T}}$.

In summary, the eigenvalues and the corresponding eigenvectors of $\mathbf{A}$ are

$$\lambda_1 = -2,\ \mathbf{x}_1 = \begin{bmatrix} 1 \\ -4 \\ 3 \end{bmatrix},\quad \lambda_2 = 0,\ \mathbf{x}_2 = \begin{bmatrix} 1 \\ 0 \\ 1 \end{bmatrix},\quad \lambda_3 = 3,\ \mathbf{x}_3 = \begin{bmatrix} 1 \\ -\frac{3}{2} \\ -2 \end{bmatrix}.$$

These three eigenvectors are nontrivial solutions of linear homogeneous algebraic equations, so they must be linearly independent. This is easily checked, because the determinant with these vectors as its columns must not vanish, and this is so because

$$\begin{vmatrix} 1 & 1 & 1 \\ -4 & 0 & -\frac{3}{2} \\ 3 & 1 & -2 \end{vmatrix} = -15 .$$

◇

As eigenvectors can be scaled arbitrarily, the essential feature of an eigenvector is not the actual values of its elements, but the fact that the ratios between elements is fixed. In calculations involving eigenvectors, the scaling of eigenvectors is often used to advantage to adjust the relative sizes of the absolute values of the elements of an eigenvector. Typically, an eigenvector of the form $\mathbf{x} = [a, b, c]^{\mathrm{T}}$ is *normalized* by setting $k = 1/\sqrt{a^2 + b^2 + c^2}$, and taking as the *normalized* eigenvector $\tilde{\mathbf{x}}$ the vector $\tilde{\mathbf{x}} = [ka,\ kb,\ kc]^{\mathrm{T}}$. In Example 5.1, in terms of this scaling, the normalized eigenvector $\mathbf{x}_1$ becomes $\tilde{\mathbf{x}}_1 = \left[\frac{1}{\sqrt{26}}, -\frac{4}{\sqrt{26}}, \frac{3}{\sqrt{26}}\right]^{\mathrm{T}}$.

The purpose of normalization is to adopt a convenient scale for the sizes of the elements in an eigenvector that ensures the magnitude of its largest element does not exceed one. This, in turn, helps limit the growth of round-off errors when an eigenvector is used repeatedly in calculations.

Another example involving the determination of eigenvalues and eigenvectors now follows. In this example two eigenvalues are repeated, though the system is such that two linearly independent eigenvectors can still be found that correspond to the single repeated eigenvalue.

Example 5.2. Find the eigenvalues and eigenvectors of

$$\mathbf{A} = \begin{bmatrix} 1 & 0 & 0 \\ 0 & 2 & -1 \\ 0 & 0 & 1 \end{bmatrix}.$$

Solution. The characteristic equation $p(\lambda) = \det[\mathbf{A} - \lambda\mathbf{I}] = 0$ becomes

$$p(\lambda) = \begin{vmatrix} 1-\lambda & 0 & 0 \\ 0 & 2-\lambda & -1 \\ 0 & 0 & 1-\lambda \end{vmatrix} = (1-\lambda)^2(2-\lambda) = 0 ,$$

so the eigenvalues of $\mathbf{A}$ are $\lambda_1 = 1$, $\lambda_2 = 1$, $\lambda_3 = 2$. In this case, the eigenvalue $\lambda = 1$ has multiplicity 2 (it is repeated twice). If possible, let us find two *different* (linearly independent) eigenvectors corresponding to $\lambda = 1$. To find these eigenvectors we must attempt to find two linearly independent solution vectors $\mathbf{x}$ of $[\mathbf{A} - \lambda\mathbf{I}]\mathbf{x} = \mathbf{0}$ when $\lambda = 1$. So setting $\lambda = \lambda_1 = 1$ the matrix equation from which the eigenvectors must be obtained becomes

$$\begin{bmatrix} 0 & 0 & 0 \\ 0 & 1 & -1 \\ 0 & 0 & 0 \end{bmatrix} \begin{bmatrix} x_1^{(1)} \\ x_2^{(1)} \\ x_3^{(1)} \end{bmatrix} = \begin{bmatrix} 0 \\ 0 \\ 0 \end{bmatrix}.$$

Only one scalar equation is obtained when this matrix equation is written out in full, and it is $x_2^{(1)} - x_3^{(1)} = 0$, in which one of the variables, say $x_3^{(1)}$, can be assigned an arbitrary value. Notice there is *no* equation involving $x_1^{(1)}$, so this variable, like $x_3^{(1)}$, can be assigned an arbitrary value.

Assigning the values $x_1^{(1)} = 0$ and $x_3^{(1)} = 1$, one of the eigenvectors corresponding to $\lambda = \lambda_1 = 1$ must be proportional to the vector $\mathbf{x}_1 = [0, 1, 1]^{\mathrm{T}}$. Next, assigning the different values $x_1^{(1)} = 1$ and $x_3^{(1)} = 0$, a second (linearly independent) eigenvector corresponding to $\lambda = \lambda_1 = 1$ will be proportional to $\boldsymbol{x}_2 = [1, 0, 0]^{\mathrm{T}}$. The third eigenvector $\mathbf{x}_3$, corresponding to setting $\lambda = \lambda_3 = 2$, follows by solving

$$\begin{bmatrix} 1-(2) & 0 & 0 \\ 0 & 2-(2) & -1 \\ 0 & 0 & 1-(2) \end{bmatrix} \begin{bmatrix} x_1^{(3)} \\ x_2^{(3)} \\ x_3^{(3)} \end{bmatrix} = \begin{bmatrix} -1 & 0 & 0 \\ 0 & 0 & -1 \\ 0 & 0 & 0-1 \end{bmatrix} \begin{bmatrix} x_1^{(3)} \\ x_2^{(3)} \\ x_3^{(3)} \end{bmatrix} = \begin{bmatrix} 0 \\ 0 \\ 0 \end{bmatrix}.$$

When written out in full, this shows that $x_1^{(3)} = x_3^{(3)} = 0$, but as there is no equation involving $x_2^{(3)}$ this may be taken to be arbitrary, so for simplicity we set $x_2^{(3)} = 1$. Thus the third eigenvector will be proportional to $\mathbf{x}_3 = [0, 1, 0]^{\mathrm{T}}$. In summary, we have shown that the eigenvalues and eigenvectors of $\mathbf{A}$ are

$$\lambda_1 = 1, \quad \mathbf{x}_1 = \begin{bmatrix} 0 \\ 1 \\ 1 \end{bmatrix}, \quad \lambda_2 = 1, \quad \mathbf{x}_2 = \begin{bmatrix} 1 \\ 0 \\ 0 \end{bmatrix}, \quad \lambda_3 = 2, \quad \mathbf{x}_3 = \begin{bmatrix} 0 \\ 1 \\ 0 \end{bmatrix}.$$

The linear independence of these eigenvectors is easily confirmed, because the value of the determinant with its columns equal to $\mathbf{x}_1$, $\boldsymbol{x}_2$ and $\mathbf{x}_3$ is 1, and so does not vanish.

◇

Not every $n \times n$ matrix $\mathbf{A}$ has n linearly independent eigenvectors. For example, the matrix

$$\mathbf{A} = \begin{bmatrix} 1 & -3 & 1 \\ -1 & -1 & 1 \\ 1 & -2 & 0 \end{bmatrix}$$

has the three eigenvalues $\lambda_1 = 2$, $\lambda_2 = \lambda_3 = -1$, which are the roots of the characteristic equation $p(\lambda) = \lambda^3 - 3\lambda - 2 = 0$. The eigenvector corresponding to the single eigenvalue $\lambda_1 = 2$ is easily shown to be proportional to $\mathbf{x}_1 = [-8, 1, -5]^{\mathrm{T}}$. However, corresponding to the repeated eigenvalue $\lambda = -1$, when the equation $[\mathbf{A} - \lambda\mathbf{I}]\mathbf{x} = \mathbf{0}$ is written out in full, it becomes the three equations $2x_1^{(2)} - 3x_2^{(2)} + x_3^{(2)} = 0$, $-x_1^{(2)} + x_3^{(2)} = 0$ and $x_1^{(2)} - 2x_2^{(2)} + x_3^{(2)} = 0$.

Discarding the third equation as redundant, the first two equations can be solved in terms of $x_3^{(2)} = k$, an arbitrary constant, when $x_1^{(2)} = k$, $x_2^{(2)} = k$, showing that

$\mathbf{x}_2 = [k,\ k,\ k]^{\mathrm{T}}$. This is the *only* possible solution when $\lambda = -1$ (twice), so setting $k = 1$ we find the single eigenvector $\boldsymbol{x}_2 = \mathbf{x}_3 = [1,\ 1,\ 1]^{\mathrm{T}}$ that corresponds to the repeated eigenvalue $\lambda = -1$. We will denote this single eigenvector by $\mathbf{x}_{(2,3)} = [1, 1, 1]^{\mathrm{T}}$, where the subscript $_{(2,3)}$ indicates that $\mathbf{x}_{(2,3)}$ corresponds to the repeated eigenvalue. The eigenvalues and normalized eigenvectors are

$$\lambda_1 = 2,\ \ \tilde{\mathbf{x}}_1 = \begin{bmatrix} -\frac{8}{3\sqrt{10}} \\ \frac{1}{3\sqrt{10}} \\ -\frac{5}{3\sqrt{10}} \end{bmatrix} \text{ and } \lambda_2 = \lambda_3 = -1,\ \ \tilde{\mathbf{x}}_{(2,3)} = \begin{bmatrix} \frac{1}{\sqrt{3}} \\ \frac{1}{\sqrt{3}} \\ \frac{1}{\sqrt{3}} \end{bmatrix}.$$

To distinguish between two different situations involving eigenvalues and eigenvectors, the number r of repeated eigenvalues is called the *algebraic multiplicity* of the eigenvalue, while the number m of linearly independent eigenvectors corresponding to a repeated eigenvalue is called the *geometric multiplicity* of the eigenvalue, and clearly $m \leq r$.

Even though a square matrix $\mathbf{A}$ with real elements leads to a characteristic equation in the form of a polynomial equation with *real* coefficients, some of the roots of the characteristic equation may be *complex*, as was seen in (4.17). We now make use of the following elementary and easily proved result from complex analysis, which we state as follows (see Exercise 34).

5.1.1 The Roots of Polynomials with Real Coefficients

The roots of a polynomial equation with *real* coefficients may be either real or complex, but if any roots are complex they must occur in *complex conjugate pairs*. Thus if the degree of a polynomial equation with real coefficients is odd, it must have at least one real root.

This result is useful when working with third-degree polynomials $p(\lambda) = 0$ with real coefficients, because in this case one root is always real.

If in simple cases the real root λ_1 can be found by trial and error, the quantity $(\lambda - \lambda_1)$ must be a factor of $p(\lambda)$. Dividing $p(\lambda)$ by the factor $(\lambda - \lambda_1)$, and equating the result to zero, will lead to a quadratic equation whose roots will be the other two eigenvalues, and these can be found from the quadratic formula (they will either be real or complex conjugates). By this method, all three roots of $p(\lambda)$ can be obtained.

If a real root of $p(\lambda)$ cannot be found by trial and error, or if the degree of $p(\lambda)$ is greater than three, it becomes necessary to use numerical methods to find the eigenvalues and eigenvectors. The numerical determination of eigenvalues and their associated eigenvectors is a special topic in numerical analysis, and it will not be discussed here.

The next example illustrates a simple case where a real root of the characteristic equation (an eigenvalue) can be found by inspection, and the result then used to find the two remaining roots (eigenvalues) that turn out to be complex conjugates.

Example 5.3. Find the eigenvalues and eigenvectors of

$$\mathbf{A} = \begin{bmatrix} 1 & 0 & 1 \\ 1 & 1 & 0 \\ 0 & 1 & 1 \end{bmatrix}.$$

Solution. The characteristic polynomial is

$$p(\lambda) = \det[\mathbf{A} - \lambda\mathbf{I}] = \begin{vmatrix} 1-\lambda & 0 & 1 \\ 1 & 1-\lambda & 0 \\ 0 & 1 & 1-\lambda \end{vmatrix} = 2 - 3\lambda + 3\lambda^2 - \lambda^3.$$

The eigenvalues of $\mathbf{A}$ are the roots of $2 - 3\lambda + 3\lambda^2 - \lambda^3 = 0$ which, for convenience, we rewrite as $\lambda^3 - 3\lambda^2 + 3\lambda - 2 = 0$. Trial and error shows $\lambda_1 = 2$ is an eigenvalue, so $(\lambda - \lambda_1) = (\lambda - 2)$ is a factor of $\lambda^3 - 3\lambda^2 + 3\lambda - 2$. Long division shows that $(\lambda^3 - 3\lambda^2 + 3\lambda - 2)/(\lambda - 2) = \lambda^2 - \lambda + 1$, so the remaining two eigenvalues are the roots of $\lambda^2 - \lambda + 1 = 0$. Applying the quadratic formula to this equation shows its roots are the complex conjugates $\frac{1}{2}(1 + i\sqrt{3})$ and $\frac{1}{2}(1 - i\sqrt{3})$. Thus $\mathbf{A}$ has the one real and the two complex conjugate eigenvalues $\lambda_1 = 2$, $\lambda_2 = \frac{1}{2}(1 + i\sqrt{3})$ and $\lambda_3 = \frac{1}{2}(1 - i\sqrt{3})$. Proceeding as usual, and setting $x_1^{(1)} = 1$, eigenvector $\mathbf{x}_1$ is found to be proportional to $\mathbf{x}_1 = [1, 1, 1]^{\mathrm{T}}$.

To determine the eigenvector $\mathbf{x}_2$ corresponding to $\lambda_2 = \frac{1}{2}(1 + i\sqrt{3})$ it is necessary to set $\lambda = \frac{1}{2}(1 + i\sqrt{3})$ in (5.2), when we obtain

$$\begin{bmatrix} 1-\frac{1}{2}(1+i\sqrt{3}) & 0 & 1 \\ 1 & 1-\frac{1}{2}(1+i\sqrt{3}) & 0 \\ 0 & 1 & 1-\frac{1}{2}(1+i\sqrt{3}) \end{bmatrix} \begin{bmatrix} x_1^{(2)} \\ x_2^{(2)} \\ x_3^{(2)} \end{bmatrix} = \begin{bmatrix} 0 \\ 0 \\ 0 \end{bmatrix}.$$

Thus the elements of $\boldsymbol{x}_2 = [x_1^{(2)}, x_2^{(2)}, x_3^{(3)}]^{\mathrm{T}}$ must satisfy the equations

$$\tfrac{1}{2}\left(1 - i\sqrt{3}\right)x_1^{(2)} + x_3^{(2)} = 0,\; x_1^{(2)} + \tfrac{1}{2}\left(1 - i\sqrt{3}\right)x_2^{(2)} = 0,\; x_2^{(2)} + \tfrac{1}{2}\left(1 + i\sqrt{3}\right)x_3^{(2)} = 0.$$

Discarding one of these equations as redundant, say the last equation, and solving the remaining equations in terms of $x_1^{(2)}$, where for convenience we set $x_1^{(2)} = 1$, we find that

$$x_1^{(2)} = 1,\; x_2^{(2)} = -\tfrac{1}{2}\left(1 + i\sqrt{3}\right),\; x_3^{(2)} = -\tfrac{1}{2}\left(1 - i\sqrt{3}\right).$$

Thus the complex eigenvector $\boldsymbol{x}_2$ corresponding to $\lambda_2 = \frac{1}{2}(1 + i\sqrt{3})$ can be taken to be

$$\mathbf{x}_2 = \begin{bmatrix} 1 \\ -\frac{1}{2}(1 + i\sqrt{3}) \\ -\frac{1}{2}(1 - i\sqrt{3}) \end{bmatrix}.$$

To find the eigenvector $\mathbf{x}_3$ it is not necessary to solve the equations with $\lambda = \lambda_3$, because λ_3 is the complex conjugate of λ_2, so the complex eigenvector $\mathbf{x}_3$ must be the complex conjugate of $\boldsymbol{x}_2$. Thus the three eigenvalues and eigenvectors of $\mathbf{A}$ are

$$\lambda_1 = 2,\ \mathbf{x}_1 = \begin{bmatrix} 1 \\ 1 \\ 1 \end{bmatrix},\ \lambda_2 = \tfrac{1}{2}\left(1 + i\sqrt{3}\right),\ \mathbf{x}_2 = \begin{bmatrix} 1 \\ -\frac{1}{2}(1+i\sqrt{3}) \\ -\frac{1}{2}(1-i\sqrt{3}) \end{bmatrix},\ \text{and}$$

$$\lambda_3 = \tfrac{1}{2}\left(1 - i\sqrt{3}\right),\ \mathbf{x}_3 = \begin{bmatrix} 1 \\ -\frac{1}{2}(1-i\sqrt{3}) \\ -\frac{1}{2}(1+i\sqrt{3}) \end{bmatrix}.$$

This has shown that, apart from working with complex numbers, the process of finding a complex eigenvector corresponding to a complex eigenvalue is the same as finding a real eigenvector that corresponds to a real eigenvalue.

◇

As an eigenvector can always be scaled by a real or complex number and still remain an eigenvector, the result of scaling by different constants can make a given eigenvector look very different. For example, if the eigenvector $\boldsymbol{x}_2$ in Example 5.3 is scaled by i, the result will still be an eigenvector associated with $\lambda_2 = \frac{1}{2}(1 + i\sqrt{3})$, though the eigenvector will look very different, because it will then become

$$\mathbf{x}_2 = \begin{bmatrix} i \\ -\frac{1}{2}(i-\sqrt{3}) \\ -\frac{1}{2}(i+\sqrt{3}) \end{bmatrix}.$$

Later, when matrices are used with systems of linear differential equations with the independent variable t, it will be found that the failure to determine eigenvectors up to a multiplicative numerical factor will make no difference to the solution of the system of equations. It will also be seen that complex conjugate eigenvectors, like $\boldsymbol{x}_2$ and $\mathbf{x}_3$ in Example 5.3, will lead to real solutions of differential equations containing terms like $e^{\alpha t}\sin \beta t$ and $e^{\alpha t}\cos \beta t$, which correspond to complex conjugate eigenvalues $\lambda = \alpha \pm i\beta$.

Theorem 5.1 *Eigenvalues and Eigenvectors of Symmetric Matrices*

A real symmetric $n \times n$ matrix $\mathbf{A}$ always has n real eigenvalues and n linearly independent eigenvectors that are mutually orthogonal.

Proof. The proof is based on the fact that if $\mathbf{A}$ is symmetric, then $\mathbf{A}^{\mathrm{T}} = \mathbf{A}$, and if $\mathbf{x}$ and $\mathbf{y}$ are n element column vectors, the product $\mathbf{y}^{\mathrm{T}}\mathbf{A}\mathbf{x}$ is a scalar quantity, so that $(\mathbf{y}^{\mathrm{T}}\mathbf{A}\mathbf{x})^{\mathrm{T}} = \mathbf{x}^{\mathrm{T}}\mathbf{A}\mathbf{y}$. It will also be necessary to use the result that an eigenvalue λ will be real if $\lambda = \bar{\lambda}$, where the overbar indicates a complex conjugate. This last result follows from the fact that if $\lambda = a + ib$, then $\bar{\lambda} = a - ib$, so $\lambda = \bar{\lambda}$ is only possible if $b = 0$ (that is if λ is purely real).

Suppose, if possible, that λ is complex, and it has associated with it a complex eigenvector $\mathbf{x}$. Then, as the characteristic equation has real coefficients, λ must have associated with it a complex conjugate eigenvalue $\bar{\lambda}$, and an associated complex conjugate eigenvector $\mathbf{x}$. Thus, λ, $\bar{\lambda}$, $\mathbf{x}$ and $\mathbf{x}$ must satisfy the equations

$$\mathbf{Ax} = \lambda\mathbf{x} \text{ and } \mathbf{Ax} = \bar{\lambda}\mathbf{x},$$

because $\mathbf{A} = \bar{\mathbf{A}}$, since $\mathbf{A}$ is real. Pre-multiplying $\mathbf{Ax} = \lambda\mathbf{x}$ by $\bar{\mathbf{x}}^{\mathrm{T}}$gives $\bar{\mathbf{x}}^{\mathrm{T}}\mathbf{Ax} = \lambda\bar{\mathbf{x}}^{\mathrm{T}}\mathbf{x}$, and pre-multiplying $A\bar{\mathbf{x}} = \bar{\lambda}\bar{\mathbf{x}}$ by $\mathbf{x}^{\mathrm{T}}$ gives $\mathbf{x}^{\mathrm{T}}A\bar{\mathbf{x}} = \bar{\lambda}\mathbf{x}^{\mathrm{T}}\bar{\mathbf{x}}$. Taking these results together, and setting $\mathbf{y} = \bar{\mathbf{x}}$ in $\mathbf{y}^{\mathrm{T}}\mathbf{Ax} = \mathbf{x}^{\mathrm{T}}\mathbf{Ay}$, which is true because $\mathbf{A}$ is symmetric, shows that $\lambda\bar{\mathbf{x}}^{\mathrm{T}}\mathbf{x} = \bar{\lambda}\mathbf{x}^{\mathrm{T}}\bar{\mathbf{x}}$, but $\bar{\mathbf{x}}^{\mathrm{T}}\mathbf{x} = \mathbf{x}^{\mathrm{T}}\bar{\mathbf{x}}$ is real, so the previous equation is only possible if $\lambda = \bar{\lambda}$, and the result is proved.

To establish the mutual orthogonality of the eigenvectors it is necessary to show that if $\mathbf{x}_i$ and $\mathbf{x}_j$ are eigenvectors of $\mathbf{A}$, corresponding to the *different* eigenvalues λ_i and λ_j, then $\mathbf{x}_i^{\mathrm{T}}\mathbf{x}_j = 0$ if $i \neq j$. Assuming the eigenvalues are all distinct (different), and using the defining equations $\mathbf{Ax}_i = \lambda_i\mathbf{x}_i$ and $\mathbf{Ax}_j = \lambda_j\mathbf{x}_j$, pre-multiplication of these, respectively, by $\mathbf{x}_i{}^{\mathrm{T}}$ and $\mathbf{x}_j{}^{\mathrm{T}}$, gives $\mathbf{x}_j{}^{\mathrm{T}}\mathbf{Ax}_i = \lambda_i\mathbf{x}_j{}^{\mathrm{T}}\,\mathbf{x}_i$ and $\mathbf{x}_i{}^{\mathrm{T}}\mathbf{Ax}_j = \lambda_j\mathbf{x}_i{}^{\mathrm{T}}\,\mathbf{x}_j$. However, from $\mathbf{y}^{\mathrm{T}}\mathbf{Ax} = \mathbf{x}^{\mathrm{T}}\mathbf{Ay}$ with $\mathbf{x} = \mathbf{x}_i$ and $\mathbf{y} = \mathbf{x}_j$, we see that $\mathbf{x}_i{}^{\mathrm{T}}\mathbf{Ax}_j = \mathbf{x}_j{}^{\mathrm{T}}\mathbf{Ax}_i$, so $\mathbf{x}_i{}^{\mathrm{T}}\mathbf{x}_j = \mathbf{x}_j{}^{\mathrm{T}}\mathbf{x}_i$. Subtracting $\mathbf{x}_j{}^{\mathrm{T}}\mathbf{Ax}_i = \lambda_i\mathbf{x}_j{}^{\mathrm{T}}\,\mathbf{x}_i$ from $\mathbf{x}_i{}^{\mathrm{T}}\mathbf{Ax}_j = \lambda_j\mathbf{x}_i{}^{\mathrm{T}}\,\mathbf{x}_j$, and using this last result, we find that $(\lambda_i - \lambda_j)\mathbf{x}_i^{\mathrm{T}}\mathbf{x}_j = 0$, but by hypothesis $\lambda_i \neq \lambda_j$ for $i \neq j$, so it must follow that $\mathbf{x}_i^{\mathrm{T}}\mathbf{x}_j = 0$ when $i \neq j$. Thus the mutual orthogonality of the eigenvectors corresponding to different eigenvalues has been proved. A similar proof, not given here, establishes the mutual orthogonality of eigenvectors when an $n \times n$ real matrix $\mathbf{A}$ has eigenvalues with algebraic multiplicities greater than 1, though there are still n linearly independent eigenvectors.

◇

Corollary 5.1. *Symmetric and Related Orthogonal Matrices Let the eigenvectors* $\mathbf{x}_1, \mathbf{x}_2, \ldots, \mathbf{x}_n$ *of a symmetric* $n \times n$ *real matrix* $\mathbf{A}$ *be normalized so that* $\mathbf{x}_i{}^T\mathbf{x}_i = 1$. *Then a matrix* $\mathbf{Q}$ *with its columns formed by the normalized eigenvectors of* $\mathbf{A}$, *arranged in any order, is an orthogonal matrix.*

Proof. The result is almost immediate, because if $\mathbf{x}_i = \left[x_1^{(i)},\ x_2^{(i)},\ \ldots\ ,x_n^{(i)}\right]^{\mathrm{T}}$, normalizing $\mathbf{x}_i$ by dividing its elements by $\left[\left(x_1^{(i)}\right)^2 + \left(x_2^{(i)}\right)^2 + \cdots + \left(x_n^{(i)}\right)^2\right]^{1/2}$, for $i = 1, 2, \ldots, n$, will produce the required normalization $\tilde{\mathbf{x}}_i$ of $\mathbf{x}_i$, and the mutual orthogonality of the vectors $\mathbf{x}_i$ and $\mathbf{x}_j$ implies the mutual orthogonality of the vectors $\tilde{\mathbf{x}}_i$ and $\tilde{\mathbf{x}}_j$ with $i \neq j$. Furthermore, by constructing a matrix $\mathbf{Q}$ with its columns the normalized eigenvectors $\tilde{\mathbf{x}}_i$, it follows that $\mathbf{Q}^{\mathrm{T}}\mathbf{Q} = \mathbf{I}$, so $\mathbf{Q}^{\mathrm{T}} = \mathbf{Q}^{-1}$ showing that $\mathbf{Q}$ is an orthogonal matrix.

◇

Example 5.4. Find the eigenvalues and eigenvectors of

$$\mathbf{A} = \begin{bmatrix} 1 & -2 & 0 \\ 0 & 1 & 2 \\ 0 & 2 & 1 \end{bmatrix},$$

and use the results to construct an orthogonal matrix $\mathbf{Q}$.

Solution. The eigenvalues and eigenvectors of **A** are easily found to be

$$\lambda_1 = 1,\ \mathbf{x}^{(1)} = \begin{bmatrix} 1 \\ 0 \\ 0 \end{bmatrix},\ \lambda_2 = 3,\ \mathbf{x}^{(2)} = \begin{bmatrix} 0 \\ 1 \\ 1 \end{bmatrix},\ \lambda_3 = -1,\ \mathbf{x}^{(3)} = \begin{bmatrix} 0 \\ -1 \\ 1 \end{bmatrix},$$

so the normalized eigenvectors are

$$\tilde{\mathbf{x}}_1 = \begin{bmatrix} 1 \\ 0 \\ 0 \end{bmatrix},\ \tilde{\mathbf{x}}_2 = \begin{bmatrix} 0 \\ \frac{1}{\sqrt{2}} \\ \frac{1}{\sqrt{2}} \end{bmatrix},\ \tilde{\mathbf{x}}_3 = \begin{bmatrix} 0 \\ -\frac{1}{\sqrt{2}} \\ \frac{1}{\sqrt{2}} \end{bmatrix}.$$

Thus an orthogonal matrix with these as its column vectors is

$$\mathbf{Q} = \begin{bmatrix} 1 & 0 & 0 \\ 0 & \frac{1}{\sqrt{2}} & -\frac{1}{\sqrt{2}} \\ 0 & \frac{1}{\sqrt{2}} & \frac{1}{\sqrt{2}} \end{bmatrix}.$$

This is easily checked, because $\mathbf{Q}\mathbf{Q}^{\mathrm{T}} = \mathbf{I}$, showing that $\mathbf{Q}^{\mathrm{T}} = \mathbf{Q}^{-1}$. Different orthogonal matrices can be formed by changing the order of the columns in **Q**. ◊

5.2 Diagonalization of Matrices

The diagonalization of general $n \times n$ (square) matrices is of fundamental importance in many applications, and especially when solving systems of linear constant coefficient differential equations, as will be seen in Chapter 6. Let us now show precisely how and when it is possible to transform a general $n \times n$ real matrix **A** into an $n \times n$ diagonal matrix **D**.

We start by considering the matrix product **AP**, where **P** is the $n \times n$ matrix with its columns the n different (linearly independent) eigenvectors $\mathbf{x}^{(i)}$ of **A** satisfying the n equations

$$\mathbf{A}\mathbf{x}^{(i)} = \lambda_i \mathbf{x}^{(i)}, i = 1, 2, \ldots, n, \tag{5.3}$$

where the matrix Eq. (5.3) is simply result (5.1) written in a different way. Now

$$\mathbf{AP} = \begin{bmatrix} a_{11} & a_{12} & \cdots & a_{1n} \\ a_{21} & a_{22} & \cdots & a_{2n} \\ \vdots & \vdots & \vdots & \vdots \\ a_{n1} & a_{n2} & \cdots & a_{nn} \end{bmatrix} \begin{bmatrix} x_1^{(1)} & x_1^{(2)} & \cdots & x_1^{(n)} \\ x_2^{(1)} & x_2^{(2)} & \cdots & x_2^{(n)} \\ \vdots & \vdots & \vdots & \vdots \\ x_n^{(1)} & x_n^{(2)} & \cdots & x_n^{(n)} \end{bmatrix} \tag{5.4}$$

$$= \left[\mathbf{A}\mathbf{x}^{(1)}, \mathbf{A}\mathbf{x}^{(2)}, \ldots, \mathbf{A}\mathbf{x}^{(n)}\right],$$

where the $\mathbf{Ax}^{(i)}$ for $i = 1, 2, \ldots, n$ are the $n \times 1$ columns of $\mathbf{AP}$. Using (5.3) allows this to be rewritten as

$$\mathbf{AP} = \left[\lambda^{(1)}\mathbf{x}^{(1)},\ \lambda_2\mathbf{x}^{(2)},\ \ldots,\ \lambda_n\mathbf{x}^{(n)}\right]. \tag{5.5}$$

Next, let $\mathbf{D}$ be the $n \times n$ diagonal matrix $\mathbf{D} = \operatorname{diag}\{\lambda_1, \lambda_2, \ldots, \lambda_n\}$ with the elements $\lambda_1, \lambda_2, \ldots, \lambda_n$ arranged along its leading diagonal in the *same* order as that of the eigenvectors $\mathbf{x}^{(i)}$ forming the columns of $\mathbf{P}$, so that

$$\mathbf{D} = \begin{bmatrix} \lambda_1 & 0 & \cdots & 0 \\ 0 & \lambda_2 & \cdots & 0 \\ \vdots & \vdots & \vdots & \vdots \\ 0 & 0 & \cdots & \lambda_n \end{bmatrix}.$$

Now form the product

$$\begin{aligned}\mathbf{PD} &= \begin{bmatrix} x_1^{(1)} & x_1^{(2)} & \cdots & x_1^{(n)} \\ x_2^{(1)} & x_2^{(2)} & \cdots & x_2^{(n)} \\ \vdots & \vdots & \vdots & \vdots \\ x_n^{(1)} & x_n^{(2)} & \cdots & x_n^{(n)} \end{bmatrix} \begin{bmatrix} \lambda_1 & 0 & \cdots & 0 \\ 0 & \lambda_2 & \cdots & 0 \\ \vdots & \vdots & \vdots & \vdots \\ 0 & 0 & \cdots & \lambda_n \end{bmatrix} \\ &= \begin{bmatrix} \lambda_1 x_1^{(1)} & \lambda_2 x_1^{(2)} & \cdots & \lambda_n x_1^{(n)} \\ \lambda_1 x_2^{(1)} & \lambda_2 x_2^{(2)} & \cdots & \lambda_n x_2^{(n)} \\ \vdots & \vdots & \vdots & \vdots \\ \lambda_1 x_n^{(1)} & \lambda_2 x_n^{(2)} & \cdots & \lambda_n x_n^{(n)} \end{bmatrix} = \left[\lambda_1\mathbf{x}^{(1)}, \lambda_2\mathbf{x}^{(2)}, \ldots, \lambda_n\mathbf{x}^{(n)}\right].\end{aligned} \tag{5.6}$$

A comparison of (5.5) and (5.6) shows that

$$\mathbf{AP} = \mathbf{PD}. \tag{5.7}$$

By hypothesis, the columns of $\mathbf{P}$ are the n linearly independent eigenvectors of $\mathbf{A}$, so $\det \mathbf{P} \neq 0$, and hence the inverse matrix $\mathbf{P}^{-1}$ will always exist. Pre-multiplying (5.7) by $\mathbf{P}^{-1}$ establishes the fundamental result that

$$\mathbf{D} = \mathbf{P}^{-1}\mathbf{AP} \text{ or, equivalently, } \mathbf{A} = \mathbf{PDP}^{-1}. \tag{5.8a}$$

Two immediate consequences of the last result in (5.8a) are that

(a) A diagonalizable matrix $\mathbf{A}$ is fully determined by its eigenvectors which form the columns of $\mathbf{P}$, and by its eigenvalues which form the elements on the diagonal of $\mathbf{D}$.
There is also the useful result that

(b)

$$\mathbf{A}^m = \mathbf{P}\mathbf{D}^m\mathbf{P}^{-1}, \tag{5.8b}$$

which follows from the result

$$\mathbf{A}^m = \underbrace{(\mathbf{P}\mathbf{D}\mathbf{P}^{-1})(\mathbf{P}\mathbf{D}\mathbf{P}^{-1})\ldots(\mathbf{P}\mathbf{D}\mathbf{P}^{-1})}_{m \text{ times}} = \mathbf{P}\mathbf{D}^m\mathbf{P}^{-1},$$

where in the products of terms all products $\mathbf{P}\mathbf{P}^{-1} = \mathbf{I}$, leaving only the result $\mathbf{P}\mathbf{D}^m\mathbf{P}^{-1}$'.

Matrix **P** is called the *diagonalizing matrix* for **A**, and the following important theorem on diagonalization has been proved.

Theorem 5.2 *Diagonalization Let an $n \times n$ real matrix* **A** *have n linearly independent eigenvectors $\mathbf{x}^{(1)}$, $\mathbf{x}^{(2)}$, ..., $\mathbf{x}^{(n)}$, with the associated eigenvalues λ_1, λ_2, ..., λ_n, some of which may be equal. Then* **A** *can always be diagonalized. To accomplish the diagonalization, let* **P** *be an $n \times n$ matrix with columns formed by the eigenvectors $\mathbf{x}^{(1)}$, $\mathbf{x}^{(2)}$, ..., $\mathbf{x}^{(n)}$ of* **A**, *and let* **D** *be an $n \times n$ diagonal matrix with the element on its leading diagonal equal to the eigenvalues of* **A** *arranged in the* same *order λ_1, λ_2, ..., λ_n as the columns of* **P**. *Then,*

$$\mathbf{D} = \mathbf{P}^{-1}\mathbf{A}\mathbf{P} \text{ or, equivalently, } \mathbf{A} = \mathbf{P}\mathbf{D}\mathbf{P}^{-1}.$$

◇

The following conclusions follow directly from Theorems 5.1 and 5.2.

5.2.1 Properties of Diagonalized Matrices

1. Diagonalization of a real $n \times n$ matrix **A** is only possible if it has n linearly independent eigenvectors.
2. The diagonalizing matrix **P** is not unique, because the columns of **P**, formed by the eigenvectors of **A**, can be arranged in any order and, furthermore, the eigenvectors of **A** can be scaled arbitrarily.
3. The order in which the eigenvectors of **A** are used to construct the columns of **P** will be the order in which the corresponding eigenvalues of **A** are arranged along the leading diagonal of **D**.
4. A real symmetric $n \times n$ matrix **A** can always be diagonalized, because it always has n linearly independent eigenvectors.

Thus a real $n \times n$ matrix will be *nondiagonalizable* if it has fewer than n linearly independent eigenvectors. So, if an eigenvalue λ_i with algebraic multiplicity m has associated with it only r linearly independent eigenvectors, with $r < m$, the matrix cannot be diagonalized. Such an eigenvalue λ_i is said to be *deficient*, and to have a *deficiency index* equal to the number $m - r$ of missing eigenvectors. The matrices in Examples 5.1 through 5.3 are examples of diagonalizable matrices, whereas a matrix like

$$\mathbf{A} = \begin{bmatrix} 1 & -3 & 1 \\ -1 & -1 & 1 \\ 1 & -2 & 0 \end{bmatrix}$$

is nondiagonalizable, because

$$\lambda_1 = 2,\ \mathbf{x}_1 = \begin{bmatrix} -8 \\ 1 \\ -5 \end{bmatrix},\ \text{while } \lambda_2 = \lambda_3 = -1 \text{ and } \mathbf{x}_2 = \mathbf{x}_3 = \begin{bmatrix} 1 \\ 1 \\ 1 \end{bmatrix},$$

so corresponding to the repeated eigenvalue $\lambda_2 = \lambda_3 = -1$ there is only a single eigenvector. Consequently the eigenvalue $\lambda = -1$ has a deficiency index $2 - 1 = 1$.

Example 5.5. Find a diagonalizing matrix **P** for

$$\mathbf{A} = \begin{bmatrix} 1 & 0 & -1 \\ -2 & -1 & 2 \\ -1 & 2 & 1 \end{bmatrix}.$$

Solution. The eigenvalues and eigenvectors of **A** were found in Example 5.1 to be

$$\lambda_1 = -2,\ \mathbf{x}_1 = \begin{bmatrix} 1 \\ -4 \\ 3 \end{bmatrix},\quad \lambda_2 = 0,\ \mathbf{x}_2 = \begin{bmatrix} 1 \\ 0 \\ 1 \end{bmatrix},\quad \lambda_3 = 3,\ \mathbf{x}_3 = \begin{bmatrix} 1 \\ -\frac{3}{2} \\ -2 \end{bmatrix}.$$

So a diagonalizing matrix **P** with columns $\mathbf{x}_1$, x_2 and $\mathbf{x}_3$ will be

$$\mathbf{P} = \begin{bmatrix} 1 & 1 & 1 \\ -4 & 0 & -\frac{3}{2} \\ 3 & 1 & -2 \end{bmatrix},$$

when the required diagonalization will be given by $\mathbf{D} = \mathbf{P}^{-1}\mathbf{A}\mathbf{P}$.

To check this, notice that the eigenvalue entries along the diagonal of **D** will be arranged in the *same* order as the eigenvectors forming the columns of **P**, so

$$\mathbf{D} = \begin{bmatrix} -2 & 0 & 0 \\ 0 & 0 & 0 \\ 0 & 0 & 3 \end{bmatrix}.$$

A simple calculation gives

$$\mathbf{P}^{-1} = \begin{bmatrix} -\frac{1}{10} & -\frac{1}{5} & \frac{1}{10} \\ \frac{5}{6} & \frac{1}{3} & \frac{1}{6} \\ \frac{4}{15} & -\frac{2}{15} & -\frac{4}{15} \end{bmatrix},$$

after which a routine matrix calculation confirms that

$$\mathbf{D}=\mathbf{P}^{-1}\mathbf{AP}=\begin{bmatrix}-\frac{1}{10} & -\frac{1}{5} & \frac{1}{10}\\ \frac{5}{6} & \frac{1}{3} & \frac{1}{6}\\ \frac{4}{15} & -\frac{2}{15} & -\frac{4}{15}\end{bmatrix}\begin{bmatrix}1 & 0 & -1\\ -2 & -1 & 2\\ -1 & 2 & 1\end{bmatrix}\begin{bmatrix}1 & 1 & 1\\ -4 & 0 & -\frac{3}{2}\\ 3 & 1 & -2\end{bmatrix}=\begin{bmatrix}-2 & 0 & 0\\ 0 & 0 & 0\\ 0 & 0 & 3\end{bmatrix}.$$

Had the eigenvectors of **A** been arranged in a different order when forming **P**, say in the order $\mathbf{x}_1$, $\mathbf{x}_3$ and $\mathbf{x}_2$, then

$$\mathbf{P}=\begin{bmatrix}1 & 1 & 1\\ -4 & -\frac{3}{2} & 0\\ 3 & -2 & 1\end{bmatrix}\text{ and }\mathbf{P}^{-1}=\begin{bmatrix}-\frac{1}{10} & -\frac{1}{5} & \frac{1}{10}\\ \frac{4}{15} & -\frac{2}{15} & -\frac{4}{15}\\ \frac{5}{6} & \frac{1}{3} & \frac{1}{6}\end{bmatrix},$$

in which case the order of the elements along the leading diagonal of **D** will now become

$$\mathbf{D}=\mathbf{P}^{-1}\mathbf{AP}=\begin{bmatrix}-2 & 0 & 0\\ 0 & 3 & 0\\ 0 & 0 & 0\end{bmatrix}.$$

This same example can be used to demonstrate that scaling eigenvectors leaves diagonalization unchanged. Suppose, for example, that when performing this last diagonalization the eigenvector $\mathbf{x}_1$ had been scaled by a factor 2 to become $\mathbf{x}_1 = [2, -8, 6]^{\mathrm{T}}$, then **P** becomes

$$\mathbf{P}=\begin{bmatrix}2 & 1 & 1\\ -8 & -\frac{3}{2} & 0\\ 6 & -2 & 1\end{bmatrix}\text{ and }\mathbf{P}^{-1}=\begin{bmatrix}-\frac{1}{20} & -\frac{1}{10} & \frac{1}{20}\\ \frac{4}{15} & -\frac{2}{15} & -\frac{4}{15}\\ \frac{5}{6} & \frac{1}{3} & \frac{1}{6}\end{bmatrix},$$

and once again

$$\mathbf{D}=\mathbf{P}^{-1}\mathbf{AP}=\begin{bmatrix}-2 & 0 & 0\\ 0 & 3 & 0\\ 0 & 0 & 0\end{bmatrix}.$$

These results also show that the arbitrariness of the scaling of an eigenvector when forming matrix **P** is removed when $\mathbf{P}^{-1}$ is used to form the matrix product $\mathbf{P}^{-1}\mathbf{AP}$. ◇

5.3 Quadratic Forms and Diagonalization

Theorem 5.2, in conjunction with Corollary 5.1, now makes it possible to show how a quadratic form $Q(\mathbf{x}) = \mathbf{x}^{\mathrm{T}}\mathbf{A}\mathbf{x}$, with **x** the column vector $\mathbf{x} = [x_1, x_2, \ldots, x_n]^{\mathrm{T}}$, can be reduced to a sum of squares. It was shown in Chapter 3 that when forming the product $\mathbf{x}^{\mathrm{T}}\mathbf{A}\mathbf{x}$, matrix **A** may always be taken to be a symmetric matrix, so the diagonalizing matrix **P** in Theorem 5.2 may be replaced by **Q**, which is an *orthogonal diagonalizing matrix* derived from the normalized eigenvectors of **A**. As $\mathbf{Q}^{-1} = \mathbf{Q}^{\mathrm{T}}$, the diagonalization result of Theorem 5.2 takes the form $\mathbf{D} = \mathbf{Q}^{\mathrm{T}}\mathbf{A}\mathbf{Q}$, so we can write $\mathbf{A} = \mathbf{Q}\mathbf{D}\mathbf{Q}^{\mathrm{T}}$. Using this expression for **A**, the quadratic form becomes $Q(\mathbf{x}) = \mathbf{x}^{\mathrm{T}}\mathbf{Q}\mathbf{D}\mathbf{Q}^{\mathrm{T}}\mathbf{x}$. If we now set $\mathbf{y} = \mathbf{Q}^{\mathrm{T}}\mathbf{x}$, with $\mathbf{y} = [y_1, y_2, \ldots, y_n]^{\mathrm{T}}$, then as $\mathbf{y}^{\mathrm{T}} = \mathbf{x}^{\mathrm{T}}\mathbf{Q}$, the quadratic form simplifies to $Q(\mathbf{y}) = \mathbf{y}^{\mathrm{T}}\mathbf{D}\mathbf{y}$. Matrix **D** is a diagonal matrix with the entries on its leading diagonal equal to the distinct eigenvalues $\lambda_1, \lambda_2, \ldots, \lambda_n$ of **A**, so $Q(\mathbf{y})$ is a sum of squares of the variables y_i given by

$$Q(\mathbf{y}) = \lambda_1 y_1^2 + \lambda_2 y_2^2 + \cdots + \lambda_n y_n^2 \tag{5.9}$$

Thus the general quadratic form in the variables in $x_1, x_2, \ldots, x_n$ has been reduced to the sum of squares of the new variables $y_1, y_2, \ldots, y_n$. The connection between the variables in **x**, in terms of the new variables in **y**, follows after premultiplication of $\mathbf{y} = \mathbf{Q}^{\mathrm{T}}\mathbf{x}$ by **Q** to obtain $\mathbf{x} = \mathbf{Q}\mathbf{y}$ after using the fact that $\mathbf{Q}^{\mathrm{T}} = \mathbf{Q}^{-1}$. If, in the quadratic form (5.9), the variables y_i are replaced by $\tilde{y}_i = y_i/\sqrt{|\lambda_i|}$, for $i = 1, 2, \ldots, n$, the quadratic form reduces to its simplest form

$$Q(\tilde{\mathbf{y}}) = \operatorname{sign}(\lambda_1)\tilde{y}_1^2 + \operatorname{sign}(\lambda_2)\tilde{y}_2^2 + \cdots + \operatorname{sign}(\lambda_n)\tilde{y}_n^2, \tag{5.10}$$

where $\operatorname{sign}(u) = 1$ if $u > 0$ and $\operatorname{sign}(u) = -1$ if $u < 0$. This result completely characterizes the original quadratic form, and however the reduction is accomplished (remember **Q** is not unique), the reduction will always be the same.

This reduction is used to classify quadratic forms according to the pattern of signs in (5.10), it being understood that when an eigenvalue λ_r is zero, the term $\tilde{y}_r^2$ in (5.10) must be omitted. In algebra, the preservation of the pattern of signs in (5.10), irrespective of the way the reduction is achieved, is known as *Sylvester's law of inertia*. The following useful and important theorem has been proved.

Theorem 5.3 *Reduction of a Quadratic Form. Let a quadratic form be* $Q(x_1, x_2, \ldots, x_n) = \mathbf{x}^T\mathbf{A}\mathbf{x}$, *with* $\mathbf{x} = [x_1, x_2, \ldots, x_n]^{\mathrm{T}}$ *and* **A** *a real* $n \times n$ *symmetric matrix so that it has distinct eigenvalues. Then the change of variable* $\mathbf{x} = \mathbf{Q}\mathbf{y}$, *with* **Q** *the orthogonal matrix in Corollary 5.1 and* $y = [y_1, y_2, \ldots, y_n]^{\mathrm{T}}$, *will reduce it to the sum of squares*

$$Q(\mathbf{y}) = \lambda_1 y_1^2 + \lambda_2 y_2^2 + \cdots + \lambda_n y_n^2,$$

where the λ_i *with* $i = 1, 2, \ldots, n$ *are the* n *distinct eigenvalues of* **A**. *The change of variable* $\tilde{y}_i = y_i/\sqrt{|\lambda_i|}$ *with* $i = 1, 2, \ldots, n$ *will reduce* $Q(\mathrm{y})$ *to the quadratic form*

$$Q(\tilde{\mathbf{y}}) = \operatorname{sign}(\lambda_1)\tilde{y}_1^2 + \operatorname{sign}(\lambda_2)\tilde{y}_2^2 + \cdots + \operatorname{sign}(\lambda_n)\tilde{y}_n^2.$$

This method of reduction will fail if multiple eigenvalues occur, even though n linearly independent eigenvectors exist.

◇

5.3.1 The Classification of Quadratic Forms

A quadratic form $Q(x_1, x_2, \dots, x_n) = \mathbf{x}^T\mathbf{A}\mathbf{x}$, with $\mathbf{x} = [x_1, x_2, \dots, x_n]^T$ and $\mathbf{A}$ a real symmetric matrix is classified follows:

1. The quadratic form Q is said to be *positive definite* if $Q(x_1, x_2, \dots, x_n) > 0$ for all $\mathbf{x} \neq \mathbf{0}$ provided all $x_1, x_2, \dots, x_n$ are not zero.
2. The quadratic form Q is said to be *negative definite* if $Q(x_1, x_2, \dots, x_n) < 0$ for all $\mathbf{x} \neq \mathbf{0}$.
3. If the quadratic form $Q(x_1, x_2, \dots, x_n) \geq 0$ for all $\mathbf{x} \neq \mathbf{0}$, the quadratic form is said to be *positive semidefinite*, while if $Q(x_1, x_2, \dots, x_n) \leq 0$ for all $\mathbf{x} \neq \mathbf{0}$ it is said to be *negative semidefinite*.
4. If the quadratic form $Q(x_1, x_2, \dots, x_n)$ can be either positive or negative for $\mathbf{x} \neq \mathbf{0}$, the quadratic form is said to be *indefinite*.

For convenience, the names *positive (negative) definite*, *positive (negative) semidefinite* and *indefinite* are also often used to describe the matrix $\mathbf{A}$ itself.

After consideration of (5.10), when expressed in words, Theorem 5.3: says that the classification of a quadratic form associated with a real symmetric matrix $\mathbf{A}$ is determined by the eigenvalues of $\mathbf{A}$. If all the eigenvalues of $\mathbf{A}$ are positive, the quadratic form will be positive definite, if they are all negative it will be negative definite. The quadratic form will be positive (negative) semidefinite if some of the eigenvalues of $\mathbf{A}$ are zero, and the remainder are positive (negative). The quadratic form will be indefinite if $\mathbf{A}$ has both positive and negative eigenvalues.

Positive and negative definite quadratic forms have many applications, so it is useful to derive the following simple test for positive or negative definiteness.

Theorem 5.4 *Determinant Test for Positive Definiteness. The quadratic form* $Q(x_1, x_2, x_3) = \mathbf{x}^T\mathbf{A}\mathbf{x}$

$$Q(x_1,x_2,x_3) = a_{11}x_1^2 + 2a_{12}x_1x_2 + a_{22}x_2^2 + 2a_{23}x_2x_3 + 2a_{13}x_1x_3 + a_{33}x_3^2\,,$$

with the coefficient matrix $\mathbf{A} = [a_{ij}]$ *will be positive definite if*

$$a_{11} > 0, \begin{vmatrix} a_{11} & a_{12} \\ a_{12} & a_{22} \end{vmatrix} > 0\,, \begin{vmatrix} a_{11} & a_{12} & a_{13} \\ a_{12} & a_{22} & a_{23} \\ a_{13} & a_{23} & a_{33} \end{vmatrix} > 0,$$

and negative definite if these inequality signs are reversed.

Proof. Using the fact that matrix $\mathbf{A}$ in a quadratic form may always be written as a symmetric matrix, we start by considering a quadratic form in two variables and write it as

$$Q(x_1,x_2) = a_{11}x_1^2 + 2a_{12}x_1x_2 + a_{22}x_2^2 = \left(x_1 + \frac{a_{12}}{a_{11}}x_2\right)^2 + \left(a_{22} - \frac{a_{12}^2}{a_{11}}\right)x_2^2\,.$$

Then if $a_{11} \neq 0$, the quadratic form $Q(x_1,x_2)$ will be strictly positive for $\mathbf{x} = [x_1, x_2]^T \neq 0$ if $a_{11} > 0$ and $a_{22} - a_{12}^2/a_{11} > 0$, but these two conditions can be written

$$a_{11}>0, \begin{vmatrix} a_{11} & a_{12} \\ a_{12} & a_{22} \end{vmatrix} >0. \tag{5.11}$$

Next, consider the quadratic form in three variables,

$$Q(x_1,x_2,x_3) = a_{11}x_1^2 + 2a_{12}x_1x_2 + a_{22}x_2^2 + 2a_{23}x_2x_3 + 2a_{13}x_1x_3 + a_{33}x_3^2. \tag{5.12}$$

Then some algebraic manipulation shows this can be written in the form

$$Q(x_1,x_2,x_3) = a_{11}\left(x_1 + \frac{a_{12}x_2 + a_{13}x_3}{a_{11}}\right)^2 + \left(a_{22} - \frac{a_{12}^2}{a_{11}}\right)x_2^2$$
$$+ 2\left(a_{23} - \frac{a_{12}a_{13}}{a_{11}}\right)x_2x_3 + \left(a_{33} - \frac{a_{13}^2}{a_{11}}\right)x_3^2.$$

Clearly, Q will be strictly positive for any $x_1 \neq 0$ if $x_2 = x_3 = 0$ and $a_{11} > 0$, so Q will be positive definite if the last three terms involving x_2 and x_3 are also positive definite. Having considered the situation when $x_1 \neq 0$, we turn now to the case when the term in x_1 (the first term) vanishes. Then for Q to be strictly positive we must have $a_{11} > 0$, and the coefficient of x_2^2 will be positive if $a_{22} - a_{12}^2/a_{11}>0$, so we have again arrived at conditions (5.11). The condition that the last two terms are strictly positive can be combined into a determinant, leading to the condition

$$\begin{vmatrix} a_{22} - \frac{a_{12}^2}{a_{11}} & a_{23} - \frac{a_{12}a_{13}}{a_{11}} \\ a_{23} - \frac{a_{12}a_{13}}{a_{11}} & a_{33} - \frac{a_{13}^2}{a_{11}} \end{vmatrix} > 0.$$

However, this last condition can be expressed as the third-order determinant

$$\begin{vmatrix} a_{11} & a_{12} & a_{13} \\ 0 & a_{22} - a_{12}^2 / a_{11} & a_{23} - a_{12}a_{13} / a_{11} \\ 0 & a_{23} - a_{12}a_{13} / a_{11} & a_{33} - a_{13}^2 / a_{11} \end{vmatrix} > 0.$$

Finally, adding suitable multiples of the first row to the second and third rows, reduces this to the result

$$\begin{vmatrix} a_{11} & a_{12} & a_{13} \\ a_{12} & a_{22} & a_{23} \\ a_{13} & a_{23} & a_{33} \end{vmatrix} > 0.$$

Combining this with the results in (5.11) establishes the positive definite part of the theorem, and a similar argument in which the inequality signs > are reversed establishes the rest of the theorem.

◇

Theorem 5.4 extends to a quadratic form in n variables called the *Routh–Hurwitz* test for a positive definite form, though the proof will not be given here.

Theorem 5.5 *The Routh–Hurwitz Test for Positive Definiteness. The quadratic form* $Q(x_1, x_2, \ldots, x_n) = \sum_{i,j=1}^{n} a_{ij}x_ix_j$ *will be positive definite if*

$$a_{11} > 0, \begin{vmatrix} a_{11} & a_{12} \\ a_{12} & a_{22} \end{vmatrix} > 0, \begin{vmatrix} a_{11} & a_{12} & a_{13} \\ a_{12} & a_{22} & a_{23} \\ a_{13} & a_{23} & a_{33} \end{vmatrix} > 0, \ldots, \begin{vmatrix} a_{11} & a_{12} & \cdots & a_{1n} \\ a_{21} & a_{22} & \cdots & a_{2n} \\ \vdots & \vdots & \vdots & \vdots \\ a_{n1} & a_{n2} & \cdots & a_{nn} \end{vmatrix} > 0.$$

◊

Theorem 5.4 has the following important implication when applied to Cartesian coordinate geometry. If the quadratic form

$$Q(x_1, x_2, x_3) = a_{11}x_1^2 + 2a_{12}x_1x_2 + a_{22}x_2^2 + 2a_{23}x_2x_3 + 2a_{13}x_1x_3 + a_{33}x_3^2$$

is positive definite, the equation $Q(x_1, x_2, x_3) =$ const. describes an ellipsoid with its origin O as the origin of a Cartesian coordinate system $O\{x_1, x_2, x_3\}$ at its center, and with its the axes oriented arbitrarily relative to the ellipsoid. Theorem 5.3 then implies it is always possible to rotate the axes into a Cartesian coordinate system $O\{X_1, X_2, X_3\}$ so that in terms of some new variables X_1, X_2, X_3 the coefficients of the product terms X_iX_j for $i \neq j$ all vanish, in which case the equation of the ellipsoid becomes

$$\tilde{a}_{11}X_1^2 + \tilde{a}_{22}X_2^2 + \tilde{a}_{33}X_3^2 = \text{constant.}$$

The new axes X_1, X_2, X_3 are symmetrical with respect to the ellipsoid, in the sense that each plane through origin O containing two of the axes cuts the ellipsoid in an ellipse, with one axis lying along its major axis and the other along its minor axis. The axes $O\{X_1, X_2, X_3\}$ are called the *principal axes* of the ellipsoid. In the simpler case of two space dimensions the quadratic form describes an ellipse, and the principal axes of the ellipse centered on the origin are its mutually perpendicular major and minor axes. For obvious reasons, in geometry a theorem equivalent to Theorem 5.4 is called the *principal axes theorem*, while in algebra and elsewhere it is known as the *orthogonal diagonalization theorem.*

Theorem 5.4 also finds various applications in differential equations, ranging from the study of coupled systems of linear differential equations describing oscillatory behavior, through to the dynamics of rotating rigid bodies where it describes the principal moments of inertia. A quite different application is found in the classification and reduction to standard forms of partial differential equations.

A typical example of an application in mechanics involving the orthogonalization of a positive definite matrix, and the significance of its associated positive definite quadratic form, arises when considering the rotation of a rigid body about an axis L passing through the origin O of an arbitrary orthogonal system of axes O $\{x_1, x_2, x_3\}$ fixed in the body. The differential equations describing the time variation of the components of the angular velocity ω of the body, called the *Euler equations*, require the introduction of a 3×3 constant matrix $\mathbf{I}_\mathrm{N}$, called the *inertia matrix* of the body relative to these axes. The three diagonal elements of $\mathbf{I}_\mathrm{N}$ are the *moments of inertia* of the body about the x_1, x_2 and x_3 axes, respectively, and the six off-diagonal elements are the *products of inertia* relative to these axes. The system of nonlinear first-order differential equations for the components of ω is complicated and it involved $\mathbf{I}_\mathrm{N}$, but it simplifies considerably if the axes are rotated about O to a new orthogonal system $\mathrm{O}\{X_1, X_2, X_3\}$ where the inertia matrix only has non-zero entries on its leading diagonal. The effect of this rotation can be understood by considering the quadratic form associated with $\mathbf{I}_\mathrm{N}$.

The quadratic form $Q(x_1, x_2, x_3)$ associated with $\mathbf{I}_\mathrm{N}$ is positive definite, and the equation $Q(x_1, x_2, x_3) = c$ with c a constant defines an ellipsoid. With a suitable choice for c it can be shown that the resulting ellipsoid has the property that the length of a radius vector from its center to the surface of the ellipsoid is equal to the reciprocal of the *radius of gyration* of the body about this radius vector. Here, the radius of gyration R_L of the body about a line L through O is given by $R_L = \sqrt{I_L/M}$, where I_L is the moment of inertia of the body about the line L through O, and M is the mass of the body. Thus, in the new coordinate system, the axes X_1, X_2, X_3 are principal axes of the ellipsoid.

Example 5.6. Reduce the quadratic form $Q(\mathbf{x}) = \mathbf{x}^\mathrm{T}\mathbf{A}\mathbf{x}$ to a sum of squares given that

$$\mathbf{A} = \begin{bmatrix} 7 & -2 & -2 \\ -2 & 1 & 4 \\ -2 & 4 & 1 \end{bmatrix}.$$

Solution. Routine calculations show the eigenvalues and eigenvectors of $\mathbf{A}$ are

$$\lambda_1 = -3,\ \mathbf{x}_1 = \begin{bmatrix} 0 \\ -1 \\ 1 \end{bmatrix},\ \lambda_2 = 3,\ \mathbf{x}_2 = \begin{bmatrix} 1 \\ 1 \\ 1 \end{bmatrix},\ \lambda_3 = 9, \mathbf{x}_3 = \begin{bmatrix} -2 \\ 1 \\ 1 \end{bmatrix},$$

so the normalized eigenvectors are

$$\tilde{\mathbf{x}}_1 = \begin{bmatrix} 0 \\ -\frac{1}{\sqrt{2}} \\ \frac{1}{\sqrt{2}} \end{bmatrix},\ \tilde{\mathbf{x}}_2 = \begin{bmatrix} \frac{1}{\sqrt{3}} \\ \frac{1}{\sqrt{3}} \\ \frac{1}{\sqrt{3}} \end{bmatrix},\ \tilde{\mathbf{x}}_3 = \begin{bmatrix} -\frac{2}{\sqrt{6}} \\ \frac{1}{\sqrt{6}} \\ \frac{1}{\sqrt{6}} \end{bmatrix}.$$

Thus the orthogonal diagonalizing matrix is

$$\mathbf{Q} = \begin{bmatrix} 0 & \frac{1}{\sqrt{3}} & -\frac{2}{\sqrt{6}} \\ -\frac{1}{\sqrt{2}} & \frac{1}{\sqrt{3}} & \frac{1}{\sqrt{6}} \\ \frac{1}{\sqrt{2}} & \frac{1}{\sqrt{3}} & \frac{1}{\sqrt{6}} \end{bmatrix},$$

and the change of variable from $\mathbf{x}$ to $\mathbf{y}$ to reduce the quadratic form to

$$Q(\mathbf{y}) = \lambda_1 y_1^2 + \lambda_2 y_2^2 + \lambda_3 y_3^2 = -3y_1^2 + 3y_2^2 + 9y_3^2$$

is given by $\mathbf{x} = \mathbf{Q}\mathbf{y}$, corresponding to

$$x_1 = \frac{1}{\sqrt{3}}y_2 - \frac{2}{\sqrt{6}}y_3,\ x_2 = -\frac{1}{\sqrt{2}}y_1 + \frac{1}{\sqrt{3}}y_2 + \frac{1}{\sqrt{6}}y_3,\ x_3 = \frac{1}{\sqrt{2}}y_1 + \frac{1}{\sqrt{3}}y_2 + \frac{1}{\sqrt{6}}y_3\ .$$

This is the reduction that was stated without proof when quadratic forms were introduced in Chapter 3, and the pattern of signs in $Q(\mathbf{y})$ show this quadratic form to be *indefinite*.

Although Theorem 5.5 does *not* show that this quadratic form is *indefinite*, it does show it is neither positive nor negative definite, because the values of the three determinants in the theorem are, respectively, 7, 6 and –19.

◇

Example 5.7. Use two different methods to show the quadratic form associated with

$$\mathbf{A} = \begin{bmatrix} 100 & 0 & -100 \\ 0 & 150 & 0 \\ -100 & 0 & 250 \end{bmatrix}$$

is positive definite.

Solution. The hardest way to establish positive definiteness is to show the eigenvalues of $\mathbf{A}$ are all positive. The characteristic equation of $\mathbf{A}$ is

$$\lambda^3 - 500\lambda^2 + 67,500\lambda - 2,250,000 = 0,$$

and after trial and error calculations one root (eigenvalue) is found to be 50. Removing the factor $(\lambda - 50)$ from the characteristic equation to leave a quadratic equation, the roots of which are the remaining roots (eigenvalues) 150 and 300. As all of the eigenvalues are positive, the quadratic form associated with $\mathbf{A}$, namely

$$\mathbf{x}^{\mathrm{T}}\mathbf{A}\mathbf{x} = 100x_1^2 - 200x_1x_3 + 150x_2^2 + 250x_3^2,$$

must be positive definite. A far simpler way to establish the positive definiteness of the quadratic form is to use Theorem 5.5 with $a_{11} = 100, a_{12} = a_{21} = 0, a_{13} = a_{31} = -100$, $a_{23} = a_{32} = 0$, $a_{22} = 150$, and $a_{33} = 250$. Evaluating the determinants in the theorem shows that

$$100 > 0, \begin{vmatrix} 100 & 0 \\ 0 & 250 \end{vmatrix} > 0 \text{ and } \begin{vmatrix} 100 & 0 & -100 \\ 0 & 150 & 0 \\ -100 & 0 & 250 \end{vmatrix} > 0,$$

so as all three determinants are positive, the quadratic form is positive definite.

◇

5.4 The Characteristic Polynomial and the Cayley–Hamilton Theorem

If $\mathbf{A}$ is an $n \times n$ matrix, the zeros $\lambda_1, \lambda_2, \ldots, \lambda_n$of the characteristic polynomial $p(\lambda) = \det[\mathbf{A} - \lambda\mathbf{I}]$ are the eigenvalues of $\mathbf{A}$. The following Theorem records an important property of the characteristic equation, and the result has many applications.

An almost trivial application of Theorem 5.6 will be found in Exercise 21.

Theorem 5.6 *The Cayley–Hamilton Theorem. Let $p(\lambda)$ be the characteristic polynomial of any $n \times n$ matrix $\mathbf{A}$. Then $\mathbf{A}$ satisfies its own matrix polynomial characteristic equation $p(\mathbf{A}) = \mathbf{0}$.*

Proof. For simplicity the theorem will only be proved for matrices $\mathbf{A}$ that are diagonalizable, though it is true for all $n \times n$ matrices with real or complex elements.

Let the characteristic polynomial of $\mathbf{A}$ be

$$p(\lambda) = (-1)^n\left(\lambda^n + c_1\lambda^{n-1} + \cdots + c_{n-1}\lambda + c_n\right).$$

If $\mathbf{A}$ is diagonalizable $\mathbf{A} = \mathbf{PDP}^{-1}$, where $\mathbf{P}$ is the matrix of n linearly independent eigenvectors of $\mathbf{A}$, and $\mathbf{D}$ is the diagonal matrix $\mathbf{D} = \text{diag}\,\{\lambda_1, \lambda_2, \ldots, \lambda_n\}$. Replacing λ in the characteristic polynomial by $\mathbf{A}$ produces the *matrix polynomial*

$$p(\mathbf{A}) = (-1)^n\{\mathbf{A}^n + c_1\mathbf{A}^{n-1} + \cdots + c_{n-1}\mathbf{A} + c_n\mathbf{I}\}.$$

However, $\mathbf{A}^2 = \left(\mathbf{PDP}^{-1}\right)\left(\mathbf{PDP}^{-1}\right) = \mathbf{PD}^2\mathbf{P}^{-1}$, $\mathbf{A}^3 = \left(\mathbf{PD}^2\mathbf{P}^{-1}\right)\mathbf{PDP}^{-1} = \mathbf{PD}^3\mathbf{P}^{-1}$, and in general $\mathbf{A}^r = \mathbf{PD}^r\mathbf{P}^{-1}$, so substituting for $\mathbf{A}^r$ into the characteristic polynomial gives

$$p(\mathbf{A}) = (-1)^n\{\mathbf{P}[\mathbf{D}^n + c_1\mathbf{D}^{n-1} + \cdots + c_{n-1}\mathbf{D} + c_n\mathbf{I}]\mathbf{P}^{-1}\}.$$

The ith row of the matrix expression in square brackets is $\lambda_i^n + c_1\lambda_i^{n-1} + \cdots + c_{n-1}\lambda_i + c_n$, which is simply $p(\lambda_i)$, and this must vanish because λ_i is a zero of the characteristic polynomial. This result is true for $i = 1, 2, \ldots, n$, so $p(\mathbf{A}) = \mathbf{P} \times \mathbf{0} \times \mathbf{P}^{-1} = \mathbf{0}$, and the theorem is proved.

◊

Example 5.8. Verify the Cayley–Hamilton theorem using the matrix

$$\mathbf{A} = \begin{bmatrix} 1 & 3 & -1 \\ 2 & 0 & 1 \\ -2 & 1 & 2 \end{bmatrix}.$$

Solution. The characteristic polynomial $p(\lambda) = \lambda^3 - 3\lambda^2 - 7\lambda + 21$, and

$$\mathbf{A}^3 = \begin{bmatrix} 13 & 27 & -7 \\ 14 & 0 & 7 \\ -26 & -5 & 14 \end{bmatrix}, \ \mathbf{A}^2 = \begin{bmatrix} 9 & 2 & 0 \\ 0 & 7 & 0 \\ -4 & -4 & 7 \end{bmatrix}.$$

Hence

$$\begin{bmatrix} 13 & 27 & -7 \\ 14 & 0 & 7 \\ -26 & -5 & 14 \end{bmatrix} - 3\begin{bmatrix} 9 & 2 & 0 \\ 0 & 7 & 0 \\ -4 & -4 & 7 \end{bmatrix} - 7\begin{bmatrix} 1 & 3 & -1 \\ 2 & 0 & 1 \\ -2 & 1 & 2 \end{bmatrix} + 21\begin{bmatrix} 1 & 0 & 0 \\ 0 & 1 & 0 \\ 0 & 0 & 1 \end{bmatrix} = [\mathbf{0}],$$

where $[\mathbf{0}]$ is the 3×3 null matrix.

◊

5.5 Eigenvalues and the Transpose Operation

The following theorem is often useful, and the result is easily proved. An illustration showing a typical application is to be found in Section 5.7.

Theorem 5.7 *The Eigenvalues of A and A^T An $n \times n$ matrix* $\mathbf{A}$ *and its transpose* $\mathbf{A}^T$ *have the same characteristic polynomial, and hence the same eigenvalues.*

Proof. The eigenvalues of $\mathbf{A}$ are the roots λ of the characteristic polynomial $\det[\mathbf{A} - \lambda\mathbf{I}] = 0$, so let us consider the matrix $[\mathbf{A} - \lambda\mathbf{I}]$. Applying result (1.13), which asserts that $[\mathbf{P} + \mathbf{Q}]^{\mathrm{T}} = \mathbf{P}^{\mathrm{T}} + \mathbf{Q}^{\mathrm{T}}$, we find that $[\mathbf{A} - \lambda\mathbf{I}]^{\mathrm{T}} = [\mathbf{A}^{\mathrm{T}} - \lambda\mathbf{I}^{\mathrm{T}}]$. However, $\mathbf{I}^{\mathrm{T}} = \mathbf{I}$, so $[\mathbf{A} - \lambda\mathbf{I}]^{\mathrm{T}} = [\mathbf{A}^{\mathrm{T}} - \lambda\mathbf{I}]$. Using this last result with Theorem 1.2 (7), which shows $\det[\mathbf{A} - \lambda\mathbf{I}] = \det[\mathbf{A} - \lambda\mathbf{I}]^{\mathrm{T}}$, it follows that $\det[\mathbf{A} - \lambda\mathbf{I}] = \det[\mathbf{A}^T - \lambda\mathbf{I}]$, so $\mathbf{A}$ and $\mathbf{A}^{\mathrm{T}}$ have the same characteristic polynomial, and hence the same eigenvalues, and the theorem is proved.

◊

5.6 Similar Matrices

Many problems in engineering, applied mathematics and physics can be formulated in terms of an $n \times n$ matrix $\mathbf{A}$, and often their solution is determined by the eigenvalues of $\mathbf{A}$. It is natural to ask if it is possible to find an $n \times n$ matrix $\mathbf{C}$ that can transform matrix $\mathbf{A}$ into an $n \times n$ matrix $\mathbf{B}$ that has the *same* eigenvalues as $\mathbf{A}$, though in a much simpler form. A simpler problem is likely to be much easier to solve, and once its solution has been found, the solution can be transformed back to give the solution of the original problem.

A typical example of this type will be considered in Chapter 6 where systems of simultaneous linear first-order constant coefficient differential equations are characterized by an $n \times n$ coefficient matrix **A**. It will be shown there that when matrix **A** can be diagonalized to a matrix **D**, with the elements on its leading diagonal the same as the eigenvalues of **A**, it becomes possible to de-couple all of the equations so they can be solved individually, after which this simplified solution can be transformed back to give the solution of the original much more complicated system of ordinary differential equations. This is just one example where it is helpful for an $n \times n$ matrix **A** to be transformed into another $n \times n$ matrix, in that case **D**, with the same eigenvalues as **A** which then enables the solution of a complicated coupled system of differential equations to be found in terms of the solution of a much simpler problem. Two matrices **A** and **B** with the property that **A** can be transformed in a special way to **B** such that the eigenvalues of **A** and **B** are the same are called *similar* matrices, the formal definition of which now follows.

5.6.1 Similar Matrices

If **A** and **B** are $n \times n$ matrices, **B** is said to be *similar* to **A** if, and only if, a non-singular $n \times n$ matrix **C** exists such that

$$\mathbf{B} = \mathbf{C}^{-1}\mathbf{A}\mathbf{C}. \tag{5.13}$$

The transformation from **B** back to **A**, where eigenvalues are again preserved is, of course, given by $\mathbf{A} = \mathbf{C}\mathbf{B}\mathbf{C}^{-1}$. An immediate consequence of (5.13) is that

$$\det \mathbf{A} = \det \mathbf{B}.$$

This follows from the first result in (3.29), because

$$\det \mathbf{B} = \det(\mathbf{C}^{-1}\mathbf{A}\mathbf{C}) = (\det \mathbf{C}^{-1})(\det \mathbf{A})(\det \mathbf{C}) = (\det \mathbf{C}^{-1})(\det \mathbf{C})(\det \mathbf{A}),$$

but $\det(\mathbf{C}^{-1}) = 1/\det \mathbf{C}$, so $\det \mathbf{B} = \det \mathbf{A}$.

The two fundamental properties of similar matrices that are useful in many applications are stated in the following Theorem.

Theorem 5.8 *Similarity and Eigenvalues.*

(a) *If the two $n \times n$ matrices* **A** *and* **B** *are similar, they each have the same characteristic polynomial, and hence the same eigenvalues.*
(b) *If* **B** *is similar to* **A**, *and* $\mathbf{A} = \mathbf{C}\mathbf{B}\mathbf{C}^{-1}$, *then* $\mathbf{x}$ *is an eigenvector of* **A** *with the corresponding eigenvalue* λ, *only if* $\mathbf{C}^{-1}\mathbf{x}$ *is an eigenvector of* **B**, *that also corresponds to the eigenvalue* λ.

Proof. (a) If $\mathbf{B}$ is similar to $\mathbf{A}$, from (5.13) we have

$$\det(\mathbf{B} - \lambda\mathbf{I}) = \det(\mathbf{C}^{-1}(\mathbf{A} - \lambda\mathbf{I})\mathbf{C}).$$

However, applying (3.29) to the expression on the right this result becomes

$$\det[\mathbf{B} - \lambda\mathbf{I}] = \det(\mathbf{C}^{-1})\det[\mathbf{A} - \lambda\mathbf{I}]\det\mathbf{C}.$$

As $\det(\mathbf{C}^{-1}) = 1/\det\mathbf{C}$ it follows that $\det[\mathbf{B} - \lambda\mathbf{I}] = \det[\mathbf{A} - \lambda\mathbf{I}]$, showing the equivalence of the two characteristic polynomials, and hence that $\mathbf{A}$ and $\mathbf{B}$ have identical eigenvalues.

(b) If $\mathbf{x}$ is an eigenvector of $\mathbf{A}$ corresponding to the eigenvalue λ, then $\mathbf{Ax} = \lambda\mathbf{x}$. However, as $\mathbf{A}$ is similar to $\mathbf{B}$ we may write $\mathbf{A} = \mathbf{CBC}^{-1}$, so $\mathbf{CBC}^{-1}\mathbf{x} = \lambda\mathbf{x}$, which can also be written in the form $\mathbf{B}(\mathbf{C}^{-1})\mathbf{x} = \lambda(\mathbf{C}^{-1}\mathbf{x})$. The result of the theorem follows by setting $\mathbf{C}^{-1}\mathbf{x} = \mathbf{y}$, when the last result becomes $(\mathbf{B} - \lambda\mathbf{I})\mathbf{y} = \mathbf{0}$, which is statement (b) in the Theorem, so the result is proved. ◊

An example of similarity has already been encountered in Theorem 5.3, because when an $n \times n$ matrix $\mathbf{A}$ can be transformed to a diagonal matrix $\mathbf{D}$, it follows that $\mathbf{A}$ and $\mathbf{D}$ have the same eigennvalues, and they are similar because a nonsingular matrix $\mathbf{P}$ exists such that $\mathbf{P}^{-1}\mathbf{AP} = \mathbf{D}$.

5.7 Left and Right Eigenvectors

So far, the definition of an eigenvector $\mathbf{x}_i$ associated with the eigenvalue λ_i is that it is a solution of $\mathbf{Ax}_i = \lambda_i\mathbf{x}_i$, or equivalently, a solution of $[\mathbf{A} - \lambda_i\mathbf{I}]\mathbf{x}_i = \mathbf{0}$. Here, $\mathbf{x}_i$ occurs on the right of this last expression, so it is appropriate to call it the *right eigenvector* associated with the eigenvalue λ_i. In certain applications it becomes necessary to consider a different type of eigenvector called a *left eigenvector* associated with the eigenvalue λ_i. To distinguish between the right and left eigenvectors, when both may arise in a calculation, the right eigenvector associated with the eigenvalue λ_i will be denoted by $\mathbf{r}_i$ and the left eigenvector by $\mathbf{l}_i$. Let us now show how $\mathbf{l}_i$ can be defined, and its relationship to $\mathbf{r}_i$.

We start from the definition of the right eigenvector $\mathbf{r}_i$ as a solution of $\mathbf{Ar}_i = \lambda_i\mathbf{r}_i$, where $\mathbf{A}$ is an $n \times n$ matrix, and take the transpose of the definition

$$(\mathbf{Ar}_i)^{\mathrm{T}} = (\lambda_i\mathbf{r}_i)^{\mathrm{T}}.$$

Using the property of the matrix transpose operation, and the fact that the eigenvalue λ_i is a scalar, the last result becomes

$$\mathbf{r}_i{}^{\mathrm{T}}\mathbf{A}^{\mathrm{T}} = \lambda_i\mathbf{r}_i{}^{\mathrm{T}}.$$

Because $\mathbf{r}_i$ is an n element column vector, its transpose $\mathbf{r}_i^{\mathrm{T}}$ is an n element row vector. Setting $\mathbf{r}_i^{\mathrm{T}} = \mathbf{l}_i$, the last result becomes

$$\mathbf{l}_i\mathbf{A}^{\mathrm{T}} = \lambda_i\mathbf{l}_i, \text{ or equivalently, } \mathbf{l}_i\left[\mathbf{A}^{\mathrm{T}} - \lambda_i\mathbf{I}\right] = \mathbf{0}.$$

Rejecting the trivial solution $\mathbf{l}_i = \mathbf{0}$, a non-trivial solution $(\mathbf{l}_i \neq 0)$ can only exist if the determinant of the expression vanishes. This is only possible if the numbers λ_i which are the eigenvalues of $\mathbf{A}$ are also the eivenvalues of $\mathbf{A}^{\mathrm{T}}$. Theorem 5.7 shows that this is indeed the case, so this last result can be replaced by

$$\mathbf{l}_i[\mathbf{A} - \lambda_i\mathbf{I}] = \mathbf{0}.$$

The position of $\mathbf{l}_i$ on the left of the expression in square brackets is the reason why $\mathbf{l}_i$ is called a *left eigenvector*, but remember that $\mathbf{l}_i$ is an n element row vector.

Let us now show the left and right eigenvectors corresponding to *different* eigenvalues are *mutually orthogonal*, by which we mean that the product $\mathbf{l}_i\mathbf{r}_j = 0$ when $i \neq j$, while $\mathbf{l}_i\mathbf{r}_i \neq 0$ for $i = 1, 2, \ldots, n$. From the definitions of $\mathbf{l}_i$ and $\mathbf{r}_j$, with $i \neq j$, we have

$$\lambda_j\mathbf{l}_i\mathbf{r}_j = (\mathbf{l}_i\mathbf{A})\mathbf{r}_j = \lambda_i\mathbf{l}_i\mathbf{r}_j,$$

and so

$$(\lambda_j - \lambda_i)\mathbf{l}_i\mathbf{r}_j = 0.$$

However, by supposition, $\lambda_i \neq \lambda_j$, so

$$\mathbf{l}_i\mathbf{r}_j = 0, \quad \text{for } i \neq j.$$

When $i = j$, corresponding elements of $\mathbf{l}_i$ and $\mathbf{r}_i$ are proportional so $\mathbf{l}_i\mathbf{r}_i$ cannot vanish, and the orthogonality is proved.

5.8 Jordan Normal Forms

The discussion that follows will be prefaced by some brief remarks to provide motivation for what at first sight might appear to be an unnecessary abstraction. Let us turn our attention to coefficient matrices that describe pairs of simultaneous linear first-order homogeneous differential equations that govern the behavior of many physical phenomena. These range from mechanical systems, to systems in electrical engineering and physics, to commercial situations involving competition for resources, and also to environmental systems where competition exists between different biological species. Because the solutions of such systems evolve with time they are often called *dynamical systems*, though in most cases the term *dynamical* is used in the sense that it refers to a continuous change with respect to time, rather than to the sense in which the term *dynamics* is used in mechanics.

Of particular interest is the question of whether a system has solutions that are *stable* or *unstable*. Here, a stable system is one whose solution remains bounded for all time, though the solution may or may not decay to zero as time increases, while an unstable system is one in which the solution grows without bound as time increases. The simplest examples of such systems are of the form

$$\frac{dx}{dt} = ax + by \text{ and } \frac{dy}{dt} = cx + dy, \tag{5.14}$$

where a, b, c and d are constants, and $x(t)$ and $y(t)$ are physical quantities that depend on the time t. A typical mechanical example, that when linearized can be reduced to a system like (5.14), is the *nonlinear pendulum equation*. This is the equation that governs the oscillations of a pendulum of length l with angle of swing from the vertical $\theta(t)$at time t, and it takes the form $d^2\theta/dt^2 + (g/l)\sin\theta = 0$, where g is the acceleration due to gravity. Provided the angle of swing is small, this equation can be linearized by replacing the nonlinear term $\sin\theta$ by θ, to obtain $d^2\theta/dt^2 + (g/l)\theta = 0$. Then, by setting $d\theta/dt = x$, this second-order equation can be written as the pair of simultaneous first-order linear equations $dx/dt = -(g/l)\theta$ and $d\theta/dt = x$ which, apart from the notation, is in the form given in (5.14).

A system like (5.14) is characterized by its real 2×2 constant coefficient matrix $\mathbf{A} = \begin{bmatrix} a & b \\ c & d \end{bmatrix}$, and the nature of its solution is determined by the eigenvalues of $\mathbf{A}$. Thus a solution will be stable and decay to zero without oscillations if both eigenvalues are real and negative, it will be oscillatory and decay to zero if the eigenvalues are complex conjugates with negative real parts, and it will be unstable if the eigenvalues are complex conjugates with positive real parts, or if the eigenvalues are real and at least one is positive. The case when the eigenvalues are purely imaginary corresponds to purely oscillatory behavior that remains bounded for all time.

Of particular interest is the way the solution of a system evolves with the passage of time from some initial conditions $x(t_0) = x_0$ and $y(t_0) = y_0$ at time t_0. Because the initial conditions describe the physical nature of the system, the quantities $x(t)$ and $y(t)$ in (5.14) describe what is referred to as the physical *state* of the system at time t. The (x, y)-plane is called the *phase-plane* of the system, and the path traced out in the phase-plane by a point $(x(t), y(t))$ as the solution of the system evolves from its initial conditions as the time t increases is called a *trajectory* in the phase-plane.

A key question that arises is how to classify the nature of all possible systems like (5.14) that are described by real 2×2 constant coefficient matrices $\mathbf{A}$. As already mentioned, the property that a solution is either *stable* or *unstable* as time increases, is determined by the eigenvalues of the coefficient matrix $\mathbf{A}$, so the answer to this question must rest with the eigenvalues of $\mathbf{A}$.

It turns out that the behavior of a system characterized by a matrix **A** is closely related to the behavior of a system in which **A** is replaced by what is called its Jordan normal form which is defined below. Consequently, the behavior of an entire class of systems can be explored by examining how a system behaves when its coefficient matrix is replaced by the Jordan normal form to which **A** is similar. This same form of analysis can be extended to examine the *local* behavior of nonlinear systems, provided they can be linearized about a state of the system that is of interest, as shown above when the nonlinear pendulum equation was linearized. Thus the identification of the types of Jordan normal form that can occur is of considerable importance. There are also many other situations where more general types of Jordan normal forms occur with $n \times n$ matrices, both in connection with physical problems and with problems in algebra and numerical analysis, though here only real 2×2 matrices will be considered.

Let us now consider similarity in the context of real 2×2 matrices **A** whose characteristic polynomials $p(\lambda)$ are quadratic polynomials in λ, and whose zeros are the two roots λ_1 and λ_2 that are the eigenvalues of **A**. An examination of all such systems reduces to the examination of the behavior of systems with a Jordan normal form as its coefficient matrix.

After some reflection, it can be seen that a 2×2 matrix **A** must belong to one of the following categories:

(i) The eigenvalues λ_1 and λ_2 of **A** are real and distinct, so that $\lambda_1 \neq \lambda_2$, in which case **A** will have two real linearly independent eigenvectors $\mathbf{x}_1$ and $\boldsymbol{x}_2$.
(ii) Matrix **A** is a diagonal matrix with a single repeated real eigenvalue λ_1 but with two real linearly independent eigenvectors $\mathbf{x}_1$ and $\boldsymbol{x}_2$.
(iii) Matrix **A** has a repeated real eigenvalue λ_1 to which there corresponds only *one* real eigenvector $\mathbf{x}_1$.
(iv) The eigenvalues of **A** are complex conjugates, in which case **A** has two linearly independent complex conjugate eigenvectors $\mathbf{x}_1$ and $\boldsymbol{x}_2$.

We now show all such real 2×2 matrices **A** must be similar to one of the following four types of *Jordan normal form*, each of which is said to be a *canonical form* for a 2×2 matrix.

Theorem 5.9 *Jordan Normal Forms for 2×2 Matrices. Every real 2×2 matrix* **A** *must be similar to just one of the following four* Jordan normal forms*:*

(a) $\mathbf{J}_1 = \begin{bmatrix} \lambda_1 & 0 \\ 0 & \lambda_2 \end{bmatrix}$, *where* **A** *is a diagonal matrix with two real eigenvalues $\lambda_1 \neq \lambda_2$, to which there correspond two real linearly independent eigenvectors $\mathbf{x}_1$ and $\boldsymbol{x}_2$.*

(b) $\mathbf{J}_2 = \begin{bmatrix} \lambda_1 & 0 \\ 0 & \lambda_1 \end{bmatrix}$, *where* **A** *is a diagonal matrix with identical elements on the leading diagonal, and a single repeated real eigenvalue λ_1, to which there correspond two real linearly independent eigenvectors $\mathbf{x}_1$ and $\boldsymbol{x}_2$.*

(c) $\mathbf{J}_3 = \begin{bmatrix} \lambda_1 & 1 \\ 0 & \lambda_1 \end{bmatrix}$, *where* $\mathbf{A}$ *has a single repeated real eigenvalue* λ_1, *but only one real eigenvector* $\mathbf{x}_1$.

(d) $\mathbf{J}_4 = \begin{bmatrix} \alpha & -\beta \\ \beta & \alpha \end{bmatrix}$ *with* $\beta > 0$, *where the eigenvalues of* $\mathbf{A}$ *are the complex conjugate complex numbers* $\lambda_{\pm} = \alpha \pm i\beta$ *corresponding to which are two (linearly independent) complex conjugate eigenvectors* $\mathbf{x}_1$ *and* $\mathbf{x}_2$, with $\mathbf{x}_2 = \mathbf{x}_1$..

Proof. The similarity of $\mathbf{A}$ with respect to the diagonal matrices $\mathbf{J}_1$ and $\mathbf{J}_2$ in (a) and (b) follows directly from the fact that in each case the matrices have two linearly independent eigenvectors, and so are diagonalizable. Notice that in case (b), because $\mathbf{A}$ is a diagonal matrix with identical elements on its leading diagonal we may write $\mathbf{A} = \lambda_1\mathbf{I}$, so if $\mathbf{C}$ is any real nonsingular 2×2 matrix, it follows that $\mathbf{C}^{-1}\mathbf{A}\mathbf{C} = \mathbf{C}^{-1}(\lambda_1\mathbf{I})\mathbf{C} = \lambda_1\mathbf{C}^{-1}\mathbf{C} = \lambda_1\mathbf{I}$, showing that $\mathbf{A}$ is similar to itself.

To prove the similarity of $\mathbf{A}$ to the Jordan matrices $\mathbf{J}_3$ and $\mathbf{J}_4$ in (c) and (d) takes a little longer, and we will start with (c). Let $\mathbf{x}_1$ be the single eigenvector corresponding to the repeated real eigenvalue λ_1, and let $\mathbf{v}$ be any nonzero two element column vector that is linearly independent of $\mathbf{x}_1$ (it is not proportional to $\mathbf{x}_1$). In the proof that follows only 2×2 matrices will be involved, so to make clear how the columns of the matrices are modified as the proof proceeds, the concept of partitioned matrices that was introduced in Chapter 3 will be used. In this notation, a 2×2 matrix $\mathbf{N}$ will be written in the form $\mathbf{N} = [\mathbf{c} \mid \mathbf{d}]$, where $\mathbf{c}$ is the first 2×1 column vector in matrix $\mathbf{N}$, and $\mathbf{d}$ is the second 2×1 column vector.

Adopting this notation, let us use the column vectors $\mathbf{x}_1$ and $\mathbf{v}$ to form the 2×2 matrix $\mathbf{C} = [\mathbf{x}_1 \mid \mathbf{v}]$. Then, because $\mathbf{x}_1$ and $\mathbf{v}$ are linearly independent, $\mathbf{C}^{-1}$ exists. The product $\mathbf{AC} = [\mathbf{Ax}_1 \mid \mathbf{Av}]$, but $\mathbf{Ax}_1 = \lambda_1\mathbf{x}_1$, so this last result becomes $\mathbf{AC} = [\lambda_1\mathbf{x}_1 \mid \mathbf{Av}]$ By defining the column vector $\mathbf{e} = [1, 0]^{\mathrm{T}}$, we can write $\mathbf{x}_1 = \mathbf{Ce}$, and using the result $\mathbf{CC}^{-1} = \mathbf{I}$ the equation becomes $\mathbf{AC} = [\lambda_1\mathbf{Ce} \mid \mathbf{CC}^{-1}\mathbf{Av}]$. Writing $\mathbf{C}$ a pre-multiplier, the expression on the right becomes $\mathbf{AC} = \mathbf{C}[\lambda_1\mathbf{e} \mid \mathbf{C}^{-1}\mathbf{Av}]$. Pre-multiplying this result by $\mathbf{C}^{-1}$ gives $\mathbf{C}^{-1}\mathbf{AC} = [\lambda_1\mathbf{e} \mid \mathbf{C}^{-1}\mathbf{Av}]$, showing that $\mathbf{A}$ is similar to $[\lambda_1\mathbf{e} \mid \mathbf{C}^{-1}\mathbf{Av}]$.

The matrix $[\lambda_1\mathbf{e} \mid \mathbf{C}^{-1}\mathbf{Av}]$ is an upper triangular matrix with λ_1 the first element on its leading diagonal. However, $\mathbf{A}$ and $\mathbf{C}^{-1}\mathbf{AC}$ must have the same eigenvalues, so $\mathbf{C}^{-1}\mathbf{AC}$ must also have a repeated eigenvalue λ_1, with the result that the upper triangular matrix $[\lambda_1\mathbf{e} \mid \mathbf{C}^{-1}\mathbf{Av}]$ must also have λ_1 as the second element on its leading diagonal. Consequently, it follows that

$$\mathbf{C}^{-1}\mathbf{AC} = \begin{bmatrix} \lambda_1 & c \\ 0 & \lambda_1 \end{bmatrix}, \tag{5.15}$$

where the constant $c \neq 0$. Result (5.15) is not yet in the standard form for $\mathbf{J}_3$, so to make $c = 1$ the matrix $\mathbf{C}$ must be scaled. The scaling is accomplished by defining the new matrix

$$\mathbf{R} = \mathbf{C}\begin{bmatrix} 1 & 0 \\ 0 & 1/c \end{bmatrix}, \tag{5.16}$$

when

$$\mathbf{J}_3 = \mathbf{R}^{-1}\mathbf{A}\mathbf{R} = \begin{bmatrix} \lambda_1 & 1 \\ 0 & \lambda_1 \end{bmatrix},$$

and the required similarity has been established.

It remains for us to consider case (d) where the complex conjugate eigenvalues $\lambda_\pm = \alpha \pm i\beta$ of $\mathbf{J}_4$ are seen to be the same as the eigenvalues of $\mathbf{A}$, so the two matrices are indeed similar. Now let us find the form taken by a matrix $\mathbf{R}$ such that $\mathbf{R}^{-1}\mathbf{A}\mathbf{R} = \mathbf{J}_4$. To do this we partition matrix $\mathbf{R}$ by setting $\mathbf{R} = [\mathbf{r}_1 \mid \mathbf{r}_2]$, and require that

$$\mathbf{A}\mathbf{R} = \mathbf{R}\begin{bmatrix} \alpha & -\beta \\ \beta & \alpha \end{bmatrix}, \tag{5.17}$$

because then $\mathbf{R}^{-1}\mathbf{A}\mathbf{R} = \mathbf{J}_4$. When this equation is expanded it becomes

$$[\mathbf{A}\mathbf{r}_1 \mid \mathbf{A}\mathbf{r}_2] = [\alpha\mathbf{r}_1 + \beta\mathbf{r}_2 \mid -\beta\mathbf{r}_1 + \alpha\mathbf{r}_2],$$

which can be rewritten as

$$[(\mathbf{A} - \alpha\mathbf{I}) - \beta\mathbf{I}\mathbf{r}_2 \mid \beta\mathbf{I}\mathbf{r}_1 + (\mathbf{A} - \alpha\mathbf{I})\mathbf{r}_2] = [\mathbf{0} \mid \mathbf{0}].$$

Using a partitioned matrix, this set of homogeneous matrix equations becomes

$$\left[\begin{array}{c|c} \mathbf{A} - \alpha\mathbf{I} & -\beta\mathbf{I} \\ \hline \beta\mathbf{I} & \mathbf{A} - \alpha\mathbf{I} \end{array}\right]\left[\begin{array}{c} \mathbf{r}_1 \\ \hline \mathbf{r}_2 \end{array}\right] = \left[\begin{array}{c} \mathbf{0} \\ \hline \mathbf{0} \end{array}\right]. \tag{5.18}$$

From Example 3.8 we have

$$\left[\begin{array}{c|c} \mathbf{A} - \alpha\mathbf{I} & -\beta\mathbf{I} \\ \hline \beta\mathbf{I} & \mathbf{A} - \alpha\mathbf{I} \end{array}\right]\left[\begin{array}{c|c} \mathbf{A} - \alpha\mathbf{I} & \beta\mathbf{I} \\ \hline -\beta\mathbf{I} & \mathbf{A} - \alpha\mathbf{I} \end{array}\right] = \left[\begin{array}{c|c} p(\mathbf{A}) & \mathbf{0} \\ \hline \mathbf{0} & p(\mathbf{A}) \end{array}\right], \tag{5.19}$$

where $p(\mathbf{A}) = \mathbf{A}^2 - 2\alpha\mathbf{A} + (\alpha^2 + \beta^2)\mathbf{I}$. However, from the Cayley–Hamilton Theorem we know that $p(\mathbf{A}) = \mathbf{0}$, so (5.19) becomes

$$\left[\begin{array}{c|c} \mathbf{A} - \alpha\mathbf{I} & -\beta\mathbf{I} \\ \hline \beta\mathbf{I} & \mathbf{A} - \alpha\mathbf{I} \end{array}\right]\left[\begin{array}{c|c} \mathbf{A} - \alpha\mathbf{I} & \beta\mathbf{I} \\ \hline -\beta\mathbf{I} & \mathbf{A} - \alpha\mathbf{I} \end{array}\right] = \left[\begin{array}{c} \mathbf{0} \\ \hline \mathbf{0} \end{array}\right]. \tag{5.20}$$

Recalling the form of matrix Eq. (5.18), we see that the columns of

$$\left[\begin{array}{c|c} \mathbf{A}-\alpha\mathbf{I} & \beta\mathbf{I} \\ \hline -\beta\mathbf{I} & \mathbf{A}-\alpha\mathbf{I} \end{array}\right] \tag{5.21}$$

must be solutions of (5.18). Using the first column of (5.21). gives $\mathbf{r}_1 = \begin{bmatrix} a_{11}-\alpha \\ a_{21} \end{bmatrix}$ and $\mathbf{r}_2 = \begin{bmatrix} -\beta \\ 0 \end{bmatrix}$, so that

$$\mathbf{R} = \begin{bmatrix} a_{11}-\alpha & -\beta \\ \beta & 0 \end{bmatrix}, \tag{5.22}$$

and we have found a matrix $\mathbf{R}$ such that

$$\mathbf{R}^{-1}\mathbf{A}\mathbf{R} = \begin{bmatrix} \alpha & -\beta \\ \beta & \alpha \end{bmatrix}. \tag{5.23}$$

Using the remaining columns in (5.21) will produce a different forms of $\mathbf{R}$, namely, $\mathbf{R}_1$, $\mathbf{R}_2$ and $\mathbf{R}_3$, though the application of each to form $\mathbf{R}_i^{-1}\mathbf{A}\mathbf{R}_i$ with $i = 1$, 2, 3 will produce the *same* reduction as the one in (5.23). For example, using the second column of (5.21) gives

$$\mathbf{R}_1 = \begin{bmatrix} a_{12} & 0 \\ a_{22} & -\beta \end{bmatrix}, \text{ but once again } \mathbf{R}_1^{-1}\mathbf{A}\mathbf{R}_1 = \begin{bmatrix} \alpha & -\beta \\ \beta & \alpha \end{bmatrix}.$$

◇

Example 5.9. Find a matrix $\mathbf{R}$ that reduces the matrix

$$\mathbf{A} = \begin{bmatrix} 5 & 2 \\ -2 & 1 \end{bmatrix}$$

to its appropriate Jordan normal form.

Solution. Matrix $\mathbf{A}$ has a single repeated eigenvalue $\lambda = 3$, to which there corresponds the single eigenvector $\mathbf{u}_1 = [1, -1]^{\mathrm{T}}$, so matrix $\mathbf{A}$ is of type (c). Construct a nonsingular matrix $\mathbf{C}$ by taking $\mathbf{u}_1$ to be its first column, and for its second column the arbitrarily chosen vector $\mathbf{u}_2 = [1, -1]^{\mathrm{T}}$, which is linearly independent of $\mathbf{u}_1$. Then

$$\mathbf{C} = \begin{bmatrix} 1 & 1 \\ -1 & 1 \end{bmatrix}, \mathbf{C}^{-1} = \begin{bmatrix} \frac{1}{2} & -\frac{1}{2} \\ \frac{1}{2} & \frac{1}{2} \end{bmatrix} \text{ and } \mathbf{C}^{-1}\mathbf{A}\mathbf{C} = \begin{bmatrix} 3 & 4 \\ 0 & 3 \end{bmatrix}.$$

To convert the element 4 to 1, as required in form (c), we see from (5.14) and (5.15) that we must set $c = 4$, when

$$\mathbf{R} = \mathbf{C}\begin{bmatrix} 1 & 0 \\ 0 & \frac{1}{4} \end{bmatrix} = \begin{bmatrix} 1 & \frac{1}{4} \\ -1 & \frac{1}{4} \end{bmatrix} \text{ and } \mathbf{R}^{-1} = \begin{bmatrix} \frac{1}{2} & -\frac{1}{2} \\ 2 & 2 \end{bmatrix}, \text{ giving } \mathbf{R}^{-1}\mathbf{AR} = \begin{bmatrix} 3 & 1 \\ 0 & 3 \end{bmatrix},$$

which is the required reduction.

◇

Example 5.10. Find the matrix that reduces

$$\mathbf{A} = \begin{bmatrix} 6 & 3 \\ -3 & 4 \end{bmatrix}$$

to its appropriate Jordan normal form.

Solution. The eigenvalues are the complex conjugates $\lambda_\pm = 5 \pm 2\sqrt{2}i$, so matrix $\mathbf{A}$ is of type (d) with $\alpha = 5$ and $\beta = 2\sqrt{2}$. Thus $\mathbf{A}$ is similar to the matrix

$$\mathbf{J}_4 = \begin{bmatrix} 5 & -2\sqrt{2} \\ 2\sqrt{2} & 5 \end{bmatrix}.$$

The matrix $\mathbf{R}$ that produces this reduction through the matrix product $\mathbf{R}^{-1}\mathbf{AR}$ given by (5.22) is

$$\mathbf{R} = \begin{bmatrix} a_{11} - \alpha & -\beta \\ a_{21} & 0 \end{bmatrix},$$

with $a_{11} = 5$, $a_{21} = -3$, $\alpha = 5$ and $\beta = 2\sqrt{2}$, so

$$\mathbf{R} = \begin{bmatrix} 1 & -2\sqrt{2} \\ -3 & 0 \end{bmatrix}.$$

To confirm this, notice that

$$\mathbf{R}^{-1}\mathbf{AR} = \begin{bmatrix} 0 & -\frac{1}{3} \\ -\frac{\sqrt{2}}{4} & -\frac{\sqrt{2}}{12} \end{bmatrix}\begin{bmatrix} 6 & 3 \\ -3 & 4 \end{bmatrix}\begin{bmatrix} 1 & -2\sqrt{2} \\ -3 & 0 \end{bmatrix} = \begin{bmatrix} 5 & -2\sqrt{2} \\ 2\sqrt{2} & 5 \end{bmatrix} = \mathbf{J}_4.$$

◇

Example 5.11. Reduce matrix

$$\mathbf{A} = \begin{bmatrix} 2 & 1 \\ -1 & 4 \end{bmatrix}$$

to its appropriate Jordan normal form.

Solution. Matrix **A** has a single repeated eigenvalue $\lambda = 3$, to which there corresponds only the single eigenvector $\mathbf{v}_1 = [1, 1]^{\mathrm{T}}$, so matrix **A** is of type (c). Construct a nonsingular matrix $\mathbf{C} = [\mathbf{v}_1, \mathbf{v}_2]$, by taking $\mathbf{v}_1$ to be its first column, and for its second column the arbitrarily vector $\mathbf{v}_2 = [1, -1]^{\mathrm{T}}$, since it is not proportional to $\mathbf{v}_1$
Then,

$$\mathbf{C} = \begin{bmatrix} 1 & 1 \\ 1 & -1 \end{bmatrix}, \mathbf{C}^{-1} = \begin{bmatrix} \frac{1}{2} & \frac{1}{2} \\ \frac{1}{2} & -\frac{1}{2} \end{bmatrix} \text{ and } \mathbf{C}^{-1}\mathbf{A}\mathbf{C} = \begin{bmatrix} 3 & -2 \\ 0 & 3 \end{bmatrix}.$$

The matrix on the right is not yet equal to $\mathbf{J}_3$, so to convert it to that form it is necessary to set $c = -\frac{1}{2}$ in (5.15), when $\mathbf{R} = \mathbf{CM} = \begin{bmatrix} 1 & 1 \\ 1 & -1 \end{bmatrix}\begin{bmatrix} 1 & 0 \\ 0 & -\frac{1}{2} \end{bmatrix}$, so that $\boldsymbol{R} = \begin{bmatrix} 1 & -\frac{1}{2} \\ 1 & \frac{1}{2} \end{bmatrix}$, and $\mathbf{R}^{-1}\mathbf{A}\mathbf{R} = \begin{bmatrix} 3 & 1 \\ 0 & 3 \end{bmatrix}$.

◊

Example 5.12. Reduce the matrix

$$\mathbf{A} = \begin{bmatrix} 0 & 4 \\ -1 & 0 \end{bmatrix}$$

to the appropriate Jordan normal form.

Solution. The eigenvectors of **A** are $\lambda_{\pm} = \pm 2i$, so the appropriate Jordan form is $\mathbf{J}_4$. Taking the positive sign for β, the Jordan normal form to which **A** is similar is found to be

$$\mathbf{J}_4 = \begin{bmatrix} 0 & -2 \\ 2 & 0 \end{bmatrix}.$$

◊

Example 5.13. Reduce matrix

$$\mathbf{A} = \begin{bmatrix} 3 & 1 \\ -2 & 1 \end{bmatrix}$$

to its appropriate Jordan normal form.

Solution. The eigenvalues of **A** are the complex conjugates $\lambda_{\pm} = 2 \pm i$, so working with the eigenvalue λ_{+}, we must set $\alpha = 2$ and $\beta = 1$. From **A** we see that $a_{11} = 3$, $a_{21} = -2$, so the general form for **C** in (d) is

$$\mathbf{C} = \begin{bmatrix} a_{11} - \alpha & -\beta \\ a_{21} & 0 \end{bmatrix} = \begin{bmatrix} 2 & -1 \\ 1 & 2 \end{bmatrix},$$

showing that $\mathbf{A}$ is similar to the matrix.

$$\mathbf{J}_4 = \begin{bmatrix} 2 & -1 \\ 1 & 2 \end{bmatrix}.$$

Had we worked with the eigenvalue $\lambda_- = \alpha - i\beta$ we would have found that $\mathbf{A}$ is similar to

$$\mathbf{J}_4 = \begin{bmatrix} 2 & 1 \\ -1 & 2 \end{bmatrix}.$$

◇

5.9 A Special Tridiagonal Matrix, Its Eigenvalues and Eigenvectors

The $n \times n$ matrices with special structures that have been considered so far have been diagonal matrices, symmetric, skew symmetric, upper triangular and lower triangular matrices. We now introduce another class of $n \times n$ matrices called *banded matrices* that occur in applications throughout engineering, physics, chemistry and numerical analysis. These are matrices in which all elements that do not lie on the leading diagonal, or on a few adjacent parallel diagonals, are zero. Symbolically, a banded matrix $\mathbf{A} = [a_{ij}]$ is one where $a_{ij} = 0$ for $j \neq i - r, i, i + s$, with r and s are small integers, and the *band width* of the matrix is equal to $r + s + 1$ (the number of diagonals that contain nonzero entries). The type of banded matrix to be considered now is a *tridiagonal matrix* with the property that $a_{ij} = 0$ for $j \neq i - 1\ , i\ ,\ i + 1,$ so its band width equal to 3 is formed by the leading diagonal and by the diagonals immediately above and below it. These two parallel diagonals are called, respectively, the *super-diagonal* and the *sub-diagonal* of the matrix. In applications a tridiagonal matrix can be very large, often containing thousands of elements, most of which are zeros. For example, an $n \times n$ tridiagonal matrix contains n^2 elements, of which only $3n - 2$ are nonzero. So if a 30×30 tridiagonal matrix is involved, which in many practical applications is rather small, the number of elements in the matrix is 900, whereas the number of nonzero elements is only 88.

The special $n \times n$ tridiagonal matrix $\mathbf{T}_n(x)$ that will concern us here has the form

$$\mathbf{T}_n(x) = \begin{bmatrix} x & -1 & 0 & 0 & \cdots & 0 & 0 & 0 \\ -1 & x & -1 & 0 & \cdots & 0 & 0 & 0 \\ 0 & -1 & x & -1 & \cdots & 0 & 0 & 0 \\ 0 & 0 & -1 & x & \cdots & 0 & 0 & 0 \\ \vdots & \vdots & \vdots & \vdots & \cdots & \vdots & \vdots & \vdots \\ 0 & 0 & 0 & 0 & \cdots & -1 & x & -1 \\ 0 & 0 & 0 & 0 & \cdots & 0 & -1 & x \end{bmatrix}, \tag{5.24}$$

and its associated determinant $D_n(x) = \det|\mathbf{T}_n(x)|$.

Our objective will be to find the value of the determinant $D_n(x)$, and the eigenvalues and eigenfunctions of matrix $\mathbf{T}_n(x)$ in terms of n and x, though before doing this it will be helpful to outline the steps that are involved. To achieve this objective we will use the Laplace expansion theorem for a determinant to deduce a recurrence (recursion) relation satisfied by $D_n(x)$ for any positive integer n. This recurrence relation turns out to be a *second-order linear difference equation*, and for any n this relates $D_n(x)$ to the values of $D_{n-1}(x)$ and $D_{n-2}(x)$, and its solution will yield a general expression for $D_n(x)$. This will then be used to determine the eigenvalues of $\mathbf{T}_n(x)$ from which, after making use of the simple structure of $\mathbf{T}_n(x)$, the eigenvectors of $\mathbf{T}_n(x)$ will be obtained.

To obtain the recurrence relation for $D_n(x)$ we will expand $D_n(x) = \det|\mathbf{T}_n(x)|$ in terms of the elements of its last column, where

$$D_n(x) = \begin{vmatrix} x & -1 & 0 & 0 & \cdots & 0 & 0 & 0 \\ -1 & x & -1 & 0 & \cdots & 0 & 0 & 0 \\ 0 & -1 & x & -1 & \cdots & 0 & 0 & 0 \\ 0 & 0 & -1 & x & \cdots & 0 & 0 & 0 \\ \vdots & \vdots & \vdots & \vdots & \cdots & \vdots & \vdots & \vdots \\ 0 & 0 & 0 & 0 & \cdots & -1 & x & -1 \\ 0 & 0 & 0 & 0 & \cdots & 0 & -1 & x \end{vmatrix}. \tag{5.25}$$

From the Laplace expansion theorem we have

$$D_n(x) = (-1)C_{n-1,n} + xC_{n,n},$$

where $C_{n-1,n}$ and C_{nn} are the cofactors of the elements -1 and x in the last column of $D_n(x)$. The cofactor $C_{n-1,n} = (-1)^{2n-1}M_{n-1,n}$, where $M_{n-1,n}$ is the corresponding minor. This minor has -1 as the only element in its last row, so when $C_{n-1,n}$ is expanded it becomes $C_{n-1,n} = (-1)(-1)^{2n-1}D_{n-2}(x) = D_{n-2}(x)$, with the result that $C_{n-1,n} = D_{n-2}(x)$. When this is multiplied by the element (-1) we find that $(-1)C_{n-1,n} = -D_{n-2}(x)$. The minor $M_{n,n} = D_{n-1}(x)$, so $C_{n,n} = (-1)^{2m}D_{n-1}(x)$ giving $xC_{n,n} = xD_{n-1}(x)$, so $D_n(x) = (-1)C_{n-1,n} + xC_{n,n}$ becomes

$$D_n(x) = xD_{n-1}(x) - D_{n-2}(x), \tag{5.26}$$

which is the required recurrence relation. This is an example of a second-order linear difference equation, and solutions of such equations are known to be of the form $D_n(x) = A\beta^n$, where A is a constant and β has to be determined by substituting $D_n(x)$ into the difference equation. The result of this substitution is

$$A\beta^{n-2}(\beta^2 - x\beta + 1) = 0. \tag{5.27}$$

Clearly A and β cannot be zero because then there is no solution, so β must be a solution of the quadratic equation

$$\beta^2 - x\beta + 1 = 0, \tag{5.28}$$

with the solution

$$\beta = \frac{x \pm \sqrt{x^2 - 4}}{2}.$$

The case of greatest practical interest occurs when $x^2 < 4$, so to simplify the subsequent analysis we will express x in terms of a parameter θ by setting $x = 2\cos\theta$, and substituting for x shows that $\beta = \cos\theta \pm i\sin\theta$, or equivalently $\beta = e^{\pm i\theta}$. Thus the general solution of the difference equation is seen to be

$$D_n(x) = A_1 e^{in\theta} + A_2 e^{-in\theta}, \tag{5.29}$$

where A_1 and A_2 are arbitrary complex constants. As $e^{in\theta}$ and $e^{-in\vartheta}$ are complex conjugates, for $D_n(x)$ to be real, as it must be because the elements of $\mathbf{T}_n(x)$ are real, it is necessary for $\mathbf{A}_1$ and A_2 to be complex conjugate constants, so we will set $\bar{A}_2 = A_1$, where the overbar signifies the complex conjugation operation.

To find an explicit solution for $D_n(x)$ it is necessary to impose initial conditions on this expression for $D_n(x)$, and to find these we compute $D_1(x)$ and $D_2(x)$ directly from $D_n(x) = \det|\mathbf{T}_n(x)|$, when we find that $D_1(x) = x$ and $D_2(x) = x^2 - 1$. Setting $n = 2$ in the difference equation gives $D_2(x) = xD_1(x) - D_0(x)$, and after substituting for $D_1(x)$ and $D_2(x)$ we obtain $x^2 - 1 = x^2 - D_0(x)$, showing that $D_0(x) = 1$. Next, setting $n = 1$ in the difference equation gives $D_1(x) = xD_0(x) - D_{-1}(x)$, and after substituting $D_1(x) = x$ and $D_0(x) = 1$ we find that $D_{-1}(x) = 0$. So two suitable initial conditions are $D_{-1} = 0, D_0 = 1$. Equivalently, we could use $D_0(x) = 1$ and $D_1(x) = x$, but the first pair of initial conditions prove to be the most convenient ones to use.

Setting $n = -1$ in $D_n(x) = A_1 e^{in\theta} + A_2 e^{-in\theta}$ gives $0 = A_1 e^{-i\theta} + A_2 e^{i\theta}$, from which it follows that $A_2 = -A_1 e^{-2i\theta}$. Setting $n = 0$ in $D_n(x) = A_1 e^{in\theta} + A_2 e^{-in\theta}$ gives $1 = A_1 + A_2$, so $1 = A_1(1 - e^{-2i\theta})$, which leads to the result $A_1 = e^{i\theta}/\,(2i\sin\theta)$. As $A_2 = \bar{A}_1$ we have $A_2 = -e^{-i\theta}/(2i\sin\theta)$, so that

$$D_n(x) = \frac{e^{i\theta}}{2i\sin\theta} e^{in\theta} - \frac{e^{-i\theta}}{2i\sin\theta} = \frac{\sin(n+1)\theta}{\sin\theta}.$$

If $D_n(x) = \det|\mathbf{T}_n(x)|$ is the determinant obtained from the tridiagonal matrix $\mathbf{T}_n$ given above, then $D_n(x)$ satisfies the recurrence relation (difference equation) $D_n(x) = xD_{n-1}(x) - D_{n-2}(x)$ with $x = 2\cos\theta$, for $n = 2, 3, \ldots$ subject to the initial conditions $D_{-1}(x) = 0$ and $D_0(x) = 1$. The explicit expression for $D_n(x)$ in terms of n is given by

$$D_n(x) = \frac{\sin(n+1)\theta}{\sin\theta}, \quad \text{for } n = 1, 2, \ldots \quad . \tag{5.30}$$

The eigenvalues λ of $\mathbf{T}_n(x)$ are the solutions of $|\mathbf{T}_n(x) - \lambda\mathbf{I}| = 0$, which are the n solutions λ of

$$\begin{vmatrix} x-\lambda & -1 & 0 & 0 & \cdots & 0 & 0 & 0 \\ -1 & x-\lambda & -1 & 0 & \cdots & 0 & 0 & 0 \\ 0 & -1 & x-\lambda & -1 & \cdots & 0 & 0 & 0 \\ 0 & 0 & -1 & x-\lambda & \cdots & 0 & 0 & 0 \\ \vdots & \vdots & \vdots & \vdots & \ddots & \vdots & \vdots & \vdots \\ 0 & 0 & 0 & 0 & \cdots & -1 & x-\lambda & -1 \\ 0 & 0 & 0 & 0 & \cdots & 0 & -1 & x-\lambda \end{vmatrix} = 0. \tag{5.31}$$

The characteristic equation for $\mathbf{T}_n$(x) follows the previous reasoning by replacing $x = 2\cos\theta$ by $x - \lambda = 2\cos\theta$. So the expression for the characteristic equation becomes

$$\frac{\sin(n+1)\theta}{\sin\theta} = 0, \tag{5.32}$$

with $x - \lambda = 2\cos\theta$. Now $\sin(n+1)\theta = 0$ when $\theta = m\pi/(n+1)$, with $n = 0$, ± 1, ± 2, $\ldots$, so $2\cos\theta = 2\cos(m\pi/(n+1))$, from which it follows that the n eigenvalues $\lambda_1, \lambda_2, \ldots, \lambda_n$ are given by

$$\lambda_m = x - 2\cos\left(\frac{m\pi}{n+1}\right), \quad m = 1,\ 2,\ \ldots,\ n. \tag{5.33}$$

The eigenvectors $\mathbf{u}_1$, $\mathbf{u}_2$, $\ldots$, $\mathbf{u}_n$ are easily found from their defining equation $[\mathbf{T}_n(x) - \lambda_m\mathbf{I}]\mathbf{u}_m = \mathbf{0}$ because of the simple structure of $\mathbf{T}_n(x)$. Matrices of this type occur in many applications so, by way of an example, we will find the eigenvalues and eigenvectors of the tridiagonal matrix $\mathbf{T}_n(x)$ when $x = 2$.

Let the eigenvector $\mathbf{u}_m = [u_1^{(m)},\ u_2^{(m)},\ \ldots,\ u_n^{(m)}]^{\mathrm{T}}$, and set $x = 2$ in $\mathbf{T}_n(x)$ in the defining matrix equation $[\mathbf{T}_n(2) - \lambda_m\mathbf{I}]\mathbf{u}_m = \mathbf{0}$. The first scalar equation obtained from this matrix equation is

$$(2 - \lambda_m)u_2^{(m)} - u_1^{(m)} = 0.$$

The $n - 2$ equations that follow become

$$-u_{i-1}^{(m)} + (2 - \lambda_m)u_i^{(m)} - u_{i+1}^{(m)} = 0 \quad \text{for } i = 2,\ 3,\ \ldots,\ n-1,$$

while the last equation becomes

$$-u_{n-1}^{(m)} + (2-\lambda_m)u_n^{(m)} = 0.$$

From the first two equations $(2-\lambda_m)u_2^{(m)} - u_1^{(m)} = 0$ and $-u_1^{(m)} + (2-\lambda_m)u_2^{(m)} - u_3^{(m)} = 0$ it follows that

$$\frac{u_1^{(m)}}{1} = \frac{u_2^{(m)}}{2-\lambda_m} = \frac{u_3^{(m)}}{(2-\lambda_m)^2 - 1}. \tag{5.34}$$

Using the fact that $2-\lambda_m = 2\cos(m\pi/(n+1))$, expanding and simplifying denominators and dividing each of these expressions by $\sin(m\pi/(n-1))$, followed by using elementary trigonometric identities like

$$2\sin(m\pi/(n+1))\cos(m\pi/(n+1)) = \sin(2m\pi/(n+1)),$$

the elements $u_1^{(m)}$, $u_2^{(m)}$ and $u_3^{(m)}$ are found to be such that

$$\frac{u_1^{(m)}}{\sin\left(\frac{m\pi}{n+1}\right)} = \frac{u_2^{(m)}}{\sin\left(\frac{2m\pi}{n+1}\right)} = \frac{u_3^{(m)}}{\sin\left(\frac{3m\pi}{n+1}\right)}.$$

So the first three elements of the eigenvector $\mathbf{u}_m$ are proportional to $\sin(2m/(n+1))$, $\sin(m\pi/(n+1))$, and $\sin(3m/(n+1))$. However, eigenvectors can be scaled arbitrarily while remaining eigenvectors, so the scale constant can be set equal to 1, when these three expressions can be taken to be the first three elements of the eigenvector $\mathbf{u}_m$. If the second and third scalar equations in $[\mathbf{T}_n(2) - \lambda_m\mathbf{I}]\mathbf{u}_m = \mathbf{0}$ are used, similar reasoning shows that $u_4^{(m)} = \sin(4m/(n+1))$, and this suggests the eigenvectors $\mathbf{u}_m$ are given by

$$\mathbf{u}_m = \left[\sin\left(\frac{m\pi}{n+1}\right), \sin\left(\frac{2m\pi}{n+1}\right), \ldots, \sin\left(\frac{nm\pi}{n+1}\right)\right]^{\mathrm{T}}, \text{ for } m = 1, 2, \ldots, n. \tag{5.35}$$

This intuitive result is correct, and it can be proved by mathematical induction, though the details of this proof are left as an exercise.

5.10 The Power Method for Eigenvalues and Eigenvectors

So far eigenvalues and eigenvectors have been found using the classical algebraic approach. This starts by finding the characteristic equation for a matrix $\mathbf{A}$ and solving it to find the eigenvalues, and then the eigenvalues are used to solve the systems of equations that determine the associated eigenvectors of $\mathbf{A}$. This method has been successful because the examples used specially constructed 3×3 matrices whose cubic characteristic equations could be solved by inspection. This usually involved finding one eigenvalue λ_1 by inspection, typically a small integer, and then factoring out the expression $(\lambda - \lambda_1)$ from the characteristic equation to arrive at a quadratic equation which was then solved for the two remaining eigenvalues using the quadratic formula.

In general, the roots of a characteristic equation cannot be found by inspection, so numerical methods must be used. This is true even when an equation as simple as a cubic is involved if a root cannot be found by inspection. When characteristic equations with degrees greater than three are involved numerical methods become a necessity. Finding eigenvectors is more difficult than finding eigenvalues, particularly when the eigenvectors correspond to eigenvalues that are complex, are repeated, or some are close together. Software programs resolve these difficulties by using a variety of special techniques to enable then to compute eigenvalues and eigenvectors accurately for an arbitrary $n \times n$ matrix. It is neither possible nor desirable to discuss these methods here, though it is appropriate to discuss a numerical approach that uses matrix methods to accurately compute some eigenvalues and eigenvectors for a fairly wide class of matrices. The method to be discussed is called the *power method*, and the computation leads to the determination of both an eigenvalue and its associated eigenvector.

The power method has its limitations, because it is only suitable for finding some real eigenvalues and their associated eigenvectors when the eigenvalues are well separated and the matrix is diagonalizable, though these properties are not known in advance. The eigenvalue with the largest absolute value is called the *dominant eigenvalue*, while the remaining eigenvalues are called the *sub-dominant eigenvalues*. The power method is an *iterative procedure* that determines the dominant eigenvalue and the elements of its associated eigenvector to a predetermined accuracy of m decimal places. This is achieved by terminating the iterative procedure after say, N iterations, when the Nth and $(N + 1)$th iterations show no change in the mth decimal place of the dominant eigenvalue and each of the elements of its associated eigenvector.

The method is based on the fact that given an $n \times n$ diagonalizable matrix $\mathbf{A}$, an arbitrary n element matrix column vector $\mathbf{v}$ can always be expressed in the form

$$\mathbf{v} = c_1\mathbf{v}_1 + c_2\mathbf{v}_2 + \cdots + c_n\mathbf{v}_n, \tag{5.36}$$

where $\mathbf{v}_1, \mathbf{v}_2, \ldots, \mathbf{v}_n$ are the n eigenvectors of $\mathbf{A}$, and the numbers $c_1, c_2, \ldots, c_n$ are suitable constants. It will be assumed that the eigenvalues are arranged in order of their absolute values with $|\lambda_1|>|\lambda_2| \geq |\lambda_3| \geq \cdots \geq |\lambda_n|$, so λ_1 is the dominant eigenvalue. Pre-multiplying $\mathbf{v}$ in (5.24) by $\mathbf{A}$, and using the fact that the eigenvectors $\mathbf{v}_i$ and the eigenvalues λ_i are related by $\mathbf{A}\mathbf{v}_i = \lambda_i\mathbf{v}_i$ for $i = 1, 2, \ldots, n$, gives

$$\mathbf{A}\mathbf{v} = c_1\mathbf{A}\mathbf{v}_1 + c_2\mathbf{A}\mathbf{v}_2 + \cdots + \mathbf{A}\mathbf{v}_n = \lambda_1\left(c_1\mathbf{v}_1 + c_2\frac{\lambda_2}{\lambda_1}\mathbf{v}_2 + \cdots + c_n\frac{\lambda_n}{\lambda_1}\mathbf{v}_n\right).$$

Iterating this result r times leads to the result

$$\mathbf{A}\mathbf{v}^r = \lambda_1^r\left\{c_1\mathbf{v}_1 + c_2\left(\frac{\lambda_2}{\lambda_1}\right)^r\mathbf{v}_2 + \cdots + c_n\left(\frac{\lambda_n}{\lambda_1}\right)^r\mathbf{v}_n\right\}. \tag{5.37}$$

The ordering of the magnitudes of the eigenvalues means that $|\lambda_r/\lambda_1| < 1$ for $r = 2, 3, \ldots$, so that when r becomes large, all terms on the right of (5.37), with the exception of $c_1\mathbf{v}_1$, will become vanishingly small, causing the expression on the right to reduce to $\lambda_1^r c_1\mathbf{v}_1$, which is a multiple of the eigenvector $\mathbf{v}_1$ corresponding to the dominant eigenvalue λ_1. If $|\lambda_1| > 1$ the scale factor multiplying $\mathbf{v}_1$ will grow rapidly as r increases, while if $|\lambda_1| < 1$ the scale factor will become vanishingly small as r increases. To overcome these difficulties it is usual to normalize the successive eigenvector approximations $\mathbf{v}_1^{(r)}$ for $\mathbf{v}_1$ at each stage of the iterative procedure by scaling successive approximations in such a way that the first element of the approximate vector is 1. As the eigenvector is unknown, the iterative process must begin by using any convenient starting approximation, which is usually taken to be the unit matrix column vector $\mathbf{v}_1^{(0)} = [1, 1, \ldots, 1]^{\mathrm{T}}$, though any other vector can be used. Once the result $\mathbf{A}\mathbf{v}^{(0)} = \mathbf{u}_1^{(1)}$ has been computed, where $\mathbf{u}_1^{(1)} = [u_1^{(1)}, u_2^{(1)}, \ldots, u_n^{(1)}]^{\mathrm{T}}$, the vector $\mathbf{u}_1^{(1)}$ is normalized by dividing each of its elements by $\beta_1 = u_1^{(1)}$ to arrive at the next approximation $\mathbf{v}_1^{(1)} = \left[1, u_2^{(1)}/u_1^{(1)}, u_3^{(1)}/u_1^{(1)}, \ldots, u_n^{(1)}/u_1^{(1)}\right]^{\mathrm{T}}$. The procedure is then repeated by computing $\mathbf{A}\mathbf{v}_1^{(1)} = \mathbf{u}_1^{(2)}$, where $\mathbf{u}_1^{(2)} = [u_1^{(2)}, u_2^{(2)}, \ldots, u_n^{(2)}]^{\mathrm{T}}$. The matrix column vector $\mathbf{u}_1^{(2)}$ is then normalized by dividing each of its elements by $\beta_2 = u_1^{(2)}$, when the next approximation for $\mathbf{v}_1$ becomes $\mathbf{v}_1^{(2)} = \left[1, u_2^{(2)}/u_1^{(2)}, u_3^{(2)}/u_1^{(2)}, \ldots, u_n^{(2)}/u_1^{(2)}\right]^{\mathrm{T}}$. As this iterative procedure is repeated, so the sequence of numbers $\{\beta_1, \beta_2, \beta_3, \ldots\}$ will converge to the dominant eigenvalue λ_1, while the sequence of vectors $\left\{\mathbf{v}_1^{(0)}, \mathbf{v}_1^{(1)}, \mathbf{v}_1^{(2)}, \ldots\right\}$ will converge to the eigenvector $\mathbf{v}_1$.

If a result is required to be accurate to m decimal places, the iterative procedure is terminated when, after N steps, the $(N+1)$th step fails to change the mth decimal place in either the approximation for λ_1, or in the elements of the approximation for the eigenvector $\mathbf{v}_1$.

Example 5.14. Use the power method to find the dominant eigenvalue and its eigenvector given that $\mathbf{A} = \begin{bmatrix} 1 & 2 & 1 \\ 1 & 0 & 1 \\ 1 & 1 & 1 \end{bmatrix}$.

Solution. In order to check the accuracy of the iterative process, notice first that the characteristic equation is $\lambda^3 - 2\lambda^2 - 3\lambda = 0$, or $\lambda(\lambda+1)(\lambda-3) = 0$, so the eigenvalues are 3, 0 and -1, so the dominant eigenvalue $\lambda_1 = 3$. A routine calculation shows the eigenvector $\mathbf{v}_1$, scaled so its first element is 1, is $\mathbf{v}_1 = \left[1, \frac{3}{5}, \frac{4}{5}\right]^{\mathrm{T}} = [1,\ 0.6,\ 0.8]^{\mathrm{T}}$.

Setting $\mathbf{v}_1^{(0)} = [1,\ 1,\ 1]^{\mathrm{T}}$, we find that

$$\mathbf{A}\mathbf{v}_1^{(0)} = \begin{bmatrix} 4 \\ 2 \\ 3 \end{bmatrix}, \text{ so } \beta_1 = 4 \text{ giving } \mathbf{v}_1^{(1)} = \begin{bmatrix} 1 \\ 0.5 \\ 0.75 \end{bmatrix}.$$

Next, using $\mathbf{v}_1^{(1)}$, we find that

$$\mathbf{A}\mathbf{v}_1^{(1)} = \begin{bmatrix} 2.75 \\ 1.75 \\ 2.25 \end{bmatrix}, \text{ so } \beta_2 = 2.75 \text{ giving } \mathbf{v}_1^{(2)} = \begin{bmatrix} 1 \\ 0.63636 \\ 0.81818 \end{bmatrix}.$$

Proceeding to the next stage of the iteration, using $\mathbf{v}_1^{(2)}$, we find that

$$\mathbf{A}\mathbf{v}_1^{(2)} = \begin{bmatrix} 3.09091 \\ 1.81818 \\ 2.45465 \end{bmatrix}, \textit{ so } \beta_1 = 3.09091 \text{ giving } \mathbf{v}_1^{(3)} = \begin{bmatrix} 1 \\ 0.58824 \\ 0.79412 \end{bmatrix}.$$

Continuing this procedure for ten iterations yields the β sequence $\beta_i = \{4, 2.75, 3.09091,\ 2.97059,\ 3.00991,\ 3.00132,\ 2.99952,\ 3.00016,\ 2.99995,\ 3.000002\}$. This is seen to be converging to the limiting value 3, in agreement with the exact value of the dominant eigenvalue $\lambda_1 = 3$ calculated at the outset. Omitting the intermediate calculations, the approximation for $\mathbf{v}_1^{(10)}$ was found to be $\mathbf{v}_1^{(10)} = [1,\ 0.59999,\ 0.79999]^{\mathrm{T}}$, which is seen to be converging to the exact result for the eigenvector $\mathbf{v}_1 = \left[1, \frac{3}{5}, \frac{4}{5}\right]^{\mathrm{T}}$.

◇

The power method can be used to find another eigenvalue and eigenvector by modifying matrix **A**. Consider a matrix **B** derived from matrix **A** by subtracting a number k from each element on the leading diagonal of **A**. Then the defining characteristic equation for matrix **B** becomes $\det[\mathbf{A} - (\lambda - k)\mathbf{I}] = \mathbf{0}$, which is simply the characteristic equation of matrix **A** with λ replaced by $(\lambda - k)$. Thus the eigenvalues of matrix **B** are those of matrix **A**, from each of which has been subtracted the number k (see Exercise 2.12). However, the eigenvectors of **B** will still be the eigenvectors of **A** that correspond to the eigenvalues λ. When applying this transformation to matrix **A**, the dominant eigenvalue of **B** will become the one closest to k.

To use this result with the power method, let λ_1 be the dominant eigenvalue of matrix **A**, then subtracting λ_1 from each element on the leading diagonal of **A** will produce a new matrix **B**, with the property that its dominant eigenvalue will now be the one closest to λ_1. Applying the power method to matrix **B** will generate an eigenvalue $\tilde{\lambda}$ which is sub-dominant to λ_1 together with its eigenvector, and its eigenvalue will be

$$\lambda = \tilde{\lambda} + \lambda_1.$$

If this method is applied to matrix **A** in Example 5.14 with the dominant eigenvalue $\lambda_1 = 3$, so matrix **B** becomes

$$\mathbf{B} = \begin{bmatrix} -2 & 2 & 1 \\ 1 & -3 & 1 \\ 1 & 1 & -2 \end{bmatrix}.$$

An application of the power method to matrix **B** will be found to converge very rapidly to the exact result $\lambda - 3 = -4$, so $\lambda = -1$, when the corresponding eigenvector will be found to be $\mathbf{v}_2 = [1, -1, 0]^{\mathrm{T}}$, though the details of the calculation are left as an exercise.

The power method can be modified so it will generate the eigenvalue of **A** with the *smallest magnitude*, together with its associated eigenvector. The modification follows from the defining relation for eigenvectors $\mathbf{A}\mathbf{x} = \lambda\mathbf{x}$. When **A** is non-singular, this result implies that

$$\mathbf{A}^{-1}\mathbf{x} = \frac{1}{\lambda}\mathbf{x}, \tag{5.38}$$

so the eigenvectors of **A** are also the eigenvectors of $\mathbf{A}^{-1}$, while the eigenvalues of $\mathbf{A}^{-1}$ are the reciprocals of the eigenvalues of **A**. So an application of the power method to $\mathbf{A}^{-1}$ will generate the required eigenvector whose eigenvalue will then be the reciprocal of the eigenvector that is required.

As the dominant eigenvalue of **A** is $\lambda_1 = 3$, matrix **B** will be obtained from **A** by subtracting 3 from each element on its leading diagonal. The power method can then be used to find the dominant eigenvalue of **B**, say $\tilde{\lambda}$, when its eigenvector will be the eigenvector of **A** corresponding, to the eigenvalue $\lambda = \tilde{\lambda} + k$.

For reference purposes, the values of the two sub-dominant eigenvalues of **A** and their eigenvectors are

$$\lambda_2 = -1,\ \mathbf{v}_2 = \begin{bmatrix} -1 \\ 1 \\ 0 \end{bmatrix} \text{ and } \lambda_3 = 0,\ \mathbf{v}_3 = \begin{bmatrix} -1 \\ 0 \\ 1 \end{bmatrix}.$$

◇

5.11 Estimating Regions Containing Eigenvalues

The eigenvalues of an $n \times n$ matrix **A** may be real or complex, and in some applications a qualitative knowledge of their location in the complex plane is useful, while in others it may even make the determination of their actual values unnecessary. In what follows, the complex plane will be called the z-plane, where $z = x + iy$ is the Cartesian representation of a complex number. The test to be described is called the *Gerschgorin circle theorem*, and although the information it provides does not identify the precise location of the eigenvalues of **A**, the test is easy to apply and it does identify either a region or regions in the z-plane that contain *all* of the eigenvalues. The theorem given here is a slight extension of the usual theorem, since it provides a little more information than the original theorem when the regions containing the eigenvalues are disjoint, in the sense that they do not overlap or have points in common. The corollary to the theorem uses matrix **A** and its transpose $\mathbf{A}^{\mathrm{T}}$ in a way that can give a better estimate of the region or regions containing the eigenvalues of **A**.

Theorem 5.10 *The Extended Gerschgorin Circle Theorem. Let* $\mathbf{A}[a_{ij}]$ be an $n \times n$ *matrix. Using matrix* **A**, *define n disks with the circular boundaries* $C_1, C_2, \ldots, C_n$ *such that their respective centers are at the points* $a_{11}, a_{22}, \ldots, a_{nn}$ *on the real axis of the z-plane, with the radius* ρ_k of the circle C_k *with its center at* $z = a_{kk}$ *given by*

$$\rho_k = \sum_{\substack{j=1 \\ j \neq k}}^{n} |a_{kj}| = |a_{k1}| + |a_{k2}| + \cdots + |a_{k,j-1}| + |a_{k,j+1}| + \cdots + |a_{kn}|.$$

Notice that when calculating the radius ρ_k *of circle* C_k *the term* $|a_{kk}|$ *is omitted from the summation.*

(i) *Then at least one eigenvalue of* **A** *will lie inside each circular disk, and the region R in the z-plane comprising the area covered by all of the circular disks will contain all of the eigenvalues of* **A**.

(ii) *If* k *circular disks form one region* R_1*, and* $n - k$ *circular disks form another region* R_2*, and regions* R_1 *and* R_2 *do not overlap or have common points (they are disjoint), then* k *eigenvalues lie in region* R_1 *and* $n - k$ *eigenvalues lie in region* R_2.

Proof (Optional). The proof of the theorem belongs to the study of complex analysis, but as the proof of part (i) of the theorem is simple an outline proof will be given here, though the proof of part (ii) which is a little more difficult will be omitted.

The rth equation in $\mathbf{Ax} = \lambda\mathbf{x}$ is

$$a_{r1}x_1 + \cdots + a_{r,r-1}x_{r-1} + (a_{rr} - \lambda)x_r + a_{r,r+1}x_{r+1} + \cdots + a_{rr}x_r = 0.$$

If this equation is solved for $(a_{rr} - \lambda)$, taking the modulus of the result and making repeated use of the triangle inequality $|a + b| \leq |a| + |b|$, leads to the inequality

$$|a_{rr} - \lambda| < \sum_{\substack{j=1 \\ j \neq r}}^{n} |a_{rj}||x_j|/|x_r|, \quad \text{for } r = 1, 2, \ldots, n.$$

Now let x_r be the element of vector **x** with the largest modulus, so that $|x_j|/|x_r| \leq 1$ for $r = 1, 2, \ldots, n$. Result (I) of the theorem follows from the inequality for $|a_{rr} - \lambda|$ when each term $|x_j|/|x_r|$is replaced by 1, after which this replacement of terms is repeated for $r = 1, 2, \ldots, n$.

◇

On occasions the Corollary that follows can be used to give a better estimate of the region or regions that contain all of the eigenvalues of **A**.

Corollary 5.9. *Finding Another Estimate For the Region In Theorem* 5.10. *Using* $\mathbf{A}^T$, *the transpose of matrix* **A** *in Theorem* 5.10, *construct the* n *Gerschgorin disks* $C_1^{\mathrm{T}}, C_2^{\mathrm{T}}, \ldots, C_n^{\mathrm{T}}$ *for matrix* $\mathbf{A}^T$, *as defined in Theorem* 5.10. *Then the eigenvalues of* **A** *all lie in the region* R_{T} *covered by these discs, which may be disjoint. Part* (ii) *of Theorem* 5.10 *is again applicable.*

Proof. The proof is almost immediate, because it follows directly from Theorem 5.7 that the eigenvalues of **A** and $\mathbf{A}^{\mathrm{T}}$ are identical.

◇

Notice that if one of the regions R and R_{T} defined in Theorem 5.10 and its Corollary lies entirely within the other region then that region is optimum, in the sense that it is the smaller of the two regions with the required property. It is, of course, possible that neither of regions R and R_{T} contains the other region, in which case neither region is optimal, while if matrix **A** is symmetrical regions R and R_{T} coincide.

◇

Example 5.15. Use Theorem 5.10 and its Corollary to find, if possible, an optimum the region or regions in the z-plane that contains all of the eigenvalues of the matrix

$$\mathbf{A} = \begin{bmatrix} 3 & 0 & -3 \\ 0 & -2 & 0 \\ 1 & 2 & -3 \end{bmatrix}.$$

Plot regions R and R_{T} in the z-plane, mark the exact positions of the eigenvalues in each region, and determine if there is an optimum region.

Solution. The Gerschgorin disks for matrix **A** are:

C_1 with its centre at $z = 3$ on the real axis and the radius $\rho_1 = |-3| = 3$;
C_2 with its centre at $z = -2$ on the real axis and the radius $\rho_2 = 0$;
C_3 with its centre at $z = -3$ on the real axis and the radius $\rho_3 = |1| + |2| = 3$.

The region R in the z-plane containing the eigenvalues of **A** is shown as the shaded area in Fig. 5.1a, where disk C_2 has degenerated to the single point where disks C_1 and C_3 meet at the origin. The characteristic equation is $\lambda^3 + 2\lambda^2 - 6\lambda - 12 = 0$, with the roots (the eigenvalues) $\lambda = -2$ and $\pm\sqrt{6}$, shown as solid dots on the real axis in Fig. 5.1a.

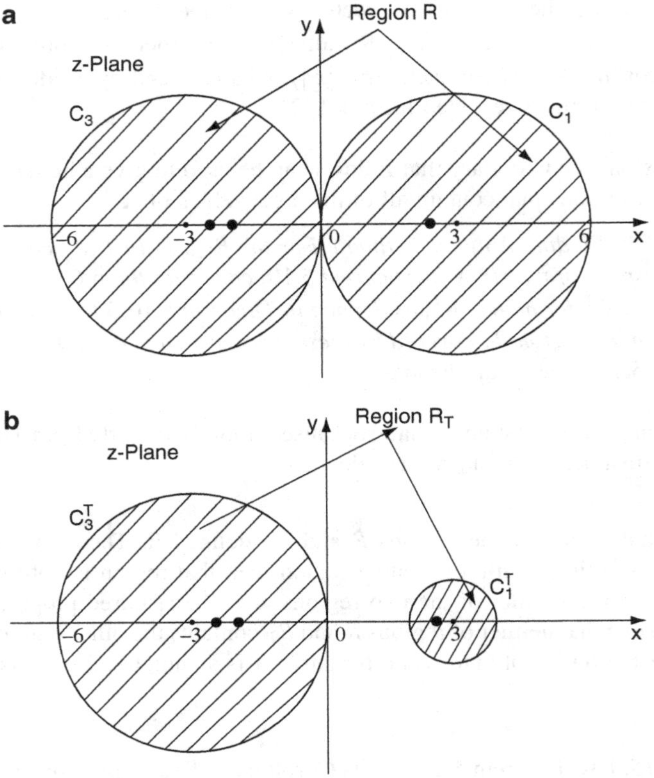

Fig. 5.1 (**a**) The region R (**b**) The region R_T

The Gerschgorin disks for matrix $\mathbf{A}^T$ are:

C_1^T with its centre at $z = 3$ on the real axis and the radius $\rho_1^T = |1| = 1$;
C_2^T with its centre at $z = -2$ on the real axis and the radius $\rho_2 = 0$;
C_3^T with its centre at $z = -3$ on the real axis and the radius $\rho_3 = |-3| = 3$.

In this case, the region R_T in the z-plane containing the eigenvalues of $\mathbf{A}$ is disjoint, and shown as the two shaded areas in Fig. 5.1b, where again the eigenvalues of $\mathbf{A}$ are shown as solid dots on the real axis. Here, again, disk C_2^T has degenerated to a point at the origin on the boundary of disk C_3^T, though disk $C_1^T$is now isolated. It follows from (ii) in Theorem 5.10 when applied to $\mathbf{A}^T$ that the Gerschgorin disk C_1^T contains one eigenvalue of $\mathbf{A}$, while Gerschgorin disk C_3^T contains two eigenvalues of $\mathbf{A}$. Disk C_2^T has a zero radius, but it is now part of disk C_3^T, so the theorem is correct when it attributes the remaining eigenvalues to the union of disks C_2^T and C_3^T. A comparison of Figs. 5.1a and b shows region $R_T$lies within region R, and so it is the optimum region, both in the sense that its Gerschgorin disks occupy the lease space in the z-plane, and because a region occupied by a single eigenvalue has been identified.

◇

5.12 The Fibonacci Sequence and Matrices

The Fibonacci sequence of numbers is $1, 1, 2, 3, 5, 8, 13, 21, 34, \ldots$. After the first two numbers, each subsequent number is generated by summing the two previous numbers in the sequence, so that $1 + 2 = 3$, $2 + 3 = 5$, $3 + 5 = 8$, and so on. The origin of this sequence dates back to 1202 when the most influential Italian algebraist of the time, Leonardo of Pisa known as *Fibonacci*, published a book on algebra in which the following famous problem was first asked and then answered:

> How many pairs of rabbits will be produced in a year, beginning with a single pair, if in every month each pair bears a new pair which then become productive from the second month on?

Fibonacci's answer was based on the two step linear difference equation

$$u_{n+2} = u_{n+1} + u_n \text{ with } n \geq 0 \text{ and } u_1 = u_0 = 1,$$

which he used sequentially to find the number of rabbits produced in a year, though he did not attempt to derive the general solution that would give the number of rabbits after n months.

It is reasonable to ask why this ancient problem and its resulting sequence should still be of interest. The answer is surprising, because sub-sequences of the Fibonacci sequence are found throughout nature, as in the spirals of sunflower heads, in pine cones, in the number of buds on the stems of different plants, in spirals found in fossils, and in patterns of veins in leaves. In mathematics, the Fibonacci sequence enters in a variety of ways, one of which occurs in the design of an optimum search algorithm for the determination of the zeros of functions and, in particular, of polynomials.

The general solution of the difference equation

$$u_n = \frac{1}{\sqrt{5}}\left[\left(\frac{1+\sqrt{5}}{2}\right)^n - \left(\frac{1-\sqrt{5}}{2}\right)^n\right]$$

can be found by the method used in Section 5.8, but the reason for considering the problem here is because the solution can also be obtained using matrix methods coupled with diagonalization.

The two-step difference equation can be transformed into a simple one-step matrix equation by setting $\mathbf{U}_n = \begin{bmatrix} u_{n+1} \\ u_n \end{bmatrix}$, when the difference equation $u_{n+2} = u_{n+1} + u_n$, together with the additional relationship $u_{n+1} = u_{n+1}$, is transformed into the simple one-step matrix equation $\mathbf{U}_{n+1} = \mathbf{A}\mathbf{U}_n$, with $\mathbf{A} = \begin{bmatrix} 1 & 1 \\ 1 & 0 \end{bmatrix}$ so that $\mathbf{U}_{n.} = \mathbf{A}^n\mathbf{U}_0$. To evaluate $\mathbf{A}^n$ we make use of diagonalization with the eigenvalues of $\mathbf{A}$ given by $\lambda_1 = \frac{1}{2}(1 - \sqrt{5})$ and $\lambda_2 = \frac{1}{2}(1 + \sqrt{5})$, and the corresponding eigenvectors

$$\mathbf{x}_1 = \begin{bmatrix} \frac{1}{2}(1-\sqrt{5}) \\ 1 \end{bmatrix}, \ \mathbf{x}_2 = \begin{bmatrix} \frac{1}{2}(1+\sqrt{5}) \\ 1 \end{bmatrix}.$$

Thus the orthogonalizing matrix **P** for matrix **A** is

$$\mathbf{P} = \begin{bmatrix} \frac{1}{2}(1-\sqrt{5}) & \frac{1}{2}(1+\sqrt{5}) \\ 1 & 1 \end{bmatrix}, \text{ while } \mathbf{P}^{-1} = \begin{bmatrix} -\frac{1}{\sqrt{5}} & \frac{1}{2\sqrt{5}}(\sqrt{5}+1) \\ \frac{1}{\sqrt{5}} & \frac{1}{2\sqrt{5}}(\sqrt{5}-1) \end{bmatrix}.$$

So setting $\mathbf{A} = \mathbf{PDP}^{-1}$, with $\mathbf{D} = \text{diag}\{\lambda_1, \lambda_2\}$ we find that $\mathbf{A}^n = \mathbf{P}\,\text{diag}\{\lambda_1^n, \lambda_2^n\}\mathbf{P}^{-1}$.

Using this result in $\mathbf{U}_{n+1} = \mathbf{A}^n\mathbf{U}_n$, and solving for u_n, gives the general solution

$$u_n = \frac{1}{\sqrt{5}}\left[\left(\frac{1+\sqrt{5}}{2}\right)^n - \left(\frac{1-\sqrt{5}}{2}\right)^n\right].$$

It is a curious fact that $\lim_{n\to\infty}(u_{n+1}/u_n) = \frac{1}{2}(1+\sqrt{5}) \approx 1.618$ is the *golden ratio* used by ancient Greek architects, and also modern ones, to ensure buildings have what is believed be the most esthetically pleasing proportions. For example architects would use a rectangle, either as a plan or as the front projection of a building, with the proportions 5:3, 8:5 or 13:8, all of which are close approximations to the golden ratio.

5.13 A Two-Point Boundary-Value Problem and a Tridiagonal Matrix

This section shows one of the ways in which a tridiagonal matrix of the type considered in Section 5.8 can arise when using a numerical method to solve a two-point boundary-value problem. A *two-point boundary-value problem* for a second-order ordinary differential equation involves finding, when possible, a solution $u(x)$ of the equation over an interval $a \le x \le b$ such that $u(a) = k_1$ and $u(b) = k_2$. So, in a two-point boundary-value problem, instead of the solution satisfying two initial conditions, the solution must satisfy one condition at $x = a$ at the lower *boundary*, and another condition at $x = b$ which is the *upper* boundary, of the interval $a \le x \le b$in which the solution is required. The case considered here is a particularly simple one, because the equation is $d^2u/dx^2 = -f(x)$ on the interval $0 \le x \le 1$, subject to the *homogeneous boundary conditions* $u(0) = u(1) = 0$. When $f(x)$ is suitably simple, this equation can be integrated analytically and the two arbitrary constants of integration chosen such that both boundary conditions are satisfied. However, although the equation is simple, when $f(x)$ cannot be integrated analytically the solution can only be found by a numerical approach.

The approach involves dividing the interval $0 \le x \le 1$ into n uniform sub-intervals of length $h = 1/n$, with the point at $x = 0$ numbered 0, and the point at $x = 1$ numbered $n + 1$, with the points at the ends of the sub-intervals called *grid points*. The first three terms of the Taylor series expansions of $u(x-h)$ and $u(x+h)$, approximate these functions by

$$u(x-h) = u(x) - hu(x) + hu''(x) + \text{an error term}$$

and

$$u(x+h) = u(x) + u'(x) + h^2u''(x) + \text{an error term.}$$

Adding these results and neglecting error terms gives $d^2u/dx^2 \approx (1/h^2)\{u(x+h) - 2u(x) + u(x-h)\}$.

In terms of this result, the equation connecting the discrete values of $u(x)$ at the grid points $j-1,\ j$ and $j+1$ becomes

$$-u_{j+1} + 2u_j - u_{j-1} = h^2 f(jh).$$

Arranging all n equations in matrix form this becomes $\mathbf{T}_n(2)\mathbf{u} = \mathbf{f}$ where the nonhomogeneous vector $\mathbf{f} = [h^2f(h),\ h^2f(2h), \ldots,\ h^2f(nh)]^{\mathrm{T}}$, so the approximate solution is given by solving $\mathbf{T}_n(2)\mathbf{u} = \mathbf{f}$ for $\mathbf{u}$. This usually involves using Gaussian elimination, though when n is small and $\mathbf{T}_n^{-1}$ is easily calculated, the solution can be found from $\mathbf{u} = \mathbf{T}_n^{-1}(2)\mathbf{f}$, where $\mathbf{T}_n(2)$ is matrix (5.24) with $x = 2$.

Setting $f(x) = 1 + e^x$ and $n = 5$ means there will be four internal grid points each separated by $h = 0.2$, so the tridiagonal matrices involved will be $\mathbf{T}_4(2)$, and for later use $\mathbf{T}_4^{-1}(2)$, given by

$$\mathbf{T}_4(2) = \begin{bmatrix} 2 & -1 & 0 & 0 \\ -1 & 2 & -1 & 0 \\ 0 & -1 & 2 & -1 \\ 0 & 0 & -1 & 2 \end{bmatrix} \text{ and } \mathbf{T}_4^{-1}(2) = \begin{bmatrix} \frac{4}{5} & \frac{3}{5} & \frac{2}{5} & \frac{1}{5} \\ \frac{3}{5} & \frac{6}{5} & \frac{4}{5} & \frac{2}{5} \\ \frac{2}{5} & \frac{4}{5} & \frac{6}{5} & \frac{3}{5} \\ \frac{1}{5} & \frac{2}{5} & \frac{3}{5} & \frac{4}{5} \end{bmatrix}.$$

The matrix $\mathbf{T}_4^{-1}(2)$ is given here because the problem is sufficiently simple that the resulting equations can be solved with the aid of this inverse matrix, the calculation of which is simplified by the symmetry of $\mathbf{T}_4(2)$ (why?). The elements of $\mathbf{f}$ are $f(jh) = h^2(1 + \exp(j/5))$ for $j = 1, 2, 3, 4$.

The exact solution is $u(x) = -1 + \left(\frac{1}{2} - e\right)x + \frac{1}{2}x^2 + e^x$, and the results that follow compare the exact and approximate solutions at the grid points.

	$u_1 = u\left(\frac{1}{5}\right)$	$u_2 = u\left(\frac{2}{5}\right)$	$u_3 = u\left(\frac{3}{5}\right)$	$u_4 = u\left(\frac{4}{5}\right)$
Exact	−0.2023	−0.3155	−0.3289	−0.2286
Approx	−0.2018	−0.3148	−0.3282	−0.2286

Increasing the value of n will improve still further the approximate solutions at the grid points at the cost of increasing to n the number of equations that need to be solved. However, problems like this do not produce matrices the size of those encountered when solving the Laplace equation in Section 3.9, if the number of grid points is increased considerably in both the x and y directions.

5.14 Matrices with Complex Elements

In certain applications, $n \times n$ matrices with complex elements occur, the most important of which are *Hermitian matrices, skew-Hermitian matrices*, and *unitary matrices*, all for which exhibit certain types of symmetry.

5.14.1 Hermitian Matrices

These are complex matrices that generalize the more familiar symmetric matrices. An *Hermitian matrix* $\mathbf{A}$ is a matrix with the property that $\bar{\mathbf{A}}^{\mathrm{T}} = \mathbf{A}$, where the overbar indicates that each element of $\mathbf{A}$ is replaced by its complex conjugate and, as usual, the superscript T indicates the matrix transpose operation. A simple example of an *Hermitian* is

$$\mathbf{A} = \begin{bmatrix} 1 & 2+i \\ 2-i & 3 \end{bmatrix}.$$

An immediate consequence of the definition of an Hermitian matrix is that the elements on its leading diagonal are all real. This follows from the fact that the ith element a_{ii} on the leading diagonal is $a_{ii} = \alpha_i + i\beta_i$, for $i = 1, 2, \ldots, n$, but $a_{ii} = \bar{a}_{ii}{}^{\mathrm{T}}$, which is only possible if $a_{ii} = \alpha_i$, showing that each element on the leading diagonal must be *real*.

The matrix $\bar{\mathbf{A}}^{\mathrm{T}}$, that is the transpose of the matrix whose elements are the complex conjugates of the corresponding elements in $\mathbf{A}$, is called the *Hermitian transpose* of $\mathbf{A}$, and it is denoted by $\mathbf{A}^{\mathrm{H}}$, so $\mathbf{A}^{\mathrm{H}} = \bar{\mathbf{A}}^{\mathrm{T}}$. In terms of this notation, an $n \times n$ matrix $\mathbf{A}$ will be *Hermitian* if

$$\mathbf{A}^{\mathrm{H}} = \mathbf{A}. \tag{5.39}$$

Notice that a *symmetric* matrix is a special case of an Hermitian matrix when all of its elements are real.

It is left as an exercise to show that the Hermitian transpose operation has the following properties similar to those of the ordinary transpose operation:

$$\left(\mathbf{A}^{\mathrm{H}}\right)^{\mathrm{H}} = \mathbf{A}, \tag{5.40}$$

and if $\mathbf{A}$ and $\mathbf{B}$ are two Hermitian matrices that are conformable for the product $\mathbf{AB}$, then

$$(\mathbf{AB})^{\mathrm{H}} = \mathbf{B}^{\mathrm{H}}\mathbf{A}^{\mathrm{H}}. \tag{5.41}$$

The definition of the norm of a vector used so far was based on a vector with real elements, and the norm is an essentially nonnegative quantity that provides a measure of the "*size*" of the vector. However, if this property is to remain true for vectors with complex elements it becomes necessary to modify the definition of the inner product of two vectors. The modification is simple, and when the elements of the n element row vector $\mathbf{x} = [x_1, x_2, \dots, x_n]$, and the n element column vector $\mathbf{y} = [y_1, y_2, \dots, y_n]^{\mathrm{T}}$ are complex, their *complex inner product* is defined as

$$\mathbf{xy} = (x_1\bar{y}_1 + x_2\bar{y}_2 + \cdots + x_n\bar{y}_n)^{1/2}. \tag{5.42}$$

It follows from this that the *norm* of a vector $\mathbf{x}$ with complex elements, written $\|\mathbf{x}\|$, defined as the inner product $(\mathbf{x}\mathbf{x}^{\mathrm{T}})^{1/2}$, is given by

$$\|\mathbf{x}\| = \left(\mathbf{x}\bar{\mathbf{x}}^{\mathrm{T}}\right)^{1/2} = (x_1\bar{x}_1 + x_2\bar{x}_2 + \cdots + x_n\bar{x}_n)^{1/2}. \tag{5.43}$$

This has the property required of a norm that it is nonnegative, as can be seen from the fact that each product $x_i\bar{x}_i$ is real and nonnegative. Notice that (5.43) is compatible with the definition of the norm of a vector $\mathbf{x}$ with *real* elements, because then (5.43) reduces to the ordinary norm $\|\mathbf{x}\| = \left(x_1^2 + x_2^2 + \cdots + x_n^2\right)^{1/2}$.

Hermitian matrices have certain properties similar to those of real symmetric matrices, and the theorem that now follows gives two of their fundamental properties.

Theorem 5.11 *Two Fundamental Properties of Hermitian Matrices. Let matrix* $\mathbf{A}$ *be Hermitian. Then:*

(i) *The eigenvalues of* $\mathbf{A}$ *are all real.*
(ii) *Eigenvectors corresponding to distinct eigenvalues are orthogonal with respect to the complex inner product.*

Proof.

(i) Let λ be any eigenvalue of the Hermitian matrix $\mathbf{A}$, with $\mathbf{x}$ the corresponding column eigenvector, then

$$\mathbf{Ax} = \lambda\mathbf{x}.$$

Then as the elements of $\mathbf{A}$ are complex, the eigenvalue λ and the eigenvector $\mathbf{x}$ may also be complex. Pre-multiplying the equation by $\bar{\mathbf{x}}^{\mathrm{T}}$ it becomes

$$\bar{\mathbf{x}}^{\mathrm{T}}\bar{\mathbf{A}}\mathbf{x} = \lambda\bar{\mathbf{x}}^{\mathrm{T}}\mathbf{x} = \lambda\|\mathbf{x}\|^2. \tag{5.44}$$

Next, taking the complex conjugate of $\mathbf{Ax} = \lambda\mathbf{x}$, and then taking the transpose of each side of the equation, we have

$$\bar{\mathbf{x}}^{\mathrm{T}}\bar{\mathbf{A}}^{\mathrm{T}} = \bar{\lambda}\bar{\mathbf{x}}^{\mathrm{T}}.$$

After post-multiplication by $\mathbf{x}$ this becomes

$$\bar{\mathbf{x}}^{\mathrm{T}}\mathbf{Ax} = \bar{\lambda}\bar{\mathbf{x}}^{\mathrm{T}}\mathbf{x} = \bar{\lambda}\|\mathbf{x}\|^2, \tag{5.45}$$

because $\mathbf{A}$ is Hermitian $\bar{\mathbf{A}}^{\mathrm{T}} = \mathbf{A}^{\mathrm{H}} = \mathbf{A}$. Subtracting (b) from (a) gives

$$(\lambda - \bar{\lambda})\|\mathbf{x}\|^2 = 0,$$

but $\|\mathbf{x}\|^2 \neq 0$, so $\lambda = \bar{\lambda}$ which is only possible if λ is real. As $\mathbf{x}$ was any eigenvector of $\mathbf{A}$, it follows that every eigenvalue of an Hermitian matrix $\mathbf{A}$ must be real.

(ii) Let $\mathbf{x}$ and $\mathbf{y}$ be eigenvectors of an Hermitian matrix $\mathbf{A}$ corresponding, respectively, to the *distinct* real eigenvalues λ and μ. Then

$$\mathbf{Ax} = \lambda\mathbf{x} \text{ and } \mathbf{Ay} = \mu\mathbf{y}.$$

Pre-multiplying the first equation by $\mathbf{y}^{\mathrm{T}}$ and the second equation by $\mathbf{x}^{\mathrm{T}}$ they become

$$\bar{\mathbf{y}}^{\mathrm{T}}\mathbf{Ax} = \lambda\bar{\mathbf{y}}^{\mathrm{T}}\mathbf{x} \tag{5.46}$$

and

$$\bar{\mathbf{x}}^{\mathrm{T}}\mathbf{Ay} = \mu\bar{\mathbf{x}}^{\mathrm{T}}\mathbf{y}. \tag{5.47}$$

Taking the complex conjugate of Eq. (5.46), followed by the transpose operation, while remembering that λ is real so that $\bar{\lambda} = \lambda$, (5.47) becomes

$$\bar{\mathbf{x}}^{\mathrm{T}}\mathbf{Ay} = \lambda\bar{\mathbf{x}}^{\mathrm{T}}\mathbf{y}, \tag{5.48}$$

because $\bar{\mathbf{A}}^{\mathrm{T}} = \mathbf{A}^{\mathrm{H}} = \mathbf{A}$. Finally, subtracting Eq. (5.47) from Eq. (5.48) gives

$$(\lambda - \mu)\bar{\mathbf{x}}^{\mathrm{T}}\mathbf{y} = 0,$$

but by hypothesis $\lambda \neq \mu$, so $\mathbf{x}^{\mathrm{T}}\mathbf{y} = 0$, confirming that the eigenvectors $\mathbf{x}$ and $\mathbf{y}$ are orthogonal with respect to the complex inner product. As $\mathbf{x}$ and $\mathbf{y}$ were any two eigenvectors of the Hermitian matrix $\mathbf{A}$ corresponding to the distinct eigenvalues λ and μ, property (ii) has been established.

◇

Example 5.14. Given the matrix

$$\mathbf{A} = \begin{bmatrix} 1 & 1+i \\ 1-i & 0 \end{bmatrix}.$$

(a) Verify that $\mathbf{A}$ is Hermitian, (b) find its eigenvalues, and (c) find its eigenvectors and verify that they are orthogonal with respect to the complex inner product.

Solution.

(a) The matrix is Hermitian because the elements on its leading diagonal are real and its off-diagonal elements are complex conjugates.

(b) The eigenvalues are the solutions of the characteristic equation.

$P(\lambda) = \begin{vmatrix} 1-\lambda & 1+i \\ 1-i & -\lambda \end{vmatrix} = 0$, corresponding to $\lambda^2 - \lambda - 2 = 0$ with the roots $\lambda = -1$ and $\lambda = 2$.

(c) Calculating the eigenvectors in the usual way shows that when $\lambda_1 = -1$ the eigenvector is proportional to

$$\mathbf{x}_1 = \begin{bmatrix} 1+i \\ -2 \end{bmatrix},$$

and when $\lambda_2 = 2$ the eigenvector is proportional to

$$\mathbf{x}_2 = \begin{bmatrix} 1+i \\ 1 \end{bmatrix}.$$

The eigenvectors $\mathbf{x}_1$ and x_2 are orthogonal with respect to the complex inner product because $(\mathbf{x}_1)^{\mathrm{T}}\bar{\mathbf{x}}_2 = 0$.

◇

5.14.2 Skew-Hermitian Matrices

A *skew-Hermitian* matrix is a generalization of an ordinary skew-symmetric matrix, and it is defined as a matrix $\mathbf{A}$ with the property that

$$\bar{\mathbf{A}}^{\mathrm{T}} = -\mathbf{A}, \text{ so the elements of } \mathbf{A} \text{ are such that } \bar{a}_{kj} = -a_{jk}. \qquad (5.49)$$

Setting $k = j$ in (5.49), and $a_{jj} = \alpha + i\beta$, it follows that $\bar{a}_{jj} = \alpha - i\beta$, so $\bar{a}_{jj} = -a_{jj}$ is only possible if $\alpha = 0$, so the elements on the leading diagonal of a skew-Hermitian matrix must either be purely imaginary or zero.

A simple example of a skew-Hermitian matrix is

$$\mathbf{A} = \begin{bmatrix} 2i & 3-i \\ -3-i & -4i \end{bmatrix}.$$

5.14.3 Unitary Matrices

A *unitary matrix* is a generalization of an orthogonal matrix, and $\mathbf{U}$ is a unitary matrix if

$$\bar{\mathbf{U}}^{\mathrm{T}} = \mathbf{U}^{-1}. \qquad (5.50)$$

A simple example of a unitary matrix is

$$\mathbf{U} = \begin{bmatrix} \frac{1}{2}\sqrt{3}i & \frac{1}{2} \\ \frac{1}{2} & \frac{1}{2}\sqrt{3}i \end{bmatrix}.$$

Theorem 5.12 *The Eigenvalues of skew-Hermitian and Unitary Matrices.*

(i) *The eigenvalues of a skew-Hermitian matrix are either purely imaginary, or zero.*
(ii) *The eigenvalues of a unitary matrix all have absolute value 1.*

Proof.

(i) If matrix $\mathbf{A}$ is skew-Hermitian, it follows from its definition that $i\mathbf{A}$ is Hermitian. The eigenvalues of $\mathbf{A}$ are the roots of the characteristic equation $|\mathbf{A} - \lambda\mathbf{I}| = 0$. Multiplying the matrix $\mathbf{A} - \lambda\mathbf{I}$ in the determinant by i will not change this result, so $|i\mathbf{A} - i\lambda\mathbf{I}| = 0$. As $i\mathbf{A}$ is Hermitian, Theorem 5.11(i) asserts that the eigenvalues $i\lambda$ are real, so the eigenvalues λ of the skew-Hermitian matrix $\mathbf{A}$ must be purely imaginary. An eigenvalue of an Hermitian matrix may be zero, so it follows that an eigenvalue of a skew-Hermitian matrix may also be zero, and the result is proved.
(ii) The proof of this result is essentially the same as the proof of result (i) in Theorem 3.1 concerning orthogonal matrices into which the complex conjugate operation has been introduced, so the details will be left as an exercise.

◊

Exercises

In Exercises 1 through 8 find the characteristic equation $p(\lambda) = \det[\mathbf{A} - \lambda\mathbf{I}] = 0$, and the eigenvalues and eigenvectors of the given matrix $\mathbf{A}$.

1.
$$\mathbf{A} = \begin{bmatrix} 1 & 0 & -1 \\ -1 & 1 & 0 \\ -1 & 0 & 1 \end{bmatrix}.$$

2.
$$\mathbf{A} = \begin{bmatrix} 2 & 0 & 1 \\ 1 & 2 & 1 \\ 0 & 1 & 1 \end{bmatrix}.$$

3.
$$\mathbf{A} = \begin{bmatrix} 1 & -1 & 0 \\ -1 & -1 & 1 \\ 0 & 1 & 1 \end{bmatrix}.$$

4.
$$\mathbf{A} = \begin{bmatrix} -1 & 1 & 0 \\ 1 & 0 & 1 \\ 0 & 1 & 1 \end{bmatrix}.$$

5.

$$\mathbf{A} = \begin{bmatrix} -3 & 1 & -1 \\ 1 & 0 & 1 \\ -1 & 2 & 1 \end{bmatrix}.$$

6.

$$\mathbf{A} = \begin{bmatrix} 2 & 1 & 0 \\ 0 & 1 & 0 \\ 1 & 1 & 1 \end{bmatrix}.$$

7.

$$\mathbf{A} = \begin{bmatrix} -1 & -2 & 2 \\ -3 & -1 & 3 \\ -3 & -2 & 4 \end{bmatrix}.$$

8.

$$\mathbf{A} = \begin{bmatrix} 1 & 0 & 0 \\ -1 & -1 & 1 \\ -1 & -2 & 2 \end{bmatrix}.$$

In Exercises 9 through 12 find a diagonalizing matrix **P** for the given matrix **A**.

9.

$$\mathbf{A} = \begin{bmatrix} -3 & 4 & 4 \\ 1 & -3 & -1 \\ -3 & 6 & 4 \end{bmatrix}.$$

10.

$$\mathbf{A} = \begin{bmatrix} 6 & 9 & 4 \\ -4 & -7 & -4 \\ -1 & -1 & 1 \end{bmatrix}.$$

11.

$$\mathbf{A} = \begin{bmatrix} -1 & 0 & -2 \\ 1 & 1 & 1 \\ -3 & -6 & -2 \end{bmatrix}.$$

12.

$$\mathbf{A} = \begin{bmatrix} 5 & 11 & 7 \\ -2 & -3 & -2 \\ 2 & 1 & 0 \end{bmatrix}.$$

In Exercises 13 and 14 find the eigenvalues and eigenvectors of matrix **A** and determine if **A** is diagonalizable.

13.

$$\mathbf{A} = \begin{bmatrix} -1 & 0 & -1 \\ -1 & 0 & -1 \\ -1 & 2 & 0 \end{bmatrix}.$$

14.

$$\mathbf{A} = \begin{bmatrix} -3 & 0 & 1 \\ 1 & 0 & 1 \\ -1 & 1 & 0 \end{bmatrix}.$$

In Exercises 15 and 16 find an orthogonal diagonalizing matrix $\mathbf{Q}$ for the symmetric matrix $\mathbf{A}$.

15.

$$\mathbf{A} = \begin{bmatrix} 1 & 0 & -1 \\ 0 & -1 & 0 \\ -1 & 0 & 1 \end{bmatrix}.$$

16.

$$\mathbf{A} = \begin{bmatrix} -3 & -2 & -2 \\ -2 & 1 & -2 \\ -2 & -2 & 1 \end{bmatrix}.$$

In Exercises 17 through 20 use a suitable orthogonal diagonalizing matrix $\mathbf{Q}$ with $\mathbf{x} = [x_1, x_2, x_3]^{\mathrm{T}}$ to reduce the quadratic form $Q(\mathbf{x}) = \mathbf{x}^{\mathrm{T}}\mathbf{A}\mathbf{x}$ to a sum of squares $Q(\mathbf{y}) = \mathbf{y}^{\mathrm{T}}\mathbf{A}\mathbf{y}$, with $\mathbf{y} = [y_1, y_2, y_3]^{\mathrm{T}}$. Find the change of variable from the x_i to the y_i to achieve the reduction, and write down and classify the quadratic form.

17. $Q(\mathbf{x}) = \mathbf{x}^{\mathrm{T}}\mathbf{A}\mathbf{x}$ with $\mathbf{A} = \begin{bmatrix} -1 & 2 & -1 \\ 2 & -1 & -1 \\ -1 & -1 & 0 \end{bmatrix}.$

18. $Q(\mathbf{x}) = \mathbf{x}^{\mathrm{T}}\mathbf{A}\mathbf{x}$ with $\mathbf{A} = \begin{bmatrix} 1 & 1 & 0 \\ 1 & 0 & 1 \\ 0 & 1 & 1 \end{bmatrix}.$

19. $Q(\mathbf{x}) = \mathbf{x}^{\mathrm{T}}\mathbf{A}\mathbf{x}$ with $\mathbf{A} = \begin{bmatrix} -1 & 0 & 2 \\ 0 & -1 & 0 \\ 2 & 0 & 2 \end{bmatrix}.$

20. $Q(\mathbf{x}) = \mathbf{x}^{\mathrm{T}}\mathbf{A}\mathbf{x}$ with $\mathbf{A} = \begin{bmatrix} 2 & -1 & 1 \\ -1 & 1 & 0 \\ 1 & 0 & 1 \end{bmatrix}.$

21. Show how when $\mathbf{A}$ is nonsingular, multiplication of the Cayley–Hamilton theorem by the inverse matrix $\mathbf{A}^{-1}$ gives a matrix polynomial that determines $\mathbf{A}^{-1}$. Use this method to find $\mathbf{A}^{-1}$, given that

$$\mathbf{A} = \begin{bmatrix} 1 & 0 & 3 \\ 2 & 1 & 1 \\ 0 & -1 & 1 \end{bmatrix},$$

and check your result by showing that $\mathbf{A}\mathbf{A}^{-1} = \mathbf{I}$. What happens if this method is used to try to find the inverse of a singular matrix $\mathbf{A}$?

In Exercises 22 through 27 classify each of the matrices and find a matrix that reduces it to the appropriate Jordan normal form.

22.
$$\mathbf{A} = \begin{bmatrix} 2 & 0 \\ 2 & 2 \end{bmatrix}.$$

23.
$$\mathbf{A} = \begin{bmatrix} 3 & 2 \\ 2 & 3 \end{bmatrix}.$$

24.
$$\mathbf{A} = \begin{bmatrix} 5 & 2 \\ -2 & 1 \end{bmatrix}.$$

25.
$$\mathbf{A} = \begin{bmatrix} 1 & 2 \\ -3 & 1 \end{bmatrix}.$$

26.
$$\mathbf{A} = \begin{bmatrix} 4 & -2 \\ 0 & 4 \end{bmatrix}.$$

27.
$$\mathbf{A} = \begin{bmatrix} -1 & 2 \\ -2 & -1 \end{bmatrix}.$$

28. Compute the determinants D_2, D_3 and D_4 associated with matrix $\mathbf{T}_n$ in Section 5.7 with x arbitrary, and confirm that $D_4 = xD_3 - D_2$.

The proof of the result in Exercise 29 that follows does not involve matrices, but it is included for the sake of completeness because it can be useful when finding the eigenvalues of a real $n \times n$ matrix.

29. Let ζ be complex zero of the nth degree polynomial $P_n(z) = z^n + a_1 z^{n-1} + a_2 z^{n-2} + \cdots + a_{n-1} z + a_n$, where the coefficients $a_1, a_2, \ldots, a_n$ are real numbers. By using the elementary properties of the complex conjugate operation show that $\bar{\zeta}$ must also be a zero of $P_n(z)$. Hence show that any pair of complex conjugate zeros of $P_n(z)$ correspond to a real quadratic factor of $P_n(z)$.

Another Hadamard Inequality for Matrices

The inequality that follows has been included because of its connection with quadratic forms, and also for general interest, though the result will not be proved here.

Hadamard's Inequality for Positive Definite Matrices

A quadratic form $Q(x_1, x_2, \ldots, x_n)$ can always be written in the form $Q(x_1, x_2, \ldots, x_n) = \mathbf{x}^{\mathrm{T}}\mathbf{A}\mathbf{x}$, with $\mathbf{x} = [x_1, x_2, \ldots, x_n]^{\mathrm{T}}$ and $\mathbf{A}$ a real $n \times n$ matrix. If $\mathbf{A}$ is positive definite (that is if the quadratic form Q is positive definite) then

$$\det \mathbf{A} = \begin{vmatrix} a_{11} & a_{12} & \cdots & a_{1n} \\ a_{21} & a_{22} & \cdots & a_{2n} \\ \vdots & \vdots & \vdots & \vdots \\ a_{n1} & a_{n2} & \cdots & a_{nn} \end{vmatrix} \leq a_{11}a_{22}\ldots a_{nn},$$

where the equality holds if and only if **A** is a diagonal matrix with positive elements.

In Exercises 30 to 32 find the matrix **A** that corresponds to the given quadratic form, and show it is positive definite by applying the Routh–Hurwitz test. The test will require finding det **A**, and use the value of det **A** to verify the Hadamard inequality.

30. $Q(x_1, x_2, x_3) = 4x_1^2 + 4x_2^2 + x_3^2 - 2x_1x_2.$

31. $Q(x_1, x_2, x_3) = \frac{5}{2}x_1^2 + x_1x_3 + x_2^2 + \frac{5}{2}x_3^2.$

32. $Q(x_1, x_2, x_3, x_4) = 2x_1^2 + 4x_1x_2 - 2x_1x_4 + 7x_2^2 + 2x_2x_4 + 3x_4^2 - 4x_2x_3 + 12x_3^2 - 2x_1x_3 - 8x_3x_4.$

The Spectral Radius

Let $\lambda_1, \lambda_2, \ldots, \lambda_n$ be the distinct eigenvalues of an $n \times n$ matrix **A**. Then the *spectral radius* $\rho(\mathbf{A})$ of matrix **A** is defined as the maximum value of the modulus $|\lambda_i|$ for $i = 1, 2, \ldots, n$, and the set of eigenvalues is called the *spectrum* of matrix **A**. The eigenvalue with the largest modulus is called the *dominant eigenvalue*. When expressed formally,

$$\rho(\mathbf{A}) = \max_i \{|\lambda_1|, |\lambda_2|, \ldots, |\lambda_n|\}.$$

As **A** may have complex eigenvalues, the interpretation of $\rho(\mathbf{A})$ in the complex plane is that $\rho(\mathbf{A})$ is the radius of the smallest circle centered on the origin that contains either in its interior or on its boundary all of the eigenvalues of **A**. The spectral radius has various applications, one of which occurs when an $n \times n$ matrix **A** is raised successively to higher powers, generating a sequence of matrices $\mathbf{A}, \mathbf{A}^2, \mathbf{A}^3, \ldots$ The spectral radius is important in this case because when $\rho(\mathbf{A}) < 1$ it can be shown that $\lim_{n\to\infty} \mathbf{A}^n = \mathbf{0}$, while if $\rho(\mathbf{A}) = 1$ *the limit* $\lim_{n\to\infty} \mathbf{A}^n$ is nonzero but bounded (the elements of **A** are all bounded). The sequence diverges if $\rho(\mathbf{A}) > 1$.

33. (a) Find the eigenvalues of matrix **A** and the spectral radius $\rho(\mathbf{A})$ if

$$\mathbf{A} = \begin{bmatrix} 0 & 1 & 2 & 1 \\ 0 & 0 & 1 & 1 \\ 0 & 0 & 0 & 1 \\ 0 & 0 & 0 & 0 \end{bmatrix}.$$

Confirm by calculation that $\lim_{n\to\infty} \mathbf{A}^n = \mathbf{0}$, in agreement with the value of $\rho(\mathbf{A})$. (b) Find the eigenvalues of matrix $\mathbf{A}$ and the spectral radius $\rho(\mathbf{A})$ if

$$\mathbf{A} = \begin{bmatrix} 1 & 0 & 0 \\ 0 & \frac{1}{3} & \frac{2}{3} \\ 0 & \frac{1}{4} & \frac{3}{4} \end{bmatrix}.$$

Confirm by calculating $\mathbf{A}^n$ for $n = 1$ to 5 that $\lim_{n\to\infty} \mathbf{A}^n$tends to a bound, and find rounded to four figures the bounds to which each of the elements of $\mathbf{A}^n$ converge.

34. Apply the power method to matrix $\mathbf{A}$ in Examples 5.1, and confirm that it yields the dominant eigenvalue and its associated eigenvector.
35. Apply Theorem 5.9 to matrices $\mathbf{A}$ in Examples 5.1 and 5.3 and, where appropriate, determine an optimum region that contains the eigenvalues. Plot the eigenvalues in your diagrams and check that they agree with the statement of the theorem.
36. Using the approach of Section 5.11, show that the solution of the difference equation $u_{n+2} = u_{n+1} + 2u_n$ with $u_0 = 0$ and $u_1 = 1$ is $u_n = \frac{1}{3}(2^n - (-1)^n)$.
37. Using the approach in Section 5.12, using $h = 0.2$, find the numerical solution of the two-point boundary-value problem $d^2u/dx^2 = -10x\cos(2\pi x)$ in the interval $0 \le x \le 1$, subject to the boundary conditions $u(0) = u(1) = 0$. Find the analytical solution and compare the analytical and numerical results.

In Exercises 38 through 41, verify matrix $\mathbf{A}$ is Hermitian, find its eigenvalues and eigenvectors, and verify that the eigenvectors are orthogonal with respect to the complex inner product.

38.
$$\mathbf{A} = \begin{bmatrix} 1 & -3i \\ 3i & 1 \end{bmatrix}.$$

39.
$$\mathbf{A} = \begin{bmatrix} 1 & i \\ -i & 1 \end{bmatrix}.$$

40.
$$\mathbf{A} = \begin{bmatrix} 0 & i & 1 \\ -i & 0 & -i \\ 1 & i & 0 \end{bmatrix}.$$

41.
$$\mathbf{A} = \begin{bmatrix} 0 & -i & 0 \\ i & 0 & -1 \\ 0 & -1 & 0 \end{bmatrix}.$$

42. Show that every $n \times n$ Hermitian matrix $\mathbf{A}$ can be written $\mathbf{A} = \mathbf{A}_1 + i\mathbf{A}_2$, where $\mathbf{A}_1$ is a real symmetric $n \times n$ matrix, and $\mathbf{A}_2$ is a real skew-symmetric $n \times n$

matrix. Give an example of this decomposition using a 3×3 Hermitian matrix of your own construction.

43. Construct a 3×3 skew-Hermitian matrix $\mathbf{A}$, and verify by direct calculation that $i\mathbf{A}$ is Hermitian.
44. Construct a 2×2 unitary matrix $\mathbf{U}$, and use it to verify that $\mathbf{U}=\left(\mathbf{U}^{\mathrm{T}}\right)^{-1}$ is an equivalent definition of a unitary matrix. Give an analytical reason why this result is true.
45. Because the modulus (absolute value) of each eigenvalue of a unitary matrix is 1, it follows that the eigenvalues must all lie on the unit circle centered on the origin in the complex plane. Verify this by (a) showing matrix $\mathbf{U}$ is unitary, where

$$\mathbf{U} = \begin{bmatrix} \frac{1}{2}(1+i) & \frac{1}{2}(1+i) & 0 \\ \frac{1}{2}(i-1) & \frac{1}{2}(1-i) & 0 \\ 0 & 0 & 1 \end{bmatrix},$$

(b) finding the characteristic equation of $\mathbf{U}$, (c) finding the eigenvalues (roots) of the characteristic equation, and (d) locating the position of each eigenvalue on the unit circle.

Chapter 6
Systems of Linear Differential Equations

6.1 Differentiation and Integration of Matrices

Before discussing the solution of systems of linear first-order constant coefficient ordinary differential equations it is necessary to develop the basic theory concerning the differentiation and integration of matrices whose elements are functions of a real variable. For the sake of completeness, the differentiation and integration of quite general matrices will be considered first, though only the simplest of these properties will be used when systems of ordinary constant coefficient differential equations are considered.

To solve linear systems of differential equations in matrix form requires differentiating matrices that are functions of a single real variable, say t. Let the $n \times 1$ column vector $\mathbf{x}(t) = [x_1(t), x_2(t), \ldots, x_n(t)]^{\mathrm{T}}$ have differentiable elements $x_i(t)$ for $i = 1, 2, \ldots, n$, and let the $m \times n$ matrix $\mathbf{G}(t) = [g_{ij}(t)]$ have differentiable elements $g_{ij}(t)$, with $i = 1, 2, \ldots, m$ and $j = 1, 2, \ldots, n$. Then the derivatives of $\mathbf{x}(t)$ and $\mathbf{G}(t)$ with respect to t are defined, respectively, as

$$\frac{d\mathbf{x}(t)}{dt} = \begin{bmatrix} dx_1/dt \\ dx_2/dt \\ \vdots \\ dx_n/dt \end{bmatrix}, \quad \frac{d\mathbf{G}(t)}{dt} = \begin{bmatrix} dg_{11}/dt & dg_2/dt & \cdots & dg_{1n}/dt \\ dg_{21}/dt & dg_{22}/dt & \cdots & dg_{2n}/dt \\ \vdots & \vdots & \vdots & \vdots \\ dg_{m1}/dt & dg_{m2}/dt & \cdots & dg_{mn}/dt \end{bmatrix}. \tag{6.1}$$

An important special case of (6.1) occurs when $\mathbf{A}$ is a constant matrix, because then $d\mathbf{A}/dt = \mathbf{0}$ and so, in particular, $d\mathbf{I}/dt = \mathbf{0}$.

Now consider the derivative of $d[AG(t)]/dt$, where $\mathbf{A} = [a_{ij}]$ is a constant $m \times n$ matrix, and $\mathbf{G}(t) = [g_{ij}(t)]$ is an $n \times r$ matrix with its elements functions of t, so that $\mathbf{AG}(t)$ is an $m \times r$ matrix. Then the element $\alpha_{ij}(t)$ in row i and column j of the matrix product $\mathbf{AG}(t)$ is

$$\alpha_{ij}(t) = a_{i1}g_{1j}(t) + a_{i2}g_{2j}(t) + \cdots + a_{in}g_{nj}(t), \quad i = 1, 2, \ldots, m, \; j = 1, 2, \ldots, r,$$

A. Jeffrey, *Matrix Operations for Engineers and Scientists*,
DOI 10.1007/978-90-481-9274-8_6, © Springer Science+Business Media B.V. 2010

and so

$$\frac{d\alpha_{ij}(t)}{dt} = a_{i1}\frac{dg_{1j}(t)}{dt} + a_{i2}\frac{dg_{2j}(t)}{dt} + \cdots + a_{in}\frac{dg_{nj}(t)}{dt}. \tag{6.2}$$

This is simply the derivative of the product $\mathbf{a}_i\mathbf{g}_j(t)$ of the vector forming the ith row $\mathbf{a}_i$ of $\mathbf{A}$ and the vector forming the jth column $\mathbf{g}_j(t)$ of $\mathbf{G}$, so from definition (6.1) and the definition of matrix multiplication, we see that

$$\frac{d[\mathbf{A}\mathbf{G}(t)]}{dt} = \mathbf{A}\frac{d\mathbf{G}(t)}{dt}. \tag{6.3}$$

When necessary, higher-order derivatives may be defined in the obvious manner:

$$\frac{d}{dt}\left(\frac{d\mathbf{G}}{dt}\right) = \frac{d^2\mathbf{G}}{dt^2},\ \frac{d}{dt}\left(\frac{d^2\mathbf{G}}{dt^2}\right) = \frac{d^3\mathbf{G}}{dt^3},\ \text{and}\ \frac{d}{dt}\left(\frac{d^n\mathbf{G}}{dt^n}\right) = \frac{d^{n+1}\mathbf{G}}{dt^{n+1}}. \tag{6.4}$$

Example 6.1. Find $d[\mathbf{A}\mathbf{G}(t)]/dt$ and $d^2[\mathbf{A}\mathbf{G}(t)]/dt^2$ if

$$\mathbf{A} = \begin{bmatrix} 1 & -3 \\ 2 & 4 \end{bmatrix} \text{ and } \mathbf{G} = \begin{bmatrix} \sin t & \cos t \\ -\cos t & \sin t \end{bmatrix}.$$

Solution. From (6.3)

$$\begin{aligned} \frac{d}{dt}[\mathbf{A}\mathbf{G}(t)] = \mathbf{A}\frac{d\mathbf{G}(t)}{dt} &= \begin{bmatrix} 1 & -3 \\ 2 & 4 \end{bmatrix}\begin{bmatrix} \cos t & -\sin t \\ \sin t & \cos t \end{bmatrix} \\ &= \begin{bmatrix} \cos t - 3\sin t & -\sin t - 3\cos t \\ 2\cos t + 4\sin t & -2\sin t + 4\cos t \end{bmatrix}, \end{aligned}$$

while from (6.4)

$$\begin{aligned} \frac{d^2}{dt^2}[\mathbf{A}\mathbf{G}(t)] = \frac{d}{dt}\left(\frac{d[\mathbf{A}\mathbf{G}(t)]}{dt}\right) &= \mathbf{A}\frac{d^2\mathbf{G}}{dt^2} \\ &= \begin{bmatrix} -\sin t - 3\cos t & -\cos t + 3\sin t \\ -2\sin t + 4\cos t & -2\cos t - 4\sin t \end{bmatrix}. \end{aligned}$$

◇

It follows directly from the definition of matrix addition and (6.1) that if $\mathbf{G}(t)$ and $\mathbf{H}(t)$ are conformable for addition, then

$$\frac{d}{dt}[\mathbf{G}(t) + \mathbf{H}(t)] = \frac{d\mathbf{G}(t)}{dt} + \frac{d\mathbf{H}(t)}{dt}. \tag{6.5}$$

Furthermore, if $\mathbf{G}(t)$ and $\mathbf{H}(t)$ are conformable for the product $\mathbf{G}(t)\mathbf{H}(t)$, then

$$\frac{d}{dt}[\mathbf{G}(t)\mathbf{H}(t)] = \frac{d\mathbf{G}(t)}{dt}\mathbf{H}(t) + \mathbf{G}(t)\frac{d\mathbf{H}(t)}{dt}. \tag{6.6}$$

To derive (6.6), consider the product $\mathbf{g}_i\mathbf{h}_j$, where $\mathbf{g}_i$ is the ith row of $\mathbf{G}(t)$ and $\mathbf{h}_j$ is the jth column of $\mathbf{H}(t)$. Then the element in the ith row and jth column of $\mathbf{G}(t)\mathbf{H}(t)$ is

$$\alpha_{ij}(t) = g_{i1}(t)h_{1j}(t) + g_{i2}(t)h_{2j}(t) + \cdots + g_{in}(t)h_{nj}(t),$$

so differentiating once with respect to t gives

$$\frac{d\alpha_{ij}(t)}{dt} = \frac{dg_{i1}(t)}{dt}h_{1j}(t) + g_{i1}(t)\frac{dh_{j1}(t)}{dt} + \cdots + \frac{dg_{in}(t)}{dt}h_{nj}(t) + g_{in}(t)\frac{dh_{nj}(t)}{dt},$$

from which (6.6) follows after the matrix $d[\mathbf{G}(t)\mathbf{H}(t)]/dt$ has been reconstructed as the sum of two products.

A less obvious result is that if $\mathbf{G}(t)$ is a nonsingular $n \times n$ matrix, then

$$\frac{d\mathbf{G}^{-1}(t)}{dt} = -\mathbf{G}^{-1}(t)\frac{d\mathbf{G}(t)}{dt}\mathbf{G}^{-1}(t). \tag{6.7}$$

This result is proved by differentiating the product $d\{\mathbf{G}(t)\mathbf{G}^{-1}(t)\}/dt$. We start from the results $\mathbf{G}(t)\mathbf{G}^{-1}(t) = \mathbf{I}$, and $d\mathbf{I}/dt = \mathbf{0}$, so it follows from (6.6) that

$$\frac{d}{dt}\left[\mathbf{G}(t)\mathbf{G}^{-1}(t)\right] = \frac{d\mathbf{G}(t)}{dt}\mathbf{G}^{-1}(t) + \mathbf{G}(t)\frac{d\mathbf{G}^{-1}(t)}{dt} = \mathbf{0},$$

and so

$$\mathbf{G}(t)\frac{d\mathbf{G}^{-1}(t)}{dt} = -\frac{d\mathbf{G}(t)}{dt}\mathbf{G}^{-1}(t). \tag{6.8}$$

Result (6.7) follows after pre-multiplication of this equation by $\mathbf{G}^{-1}(t)$.

Example 6.2. Find $d\mathbf{G}^{-1}(t)/dt$ if

$$\mathbf{G}(t) = \begin{bmatrix} \cos t & \sin t \\ -\sin t & \cos t \end{bmatrix}.$$

Solution. There are several ways of finding $d\mathbf{G}^{-1}(t)/dt$, the most elementary and in this case the simplest, being to compute $\mathbf{G}^{-1}(t)$, and then to differentiate it. A routine calculation shows that

$$\mathbf{G}^{-1}(t) = \begin{bmatrix} \cos t & -\sin t \\ \sin t & \cos t \end{bmatrix},$$

so from (6.1)

$$\frac{d\mathbf{G}^{-1}(t)}{dt} = \begin{bmatrix} -\sin t & -\cos t \\ \cos t & -\sin t \end{bmatrix}.$$

A different way of finding $d\mathbf{G}^{-1}(t)/dt$ makes use of (6.7). We have

$$\frac{d\mathbf{G}(t)}{dt} = \begin{bmatrix} -\sin t & \cos t \\ -\cos t & -\sin t \end{bmatrix},$$

so using the above expressions for $\mathbf{G}^{-1}(t)$ and $d\mathbf{G}(t)/dt$ in (6.7) and simplifying the result gives the expected result

$$\frac{d\mathbf{G}^{-1}(t)}{dt} = -\mathbf{G}^{-1}(t)\frac{d\mathbf{G}(t)}{dt}\mathbf{G}^{-1}(t) = \begin{bmatrix} -\sin t & -\cos t \\ \cos t & -\sin t \end{bmatrix}.$$

◊

By definition, if $\mathbf{A}(t) = [a_{ij}(t)]$ is an $m \times n$ matrix, with $i = 1, 2, \ldots, m$ and $j = 1, 2, \ldots, n$, then the indefinite integral of the element in the ith row and jth column of $\mathbf{A}(t)$ is $\int a_{ij}(t)dt$, so the indefinite integral of $\mathbf{A}(t)$ is defined as $\left[\int a_{ij}(t)dt\right]$, so

$$\int \mathbf{A}(t)dt = \left[\int a_{ij}(t)dt\right], \tag{6.9}$$

where, of course, an arbitrary constant matrix must be added after the integration has been performed on each element $a_{ij}(t)$. Similarly, the definite integral of $\mathbf{A}(t)$ between the limits $t = a$ and $t = b$ is defined as the $m \times n$ matrix with the element in its ith row and jth column equal to $\int_a^b a_{ij}(t)dt$, so that

$$\int_a^b \mathbf{A}(t)dt = \left[\int_a^b a_{ij}(t)dt\right]. \tag{6.10}$$

Example 6.3. Find (a) $\int A(t)dt$ and (b) $\int_0^{\pi/2} \mathbf{A}(t)dt$, if $\mathbf{A}(t) = \begin{bmatrix} 2\sin t & \cos t \\ -3\cos t & \sin t \end{bmatrix}$.

Solution.

$$\text{(a)} \quad \int \mathbf{A}(t)dt = \begin{bmatrix} -2\cos t + C_1 & \sin t + C_2 \\ -3\sin t + C_3 & -\cos t + C_4 \end{bmatrix},$$

so $\int \mathbf{A}(t)dt = \begin{bmatrix} -2\cos t & \sin t \\ -3\sin t & -\cos t \end{bmatrix} + \mathbf{C}$, where $\mathbf{C} = \begin{bmatrix} C_1 & C_2 \\ C_3 & C_4 \end{bmatrix}$ is an arbitrary constant matrix.

$$\text{(b)} \quad \int_0^{\pi/2} \mathbf{A}(t)dt = \begin{bmatrix} 2 & 1 \\ -3 & 2 \end{bmatrix}.$$

◊

6.2 Systems of Homogeneous Constant Coefficient Differential Equations

In what follows, some elementary knowledge concerning the integration of constant coefficient ordinary differential equations will be assumed. The development of the theory of systems of differential equations presented here will be confined to the solution of systems of n linear first-order constant coefficient equations in n unknowns. The most general first-order system of this kind involving the n unknown functions $x_1(t), x_2(t), \ldots, x_n(t)$ of the independent variable t is

$$
\begin{aligned}
b_{11}\frac{dx_1}{dt} + b_{12}\frac{dx_2}{dt} + \cdots + b_{1n}\frac{dx_n}{dt} &= c_{11}x_1 + c_{12}x_2 + \cdots + c_{1n}x_n + h_1(t),\\
b_{21}\frac{dx_1}{dt} + b_{22}\frac{dx_2}{dt} + \cdots + b_{2n}\frac{dx_n}{dt} &= c_{21}x_1 + c_{22}x_2 + \cdots + c_{2n}x_n + h_2(t),\\
&\cdots\cdots\cdots\cdots\cdots\\
b_{n1}\frac{dx_1}{dt} + b_{n2}\frac{dx_2}{dt} + \cdots + b_{nn}\frac{dx_n}{dt} &= c_{n1}x_1 + c_{n2}x_2 + \cdots + c_{nn}x_n + h_n(t),
\end{aligned}
\tag{6.11}
$$

where the coefficients b_{ij} and c_{ij} are constants, and the $h_i(t)$ are arbitrary functions of t. Subsequently, it will be assumed that the n equations in (6.11) are linearly independent, so no equation in (6.11) is a linear combination of the other equations.

By defining the $n \times n$ constant matrices $\mathbf{B} = [b_{ij}]$, $\mathbf{C} = [c_{ij}]$, the variable $n \times 1$ column matrices $\mathbf{x}(t) = [x_1(t), x_2(t), \ldots, x_n(t)]^{\mathrm{T}}$, and $\mathbf{h}(t) = [h_1(t), h_2(t), \ldots, h_n(t)]^{\mathrm{T}}$, system (6.11) can be written more concisely as

$$\mathbf{B}\frac{d\mathbf{x}}{dt} = \mathbf{C}\mathbf{x} + \mathbf{h}(t). \tag{6.12}$$

By hypothesis, the equations in system (6.11) are linearly independent, so the coefficient matrix $\mathbf{B}$ has an inverse $\mathbf{B}^{-1}$, and after pre-multiplying (6.12) by $\mathbf{B}^{-1}$ it becomes

$$\frac{d\mathbf{x}}{dt} = \mathbf{B}^{-1}\mathbf{C}\mathbf{x} + \mathbf{B}^{-1}\mathbf{h}(t).$$

Defining the $n \times n$ constant matrix, $\mathbf{A} = \mathbf{B}^{-1}\mathbf{C}$, and the variable $n \times 1$ column matrix $\mathbf{f}(t) = \mathbf{B}^{-1}\mathbf{h}(t)$, shows system (6.11) can always be reduced to the standard form

$$\frac{d\mathbf{x}}{dt} = \mathbf{A}\mathbf{x} + \mathbf{f}(t). \tag{6.13}$$

In what follows, only systems of this type will be considered. System (6.13), equivalently system (6.11), is *nonhomogeneous* when the vector $\mathbf{f}(t) \neq \mathbf{0}$, otherwise it is *homogeneous*.

6.2.1 The Homogeneous System

In this section we will consider the *homogeneous* system

$$\frac{d\mathbf{x}}{dt} = \mathbf{A}\mathbf{x}, \tag{6.14}$$

and establish a connection between its general solution, and the eigenvectors of **A**. Our concern will be to find both the general solution of system (6.14), and then the solution of an *initial-value problem* for the system. That is, finding a solution of system (6.14) subject to a set of *n initial conditions* of the form $x_i(t_1) = k_i$, with $i = 1, 2, \ldots, n$, where the constants k_i are the values the functions $x_i(t)$ are required to satisfy initially when $t = t_1$.

Modeling our approach on the elementary one used when solving a single constant coefficient linear differential equation, we will attempt to find solutions of (6.14) of the form

$$\mathbf{x}(t) = \tilde{\mathbf{x}}e^{\lambda t}, \tag{6.15}$$

where $\tilde{\mathbf{x}}$ is a constant $n \times 1$ column vector. Substituting (6.15) into (6.14) gives

$$\lambda e^{\lambda t}\tilde{\mathbf{x}} = e^{\lambda t}\mathbf{A}\tilde{\mathbf{x}}, \tag{6.16}$$

and after cancellation of the nonvanishing scalar factor $e^{\lambda t}$, followed by some re-arrangement of terms, we find that λ must be a solution of the system of matrix equations

$$[\mathbf{A} - \lambda\mathbf{I}]\tilde{\mathbf{x}} = \mathbf{0}. \tag{6.17}$$

This shows that the permissible values of λ in (6.15) are the eigenvalues λ_1, λ_2, $\ldots$, λ_n of **A**, while the associated constant column vectors $\tilde{\mathbf{x}}_1, \tilde{\mathbf{x}}_2, \ldots, \tilde{\mathbf{x}}_n$ are the corresponding eigenvectors of **A**. When **A** has a full set of n linearly independent eigenvectors, the linearly independent solutions of (6.14) are $\mathbf{x}_i(t) = \tilde{\mathbf{x}}_i e^{\lambda_i t}$, for $i = 1, 2, \ldots, n$. An $n \times n$ matrix $\mathbf{\Phi}(t) = [\mathbf{x}_1(t), \mathbf{x}_2(t), \ldots, \mathbf{x}_n(t)]$, with its columns the solution vectors $\mathbf{x}_i(t)$, is called a *fundamental matrix* for system (6.14).

The general solution $\mathbf{x}(t)$ of (6.14) will be an arbitrary linear combination of the n linearly independent eigenvectors $\mathbf{x}_i(t)$ of the form

$$\mathbf{x}(t) = C_1\mathbf{x}_1(t) + C_2\mathbf{x}_2(t) + \cdots + C_n\mathbf{x}_n(t), \tag{6.18}$$

where the C_i are arbitrary constants. In terms of the fundamental matrix $\mathbf{\Phi}(t)$, the general solution of (6.14) becomes

$$\mathbf{x}(t) = \Phi(t)\mathbf{C}, \tag{6.19}$$

where **C** is the column matrix $\mathbf{C} = [C_1, C_2, \ldots, C_n]^{\mathrm{T}}$.

A fundamental matrix is not unique, because the eigenvectors forming its columns can be arranged in different orders, and each eigenvector can be multiplied by a constant factor and still remain an eigenvector. This nonuniqueness of the fundamental matrix can cause the arbitrary constants in general solutions to appear differently, depending on how the fundamental matrix has been constructed. However, these different forms of the general solution of (6.14) are unimportant, because the solution of a corresponding initial-value problem is unique, so when the arbitrary constants are chosen to make the $x_i(t)$ satisfy the n initial conditions, all forms of general solution in which arbitrary constants may appear differently will give rise to the same unique solution of the initial value problem.

Example 6.4. Find the general solution of the system of equations

$$\frac{dx_1}{dt} = x_2, \qquad \frac{dx_2}{dt} = x_1 .$$

Solution. In matrix form the system becomes $d\mathbf{x}/dt = \mathbf{A}\mathbf{x}$, where $\mathbf{x} = [x_1, x_2]^{\mathrm{T}}$ and

$$\mathbf{A} = \begin{bmatrix} 0 & 1 \\ 1 & 0 \end{bmatrix}.$$

The eigenvalues and eigenvectors of $\mathbf{A}$ are

$$\lambda_1 = 1\,, \quad \mathbf{x}_1 = \begin{bmatrix} 1 \\ 1 \end{bmatrix}, \quad \lambda_2 = -1, \quad \mathbf{x}_2 = \begin{bmatrix} -1 \\ 1 \end{bmatrix}.$$

As the vectors $e^{-\lambda_i t}\mathbf{x}_i$, with $i = 1, 2$ are solutions of the system, we may take the fundamental matrix to be

$$\boldsymbol{\Phi}(t) = \begin{bmatrix} e^t & -e^{-t} \\ e^t & e^{-t} \end{bmatrix}.$$

Setting $\mathbf{C} = [C_1, C_2]^{\mathrm{T}}$, with C_1 and C_2 arbitrary constants, the general solution of the system $\mathbf{x}(t) = \boldsymbol{\Phi}(t)\mathbf{C}$ becomes

$$\mathbf{x}(t) = \begin{bmatrix} e^t & -e^{-t} \\ e^t & e^{-t} \end{bmatrix} \begin{bmatrix} C_1 \\ C_2 \end{bmatrix} = \begin{bmatrix} C_1 e^t - C_2 e^{-t} \\ C_1 e^t + C_2 e^{-t} \end{bmatrix}.$$

In scalar form the solution is

$$x_1(t) = C_1 e^t - C_2 e^{-t} \quad \text{and} \quad x_2(t) = C_1 e^t + C_2 e^{-t}.$$

◇

The next example shows how to deal with the case of complex eigenvalues and eigenvectors. It also illustrates how, unlike the case of the single scalar equation $dx/dt = ax$ with only the exponential solution $x(t) = Ce^{at}$, a linear first-order system of differential equations can have trigonometric functions occurring in its general solution, as well as exponential functions.

Example 6.5. Find the general solution of the system of equations

$$\frac{dx_1}{dt} = x_1 + x_2, \quad \frac{dx_2}{dt} = x_2 - x_1 .$$

Use the result to solve the initial-value problem $x_1(\pi/2) = 1$, $x_2(\pi/2) = 2$.

Solution. In matrix form the system becomes $d\mathbf{x}/dt = \mathbf{A}\mathbf{x}$, with $\mathbf{x} = [x_1, x_2]^{\mathrm{T}}$, where

$$\mathbf{A} = \begin{bmatrix} 1 & 1 \\ -1 & 1 \end{bmatrix}.$$

The eigenvalues and eigenvectors of **A** are

$$\lambda_1 = 1 + i\,, \quad \mathbf{x}_1 = \begin{bmatrix} -i \\ 1 \end{bmatrix}, \quad \lambda_2 = 1 - i\,, \quad \mathbf{x}_2 = \begin{bmatrix} i \\ 1 \end{bmatrix}.$$

So, as the vectors $e^{-\lambda_i t}\mathbf{x}_i$ with $i = 1, 2$ are linearly independent solutions, a fundamental matrix is

$$\Phi(t) = \begin{bmatrix} -ie^{(1+i)t} & ie^{(1-i)t} \\ e^{(1+i)t} & e^{(1-i)t} \end{bmatrix}.$$

As the elements of the fundamental matrix are complex, the arbitrary constants in the matrix **C** must also be complex. Thus the general solution $\mathbf{x}(t) = \mathbf{\Phi}(t)\mathbf{C}$ becomes

$$\mathbf{x}(t) = \begin{bmatrix} -ie^{(1+i)t} & ie^{(1-i)t} \\ e^{(1+i)t} & e^{(1-i)t} \end{bmatrix} \begin{bmatrix} C_1 \\ C_2 \end{bmatrix} = \begin{bmatrix} -iC_1e^{(1+i)t} + iC_2e^{(1-i)t} \\ C_1e^{(1+i)t} + C_2e^{(1-i)t} \end{bmatrix},$$

where C_1 and C_2 are *complex* constants.

For the solution $\mathbf{x}(t)$ to be real, the two terms in each row of the solution vector on the right must be complex conjugates to allow their imaginary parts to cancel and, furthermore, the general solution of the original first-order system can only contain *two* real arbitrary constants The exponential factors $e^{(1+i)t}$ and $e^{(1-i)t}$ are already complex conjugates, as are the factors $-i$ and i, so to make the terms real it is necessary that the complex constants C_1 and C_2 are also complex conjugates, so let us set $C_1 = a + ib$ and $C_2 = a - ib$, then after simplification the general solution becomes

$$\mathbf{x}(t) = \begin{bmatrix} 2ae^t \sin t + 2be^t \cos t \\ 2ae^t \cos t - 2be^t \sin t \end{bmatrix}.$$

Both a and b are arbitrary constants, so to simplify this result we set $k_1 = 2a$ and $k_2 = 2b$, when the general solution becomes

$$x_1(t) = e^t(k_1 \sin t + k_2 \cos t) \text{ and } x_2(t) = e^t(k_1 \cos t - k_2 \sin t).$$

To satisfy the initial conditions $x_1(\pi/2) = 1, x_2(\pi/2) = 0$ we set $t = \pi/2$ in the general solution and then impose the initial conditions to obtain: (initial condition $x_1(\pi/2) = 1$): $1 = e^{\pi/2}\ k_1$ (initial condition $x_1(\pi/2) = 2$): $2 = -e^{\pi/2}k_2$, showing $k_1 = e^{-\pi/2}$ and $k_2 = -2e^{-\pi/2}$. Thus the solution of the initial-value problem is found to be

$$x_1(t) = e^{(t-\pi/2)}(\sin t - 2\cos t),\quad x_2(t) = e^{(t-\pi/2)}(\cos t + 2\sin t),\quad t \geq \pi/2.$$

◊

This method of finding a general solution for a system of homogeneous linear first-order constant coefficient equations extends to the solution of a single higher-order equation, and to systems of higher-order equations. In this case, all that is necessary is to introduce higher-order derivatives as new unknowns when, for example, a single nth-order equation can be replaced by an equivalent set of n first-order equations. This approach is most easily illustrated by example.

Example 6.6. Find the general solution of the following third-order differential equation by converting it to a first-order system:

$$\frac{d^3y}{dt^3} + \frac{d^2y}{dt^2} + \frac{dy}{dt} + y = 0\,.$$

Solution. Introduce the two new dependent variables z_1 and z_2 by setting

$$\frac{dy}{dt} = z_1 \text{ and } \frac{d^2y}{dt^2} = \frac{dz_1}{dt} = z_2\,.$$

The third-order equation can now be replaced by the equivalent first-order system

$$\frac{dy}{dt} = z_1,\quad \frac{dz_1}{dt} = z_2\quad \frac{dz_2}{dt} + z_2 + z_1 + y = 0\,.$$

When written in matrix form, this system becomes

$$\frac{d\mathbf{z}}{dt} = \mathbf{Az} \text{ with } \mathbf{z} = \begin{bmatrix} y(t) \\ z_1(t) \\ z_2(t) \end{bmatrix},\quad \mathbf{A} = \begin{bmatrix} 0 & 1 & 0 \\ 0 & 0 & 1 \\ -1 & -1 & -1 \end{bmatrix}.$$

The eigenvalues and eigenvectors of $\mathbf{A}$ are

$$\lambda_1 = -1\,,\ \mathbf{x}_1 = \begin{bmatrix} 1 \\ -1 \\ 1 \end{bmatrix},\quad \lambda_2 = i\,,\ \mathbf{x}_2 = \begin{bmatrix} -1 \\ -i \\ 1 \end{bmatrix},\quad \lambda_3 = -i\,,\ \mathbf{x}_3 = \begin{bmatrix} -1 \\ i \\ 1 \end{bmatrix}.$$

As the vectors $e^{-\lambda_i t}\mathbf{x}_i$ with $i = 1, 2, 3$ are solutions, a fundamental matrix is

$$\Phi(t) = \begin{bmatrix} e^{-t} & -e^{it} & -e^{-it} \\ -e^{-t} & -ie^{it} & ie^{-it} \\ e^{-t} & e^{it} & e^{-it} \end{bmatrix},$$

when the general solution becomes

$$\begin{bmatrix} y \\ z_1 \\ z_2 \end{bmatrix} = \begin{bmatrix} e^{-t} & -e^{it} & -e^{-it} \\ -e^{-t} & -ie^{it} & ie^{-it} \\ e^{-t} & e^{it} & e^{-it} \end{bmatrix} \begin{bmatrix} C_1 \\ C_2 \\ C_3 \end{bmatrix},$$

where for the solution to be real, the arbitrary constants C_1, C_2 and C_3 must be complex numbers.

As the solution $y(t)$ of the original third-order differential equation is needed, it is only necessary to extract this solution from the first row of this matrix equation, from which we find that

$$y(t) = C_1 e^{-t} + C_2 e^{it} + C_3 e^{-it}.$$

A real solution is required, so reasoning as in Example 6.5, we see that the arbitrary constants C_2 and C_3 must be complex conjugates, so setting $C_2 = a + ib$ and $C_3 = a - ib$, with a and b arbitrary real constants, leads to the result

$$y(t) = C_1 e^{-t} + 2a\cos t - 2b\sin t.$$

For convenience, writing C_2 in place of $2a$ and C_3 in place of $-2b$ (not the original C_2 and C_3) we arrive at the general solution

$$y(t) = C_1 e^{-t} + C_2\cos t + C_3\sin t.$$

Solving for z_1 and z_2 will give dy/dt and d^2y/dt^2, though these solutions are not required. If needed, the simplest way to determine dy/dt and d^2y/dt^2 is by differentiation of $y(t)$.

◇

The approach used in Example 6.6 extends immediately to the homogeneous nth-order constant coefficient equation

$$\frac{d^n y}{dt^n} + a_{n-1}\frac{d^{n-1}y}{dt^{n-1}} + a_{n-2}\frac{d^{n-2}y}{dt^{n-2}} + \cdots + a_1\frac{dy}{dt} + a_0 y = 0.$$

This can be replaced by the equivalent $n \times n$ first-order matrix system

$$\frac{d\mathbf{z}}{dt} = \mathbf{A}\mathbf{z} \text{ with } \mathbf{z} = \begin{bmatrix} y \\ z_1 \\ z_2 \\ \vdots \\ z_{n-1} \end{bmatrix}, \quad \mathbf{A} = \begin{bmatrix} 0 & 1 & 0 & \cdots & 0 \\ 0 & 0 & 1 & \cdots & 0 \\ \vdots & \vdots & \vdots & \vdots & \vdots \\ 0 & 0 & 0 & \cdots & 1 \\ -a_0 & -a_1 & -a_2 & \cdots & -a_{n-1} \end{bmatrix}, \tag{6.20}$$

where

$$\frac{dy}{dt} = z_1, \quad \frac{d^2y}{dt^2} = \frac{dz_1}{dt} = z_2, \quad \frac{dz_2}{dt} = z_3, \quad \ldots, \quad \frac{d^{n-1}z_{n-2}}{dt^{n-1}} = z_{n-1}. \tag{6.21}$$

This matrix system can now be solved as in the previous examples.

We mention here that an nth-order system can be reduced to a set of n first-order equations in more than one way, though the method of reduction used here is usually the simplest. For an example of a different way of reducing a higher-order equation to a system see the remarks at the end of the next section.

The matrix approach to be adopted when solving the *nonhomogeneous* system (6.13) cannot make direct use of the fundamental matrix associated with a homogeneous system. This is because solutions of nonhomogeneous systems *do not* possess the linear superposition property of the homogeneous systems of equations used in (6.18).

6.3 An Application of Diagonalization

Before discussing the solution of nonhomogeneous systems we first describe a different approach to the solution of homogeneous systems that extends easily to the nonhomogeneous case, and to do this we first examine the specially simple homogeneous system

$$\frac{d\mathbf{x}}{dt} = \mathbf{D}\mathbf{x}, \tag{6.22}$$

where the coefficient matrix $\mathbf{D}$ is the diagonal matrix $\mathbf{D} = \text{diag}\{\alpha_1, \alpha_2, \ldots, \alpha_n\}$. When written out in full (6.22) becomes

$$\begin{bmatrix} dx_1/dt \\ dx_2/dt \\ \vdots \\ dx_n/dt \end{bmatrix} = \begin{bmatrix} \alpha_1 & 0 & 0 & 0 \\ 0 & \alpha_2 & 0 & 0 \\ \vdots & \vdots & \vdots & \vdots \\ 0 & 0 & \cdots & \alpha_n \end{bmatrix} \begin{bmatrix} x_1 \\ x_2 \\ \vdots \\ x_n \end{bmatrix}. \tag{6.23}$$

This is just a set of n separate (not simultaneous) first-order linear differential equations $dx_i/dt = \alpha_i x_i$, each with the general solution $x_i(t) = C_i \exp(\alpha_i t)$ for $i = 1, 2, \ldots, n$, where the C_i are arbitrary constants. In matrix form the general solution of (6.23) becomes

$$\mathbf{x}(t) = [C_1 e^{d_1 t}, C_2 e^{d_2 t}, \ldots, C_n e^{d_n t}]^{\mathrm{T}}. \tag{6.24}$$

This suggests a different approach when solving a general homogeneous system

$$\frac{d\mathbf{x}}{dt} = \mathbf{A}\mathbf{x}, \tag{6.25}$$

when $\mathbf{A}$ is a general $n \times n$ matrix. The idea is to try to find how to change the dependent variable column vector $\mathbf{x}(t)$ to a new dependent variable column vector $\mathbf{z}(t)$ in such a way that $\mathbf{A}$ is replaced by a diagonal matrix $\mathbf{D}$. If this can be done, the general solution for $\mathbf{z}(t)$ follows at once as in (6.24). Changing back from $\mathbf{z}(t)$ to $\mathbf{x}(t)$ will then give the required general solution $\mathbf{x}(t)$ of (6.25).

To obtain such a simplification we will make use of the diagonalization process described in Chapter 5. There the diagonalization of an $n \times n$ matrix $\mathbf{A}$ was found to be possible subject to the condition that $\mathbf{A}$ has a full set of n linearly independent eigenvectors. It was shown that if the n linearly independent eigenvectors of $\mathbf{A}$ are $\mathbf{x}_1, \mathbf{x}_2, \ldots, \mathbf{x}_n$, corresponding to the n eigenvalues $\lambda_1, \lambda_2, \ldots, \lambda_n$, the diagonal matrix $\mathbf{D} = \mathrm{diag}\{\lambda_1, \lambda_2, \ldots, \lambda_n\}$ can be written in the form $\mathbf{D} = \mathbf{P}^{-1}\mathbf{A}\mathbf{P}$, where the columns of $\mathbf{P}$ are the eigenvectors $\mathbf{x}_i$, and the eigenvectors in $\mathbf{P}$ are arranged in the same order as the eigenvalues λ_i in $\mathbf{D}$. Pre-multiplying $\mathbf{D} = \mathbf{P}^{-1}\mathbf{A}\mathbf{P}$ by $\mathbf{P}$, and post-multiplying the result by $\mathbf{P}^{-1}$, gives $\mathbf{A} = \mathbf{P}\mathbf{D}\mathbf{P}^{-1}$.

Substituting this expression for $\mathbf{A}$ in (6.25) it becomes

$$\frac{d\mathbf{x}}{dt} = \mathbf{P}\mathbf{D}\mathbf{P}^{-1}\mathbf{x}, \tag{6.26}$$

after which pre-multiplication by $\mathbf{P}^{-1}$ we then find that

$$\mathbf{P}^{-1}\frac{d\mathbf{x}}{dt} = \mathbf{D}\mathbf{P}^{-1}\mathbf{x}. \tag{6.27}$$

As $\mathbf{P}$ is a constant matrix it follows that, $\mathbf{P}^{-1}(d\mathbf{x}/dt) = d(\mathbf{P}^{-1}\mathbf{x})/dt$, so (6.27) simplifies to

$$\frac{d}{dt}\left(\mathbf{P}^{-1}\mathbf{x}\right) = \mathbf{D}\mathbf{P}^{-1}\mathbf{x}. \tag{6.28}$$

The required reduction is now almost complete, because defining the new column vector $\mathbf{z} = \mathbf{P}^{-1}\mathbf{x}$, transforms (6.28) into

$$\frac{d\mathbf{z}}{dt} = \mathbf{D}\mathbf{z}, \tag{6.29}$$

which is precisely the form given in (6.23). Hence the general solution for each new variable $z_1(t), z_2(t), \ldots, z_n(t)$ in (6.29) is $z_i(t) = C_i\exp(\lambda_i t)$, for $i = 1, 2, \ldots, n$, with $C_1, C_2, \ldots, C_n$ arbitrary constants. The required solution vector $\mathbf{x}(t)$ is recovered from the vector $\mathbf{z}(t) = \mathbf{P}^{-1}\mathbf{x}(t)$ with elements $z_i(t)$ by pre-multiplication of $\mathbf{z}$ by $\mathbf{P}$, to give $\mathbf{x}(t) = \mathbf{P}\mathbf{z}(t)$.

Example 6.7. Use diagonalization to find the general solution of

$$\frac{d\mathbf{x}}{dt} = \mathbf{A}\mathbf{x} \quad \text{where } \mathbf{x}(t) = \begin{bmatrix} x_1(t) \\ x_2(t) \\ x_3(t) \end{bmatrix}, \ \mathbf{A} = \begin{bmatrix} 1 & 0 & -1 \\ -2 & -1 & 2 \\ -1 & 2 & 1 \end{bmatrix}.$$

Solution. It was shown in Example 5.5 that the eigenvalues of $\mathbf{A}$ are $\lambda_1 = -2$, $\lambda_2 = 0$, $\lambda_3 = -3$, and the diagonalizing matrix $\mathbf{P}$ in $\mathbf{A} = \mathbf{P}\mathbf{D}\mathbf{P}^{-1}$ is

$$\mathbf{P} = \begin{bmatrix} 1 & 1 & 1 \\ -4 & 0 & -\frac{3}{2} \\ 3 & 1 & -2 \end{bmatrix}, \quad \text{with } \mathbf{D} = \begin{bmatrix} -2 & 0 & 0 \\ 0 & 0 & 0 \\ 0 & 0 & 3 \end{bmatrix}.$$

Setting $\mathbf{z}(t) = [z_1(t), z_2(t), z_3(t)]^{\mathrm{T}}$, it follows from the diagonalized system $d\mathbf{z}/dt = \mathbf{D}\mathbf{z}$, corresponding to (6.23), that $dz_1/dt = -2z_1$, $dz_2/dt = 0$ and $dz_3/dt = 3z_3$, so $z_1(t) = C_1e^{-2t}$, $z_2(t) = C_2$ and $z_3(t) = C_3e^{3t}$, with C_1, C_2 and C_3 arbitrary constants. The solution vector $\mathbf{x}(t)$ obtained from $\mathbf{x}(t) = \mathbf{P}\mathbf{z}(t)$ then becomes

$$\begin{bmatrix} x_1(t) \\ x_2(t) \\ x_3(t) \end{bmatrix} = \begin{bmatrix} 1 & 1 & 1 \\ -4 & 0 & -\frac{3}{2} \\ 3 & 1 & -2 \end{bmatrix} \begin{bmatrix} C_1e^{-2t} \\ C_2 \\ C_3e^{3t} \end{bmatrix} = \begin{bmatrix} C_1e^{-2t} + C_2 + C_3e^{3t} \\ -4C_1e^{-2t} - \frac{3}{2}C_3e^{3t} \\ 3C_1e^{-2t} + C_2 - 2C_3e^{3t} \end{bmatrix},$$

so in scalar form

$$x_1(t) = C_1e^{-2t} + C_2 + C_3e^{3t}, \ x_2(t) = -4C_1e^{-2t} - \tfrac{3}{2}C_3e^{3t}, \ x_3(t) = 3C_1e^{-2t} + C_2 - 2C_3e^{3t}.$$

◇

The last example in this section shows how diagonalization can be used to solve an initial-value problem for a special system of linear homogeneous second-order equations.

Example 6.8. Use diagonalization to solve the system of linear second-order equations $d^2\mathbf{u}/dt^2 = \mathbf{A}\mathbf{u}$, where $\mathbf{A} = \begin{bmatrix} 2 & -1 \\ -1 & 2 \end{bmatrix}$ and $\mathbf{u} = \begin{bmatrix} u_1 \\ u_2 \end{bmatrix}$, subject to the initial conditions

$$u_1(0) = 1, \ u_1'(0) = 1, \ u_2(0) = 0 \text{ and } u_2'(0) = 1.$$

Solution. The eigenvalues and eigenvectors of $\mathbf{A}$ are$\lambda_1 = 1, \mathbf{x}_1 = \begin{bmatrix} 1 \\ 1 \end{bmatrix}$, $\lambda_2 = \begin{bmatrix} 1 \\ -1 \end{bmatrix}$, so as there are two linearly independent eigenvectors $\mathbf{A}$ can be diagonalized by $\mathbf{P} = \begin{bmatrix} 1 & 1 \\ 1 & -1 \end{bmatrix}$. Thus $\mathbf{A} = \mathbf{PDP}^{-1}$, where $\mathbf{D} = \begin{bmatrix} 1 & 0 \\ 0 & 3. \end{bmatrix}$ and $\mathbf{P}^{-1} = \begin{bmatrix} \frac{1}{2} & \frac{1}{2} \\ \frac{1}{2} & -\frac{1}{2} \end{bmatrix}$, so the matrix differential equation becomes $d^2\mathbf{u}/dt^2 = \mathbf{PDP}^{-1}\mathbf{u}$. Pre-multiplying this equation by $\mathbf{P}^{-1}$ it becomes $\mathbf{P}^{-1}d^2\mathbf{u}/dt^2 = \mathbf{DP}^{-1}\mathbf{u}$, but $\mathbf{P}^{-1}$ is a constant matrix so it can be taken under the differentiation sign when the equation reduces to $d^2(\mathbf{P}^{-1}\mathbf{u})/dt^2 = \mathbf{DP}^{-1}\mathbf{u}$. Setting $\mathbf{v} = \mathbf{P}^{-1}\mathbf{u}$ the equation becomes $d^2\mathbf{v}/dt^2 = \mathbf{Dv}$, where now the elements of $\mathbf{v} = [v_1, v_2]^{\mathrm{T}}$ have been separated, because the equation splits into the two scalar equations

$$\frac{d^2v_1}{dt^2} = v_1 \text{ and } \frac{d^2v_2}{dt^2} = 3v_2.$$

The general solutions of these two equations are easily shown to be

$$\mathbf{v}_1 = B_1e^x + B_2e^{-x} \text{ and } \mathbf{v}_2 = C_1e^{\sqrt{3}x} + C_2e^{-\sqrt{3}x}.$$

To determine the arbitrary constants B_1, B_2, C_1 and C_2 it is necessary to have initial conditions for $\mathbf{v}_1$ and $\mathbf{v}_2$, but the initial conditions have been given for u_1 and u_2. To find the initial conditions for $\mathbf{v}$ use must be made of $\mathbf{v} = \mathbf{P}^{-1}\mathbf{u}$, so $\mathbf{v}(0) = \mathbf{P}^{-1}\mathbf{u}(0)$, and $\mathbf{v}'(0) = \mathbf{P}^{-1}\mathbf{u}'(0)$. Substituting the initial conditions for $\mathbf{u}$ shows $v_1(0) = \frac{1}{2}, v_1'(0) = 1, v_2(0) = \frac{1}{2}$ and $v_2'(0) = 0$. When these conditions are used with v_1 and v_2 the following solutions are obtained

$$v_1 = \frac{3}{4}e^x - \frac{1}{4}e^{-x} \text{ and } v_2 = \frac{1}{4}e^{\sqrt{3}x} + \frac{1}{4}e^{-\sqrt{3}x}.$$

Finally, to find u_1 and u_2, we must use the result $\mathbf{u} = \mathbf{Pv}$. Substituting for $\mathbf{P}$ and $\mathbf{v} = [v_1, v_2]^{\mathrm{T}}$, and combining terms, gives

$$u_1 = \tfrac{1}{2}e^x - \tfrac{1}{2}\sinh(x) + \tfrac{1}{2}\cosh(\sqrt{3}x), \quad u_2 = \tfrac{1}{2}e^x - \tfrac{1}{2}\sinh(x) - \tfrac{1}{2}\cosh(\sqrt{3}x).$$

◇

6.4 The Nonhomogeneous Case

When matrix $\mathbf{A}$ can be diagonalized, only a small additional step is required to solve the nonhomogeneous system

$$\frac{d\mathbf{x}}{dt} = \mathbf{Ax} + \mathbf{f}(t). \tag{6.30}$$

Recalling from (6.13) that $\mathbf{f}(t)$ is a column vector $\mathbf{f}(t) = [f_1(t), f_2(t), \ldots, f_n(t)]^T$, and setting $\mathbf{A} = \mathbf{PDP}^{-1}$ in (6.30), where $\mathbf{P}$ diagonalizes $\mathbf{A}$, the system becomes

$$\frac{d\mathbf{x}}{dt} = \mathbf{PDP}^{-1}\mathbf{x} + \mathbf{f}(t).$$

Pre-multiplication by the constant matrix $\mathbf{P}^{-1}$ reduces the system to

$$\frac{d(\mathbf{P}^{-1}\mathbf{x})}{dt} = \mathbf{DP}^{-1}\mathbf{x} + \mathbf{P}^{-1}\mathbf{f}(t),$$

so setting $\mathbf{z}(t) = \mathbf{P}^{-1}\mathbf{x}(t)$ this becomes

$$\frac{d\mathbf{z}}{dt} = \mathbf{Dz} + \mathbf{P}^{-1}\mathbf{f}(t). \tag{6.31}$$

Writing $\mathbf{g}(t) = \mathbf{P}^{-1}\mathbf{f}(t)$, with $\mathbf{g}(t) = [g_1(t), g_2(t), \ldots, g_n(t)]^T$, where the functions $g_i(t)$ are known in terms of the elements of the nonhomogeneous vector $\mathbf{f}(t)$, result (6.31) becomes

$$\frac{d\mathbf{z}}{dt} = \mathbf{Dz} + \mathbf{g}(t). \tag{6.32}$$

The solution of (6.32) now simplifies to the solution of the n separate nonhomogeneous equations $dz_i/dt = \lambda_i z_i + g_i(t)$ for $i = 1, 2, \ldots, n$, whereas before the elements of the diagonal matrix $\mathbf{D} = \text{diag}\{\lambda_1, \lambda_2, \ldots, \lambda_n\}$ are the eigenvalues of $\mathbf{A}$ corresponding to it eigenvectors $\mathbf{x}_1, \mathbf{x}_2, \ldots, \mathbf{x}_n$ occurring in the diagonalizing matrix $\mathbf{P}$. Once the vector $\mathbf{z}(t)$ has been found, the solution of the nonhomogeneous system (6.30) follows from the result $\mathbf{x}(t) = \mathbf{Pz}(t)$.

When no initial conditions are specified, each element of $\mathbf{x}(t)$ will be the sum of the general solution of the corresponding equation in the homogeneous system, to which is added a particular integral produced by the nonhomogeneous term $\mathbf{f}(t)$.

To solve an initial-value problem it is first necessary to find the general solution for $\mathbf{x}(t)$, and then to match the arbitrary constants involved to the initial conditions.

Example 6.9. Use diagonalization to find the solution of the nonhomogeneous system

$$\frac{d\mathbf{x}}{dt} = \mathbf{Ax} + \mathbf{f}(t) \text{ when } \mathbf{A} = \begin{bmatrix} 1 & 2 \\ 2 & 1 \end{bmatrix}, \ \mathbf{x} = \begin{bmatrix} x_1 \\ x_2 \end{bmatrix} \text{ and } \mathbf{f}(t) = \begin{bmatrix} 1+t \\ \cos t \end{bmatrix},$$

given that $x_1(0) = 1$ and $x_2(0) = 2$.

Solution. The eigenvalues and eigenvectors of $\mathbf{A}$ are

$$\lambda_1 = -1, \ \mathbf{x}_1 = \begin{bmatrix} -1 \\ 1 \end{bmatrix}, \ \lambda_2 = 3, \ \mathbf{x}_2 = \begin{bmatrix} 1 \\ 1 \end{bmatrix},$$

so the diagonalizing matrix is

$$\mathbf{P} = \begin{bmatrix} -1 & 1 \\ 1 & 1 \end{bmatrix} \text{ with } \mathbf{D} = \begin{bmatrix} -1 & 0 \\ 0 & 3 \end{bmatrix}, \text{ and } \mathbf{P}^{-1} = \begin{bmatrix} -\frac{1}{2} & \frac{1}{2} \\ \frac{1}{2} & \frac{1}{2} \end{bmatrix}$$

Using these results, Eq. (6.31) becomes

$$\begin{bmatrix} dz_1/dt \\ dz_2/dt \end{bmatrix} = \begin{bmatrix} -1 & 0 \\ 0 & 3 \end{bmatrix} \begin{bmatrix} z_1 \\ z_2 \end{bmatrix} + \begin{bmatrix} -\frac{1}{2} & \frac{1}{2} \\ \frac{1}{2} & \frac{1}{2} \end{bmatrix} \begin{bmatrix} 1+t \\ \cos t \end{bmatrix}.$$

The variables $z_i(t)$ are now *separated*, and from this last result we find that

$$\frac{dz_1}{dt} = -z_1 + \tfrac{1}{2}(\cos t - 1 - t) \text{ and } \frac{dz_2}{dt} = 3z_2 + \tfrac{1}{2}(\cos t + 1 + t).$$

For convenience in what follows, the method of solution of a general linear first-order differential equation by means of an integrating factor is reviewed in Appendix 1 at the end of this chapter. Solving these linear first-order equations gives

$$z_1(t) = \tfrac{1}{4}\cos t + \tfrac{1}{4}\sin t - \tfrac{1}{2}t + C_1 e^{-t} \text{ and}$$
$$z_2(t) = -\tfrac{3}{20}\cos t + \tfrac{1}{20}\sin t - \tfrac{2}{9} - \tfrac{1}{6}t + C_2 e^{3t}.$$

Using these as the elements of $\mathbf{z}(t) = [z_1(t), z_2(t)]^{\mathrm{T}}$ in $\mathbf{x}(t) = \mathbf{P}\mathbf{z}(t)$ shows the required general solution to be

$$x_1(t) = -\tfrac{2}{5}\cos t - \tfrac{1}{5}\sin t + \tfrac{1}{3}t - \tfrac{2}{9} - C_1 e^{-t} + C_2 e^{3t},$$

$$x_2(t) = \tfrac{1}{10}\cos t + \tfrac{3}{10}\sin t - \tfrac{2}{3}t - \tfrac{2}{9} + C_1 e^{-t} + C_2 e^{3t}.$$

In each of these general solutions, the first four terms on the right represent the particular integral, while the last two terms containing the arbitrary constants C_1 and C_2 are the solution of the homogeneous form of the equation, usually called the *complementary function*. Using the initial conditions $x_1(0) = 1$ and $x_2(0) = 2$, some simple calculations show that $C_1 = \frac{1}{4}$, $C_2 = \frac{337}{180}$, so the solution of the initial-value problem becomes

$$x_1(t) = -\tfrac{2}{5}\cos t - \tfrac{1}{5}\sin t + \tfrac{1}{3}t - \tfrac{2}{9} - \tfrac{1}{4}e^{-t} + \tfrac{337}{180}e^{3t},$$

$$x_2(t) = \tfrac{1}{10}\cos t + \tfrac{3}{10}\sin t - \tfrac{2}{3}t - \tfrac{2}{9} + \tfrac{1}{4}e^{-t} + \tfrac{337}{180}e^{3t}.$$

◇

6.5 Matrix Methods and the Laplace Transform

This section shows how the solution of an initial-value problem for a nonhomogeneous linear system of differential equations with initial conditions specified at $t = 0$ can be found by using the Laplace transform in conjunction with a matrix approach. It will, however, be seen that finding the inverse transform requires some algebraic manipulation, though when the system is complicated the effort required to find the inverse transform is not very different from the effort required when making a direct application of the Laplace transform.

We have seen how the general solution of both homogeneous and nonhomogeneous linear constant coefficient first-order systems of equations can be found when the coefficient matrix can be diagonalized. In particular, because the methods described lead to general solutions, it allows initial-value problems to be imposed for any value $t = t_0$ of the independent variable. It is, however, a familiar fact that the Laplace transform method can only be used to solve initial-value problems for linear differential equations when initial conditions are imposed at $t = 0$. So, unlike the previous methods, the Laplace transform approach only solves initial-value problems, and does *not* lead to general solutions. However, because of the importance and wide use of the Laplace transform, mention must be made of its use with matrix systems of linear differential equations.

In first accounts of differential equations, the Laplace transform method is usually only applied to scalar equations, though the approach is easily extended to solve initial-value problems for first-order systems of constant coefficient matrix differential equations. As the Laplace transform method does not depend on the eigenvalues and eigenvectors of the coefficient matrix **A**, it has the advantage that it does not require knowledge of the eigenvalues of matrix **A** nor, as the eigenvectors of **A** are not used, is it necessary for the coefficient matrix **A** to have a complete set of eigenvectors.

Before proceeding further, we recall that the Laplace transform $X(s)$ of $x(t)$, denoted by writing $X(s) = \mathcal{L}\{x(t)\}$, is defined as

$$X(s) = \mathcal{L}\{x(t)\} = \int_0^\infty e^{-st}x(t)dt, \tag{6.33}$$

where s is the Laplace transform variable, and the functions $x(t)$ are restricted to those for which the improper integral on the right of (6.33) exists. The *inversion process*, that is finding $x(t)$ from its Laplace transform $X(s)$, will be denoted by $x(t) = \mathcal{L}^{-1}\{X(s)\}$, and in all straightforward cases it is performed using tables of Laplace transform pairs coupled with the use of some simple rules. An outline of the essential details of the Laplace transform is given in Appendix 2 at the end of this chapter, where a a short table of transform pairs is also given.

A Laplace transform pair is a function $x(t)$ and its associated Laplace transform $X(s)$. Then, given $x(t)$, its Laplace transform $X(s)$ can be found from the table and, conversely, given $X(s)$, the inverse Laplace transform $x(t) = \mathcal{L}^{-1}\{X(s)\}$ can be found by using the table in reverse, usually with the help of some simple rules that extend the table of transform pairs. For example, if $x(t) = \cos at$, then $\mathcal{L}\{\cos at\} = s/(s^2 + a^2)$, so that $\cos at$ and $s/(s^2 + a^2)$ is a typical Laplace transform pair. Then, given $\cos at$, its Laplace transform $s/(s^2 + a^2)$ follows from a table of Laplace transform pairs and, conversely, when the transform $s/(s^2 + a^2)$ is obtained in a calculation, using the table of Laplace transform pairs in reverse it follows that the *inverse* Laplace transform of $s/(s^2 + a^2)$ is $\cos at$.

The adaptation of the Laplace transform approach to the solution of systems of linear differential equations is illustrated by the following examples that show how the approach also extends in a natural way to higher-order systems. However, to limit the length of the examples, it will be assumed that the reader is familiar with the elements of Laplace transform theory. In particular, familiarity will be assumed with the technique of partial fraction expansion used to simplify the transformed solution, and also with the standard results needed to interpret the partial fractions as functions of t.

Consider the initial-value problem for the system of n linear first-order constant coefficient equations

$$\frac{d\mathbf{x}}{dt} = \mathbf{A}\mathbf{x} + \mathbf{f}(t)\,, \tag{6.34}$$

where

$$\mathbf{x}(t) = \begin{bmatrix} x_1(t) \\ x_2(t) \\ \vdots \\ x_n(t) \end{bmatrix}, \quad \mathbf{A} = \begin{bmatrix} a_{11} & a_{12} & \cdots & a_{1n} \\ a_{21} & a_{22} & \cdots & a_{2n} \\ \vdots & \vdots & \vdots & \vdots \\ a_{n1} & a_{n2} & \cdots & a_{nn} \end{bmatrix}, \quad \mathbf{f}(t) = \begin{bmatrix} f_1(t) \\ f_2(t) \\ \vdots \\ f_n(t) \end{bmatrix},$$

subject to the initial conditions $\mathbf{x}(0) = [k_1, k_2, \ldots, k_n]^{\mathrm{T}}$, with $k_1, k_2, \ldots, k_n$ the arbitrary *initial values*.

Using the familiar property of the Laplace transform of a derivative, that $\mathcal{L}\{dx_i(t)/dt\} = sX_i(s) - x_i(0)$, so that $\mathcal{L}\{dx_i(t)/dt\} = sX_i(s) - k_i$, for $i = 1, 2, \ldots, n$, the result of taking the Laplace transform of system (6.34) is the matrix system involving the transformed variables $X_i(s) = \mathcal{L}\{x_i(t)\}$, for $i = 1, 2, \ldots, n$

$$\begin{bmatrix} sX_1(s) - k_1 \\ sX_2(s) - k_2 \\ \vdots \\ sX_n(s) - k_n \end{bmatrix} = \mathbf{A} \begin{bmatrix} X_1(s) \\ X_2(s) \\ \vdots \\ X_n(s) \end{bmatrix} + \begin{bmatrix} F_1(s) \\ F_2(s) \\ \vdots \\ F_n(s) \end{bmatrix}, \tag{6.35}$$

where $\mathcal{L}\{f_i(t)\} = F_i(s)$. After rearrangement, this becomes

$$[s\mathbf{I} - \mathbf{A}]\mathbf{Z}(s) = \mathbf{x}(0) + \mathbf{F}(s), \tag{6.36}$$

where $\mathbf{Z}(s) = [\mathcal{L}\{x_1(t)\}, \mathcal{L}\{x_2(t)\}, \ldots, \mathcal{L}\{x_n(t)\}]^{\mathrm{T}}$. Pre-multiplication by $[s\mathbf{I} - \mathbf{A}]^{-1}$ then gives

$$\mathbf{Z}(s) = [s\mathbf{I} - \mathbf{A}]^{-1}[\mathbf{x}(0) + \mathbf{F}(s)], \tag{6.37}$$

and so

$$\mathbf{x}(t) = \mathcal{L}^{-1}\Big\{[s\mathbf{I} - \mathbf{A}]^{-1}[\mathbf{x}(0) + \mathbf{F}(s)]\Big\}. \tag{6.38}$$

The advantage this approach has over the ones in previous sections is that it does not require the determination of the eigenvalues or the eigenvectors of $\mathbf{A}$, so the method is applicable irrespective of whether or not $\mathbf{A}$ has a full set of eigenvectors. It also has the advantage that it avoids dealing with any complex eigenvalues and eigenvectors that might arise. The disadvantage of the method is that it only solves initial-value problems for $\mathbf{x}(t)$, and in addition the algebraic complexity of the computation required when finding and then inverting the transform $\mathbf{Z}(s) = [s\mathbf{I} - \mathbf{A}]^{-1}[\mathbf{x}(0) + \mathbf{F}(s)]$ can be tiresome.

Example 6.10. Solve the initial-value problem

$$\frac{dx_1}{dt} = x_1 - x_2 + 2t\,, \quad \frac{dx_2}{dt} = x_2 - 4x_1 + 1\,, \quad x_1(0) = 1, x_2(0) = 0.$$

Solution. Using the notation introduced previously,

$$\mathbf{A} = \begin{bmatrix} 1 & -1 \\ -4 & 1 \end{bmatrix}, \text{ so } s\mathbf{I} - \mathbf{A} = \begin{bmatrix} s-1 & 1 \\ 4 & s-1 \end{bmatrix}, \text{ and } \mathbf{x}(0) = \begin{bmatrix} 1 \\ 0 \end{bmatrix}.$$

$$\mathcal{L}\{2t\} = 2/s^2, \quad \text{and } \mathcal{L}\{1\} = 1/s, \text{ so } \mathbf{F}(s) = \left[2/s^2,\ 1/s\right]^{\mathrm{T}}.$$

Routine calculations then give

$$[s\mathbf{I} - \mathbf{A}]^{-1} = \frac{1}{(s^2 - 2s - 3)} \begin{bmatrix} s-1 & -1 \\ -4 & s-1 \end{bmatrix}, \mathbf{F}(s) = \begin{bmatrix} 1/(s-3) \\ -2/s^2 \end{bmatrix}, \text{ and } \mathbf{x}(0) = \begin{bmatrix} 0 \\ 0 \end{bmatrix},$$

so

$$\mathbf{Z}(s) = \begin{bmatrix} Z_1(s) \\ Z_2(s) \end{bmatrix} = [s\mathbf{I} - \mathbf{A}]^{-1}[\mathbf{x}(0) + \mathbf{F}(s)] = \frac{1}{s^2(s^2 - 2s - 3)} \begin{bmatrix} s^3 - s^2 + s - 2 \\ -3s^2 - s - 8 \end{bmatrix}.$$

As $Z_1(s) = \mathcal{L}\{x_1(t)\}$ and $Z_2(s) = \mathcal{L}\{x_2(t)\}$, we see that

$$x_1(t) = \mathcal{L}^{-1}\left\{\frac{s^3 - s^2 + s - 2}{s^2(s^2 - 2s - 3)}\right\} \text{ and } x_2(t) = \mathcal{L}^{-1}\left\{\frac{-3s^2 - s - 8}{s^2(s^2 - 2s - 3)}\right\}.$$

Simplifying these transformed solutions by means of partial fractions, and then using tables of Laplace transform pairs to express the result in terms of t gives

$$x_1(t) = -\frac{7}{9} + \frac{2}{3}t + \frac{5}{4}e^{-t} + \frac{19}{36}e^{3t} \text{ and } x_2(t) = -\frac{13}{9} + \frac{8}{3}t + \frac{5}{2}e^{-t} - \frac{19}{18}e^{3t}.$$

The details of the partial fraction expansion and the use of a table of Laplace transform pairs to arrive at $x_1(t)$ and $x_2(t)$ are left as an exercise. Having reached the stage of finding $Z_1(s)$ and $Z_2(s)$, the work required to invert the transforms of the solutions is precisely the same as would have been involved had the Laplace transform been applied directly, without the use of matrices.

Although the eigenvalues and eigenvectors of **A** were *not* used in these calculations, we mention that they are

$$\lambda_1 = 3\,,\ \mathbf{x}_1 = \begin{bmatrix} 1 \\ 2 \end{bmatrix} \quad \text{and} \quad \lambda_2 = -2\,,\ \mathbf{x}_2 = \begin{bmatrix} 1 \\ -3 \end{bmatrix}.$$

This shows that in this case, because **A** has a full set of eigenvectors, this same solution could have been obtained by diagonalizing **A** to find the general solution, and then imposing the initial conditions to determine the values of the constants of integration.

◇

Example 6.11. Solve the initial-value problem

$$\frac{dx_1}{dt} = 2x_1 + 4x_2 - 2x_3 + 1\,,\quad \frac{dx_2}{dt} = -x_2 - x_3 + \sin t,$$
$$\frac{dx_3}{dt} = x_2 + x_3,\ x_1(0) = 1, x_2(0) = 0\,,\ x_3(0) = 0.$$

Solution.

$$\mathbf{x}(t) = \begin{bmatrix} x_1(t) \\ x_2(t) \\ x_3(t) \end{bmatrix},\ \mathbf{A} = \begin{bmatrix} 2 & 4 & -2 \\ 0 & -1 & -1 \\ 0 & 1 & 1 \end{bmatrix},\ \mathbf{f}(t) = \begin{bmatrix} 1 \\ \sin t \\ 0 \end{bmatrix},\ \text{and } \mathbf{x}(0) = \begin{bmatrix} 1 \\ 0 \\ 0 \end{bmatrix}.$$

Thus

$$[s\mathbf{I} - \mathbf{A}] = \begin{bmatrix} s-2 & -4 & 2 \\ 0 & s+1 & 1 \\ 0 & -1 & s-1 \end{bmatrix},$$

$$[s\mathbf{I} - \mathbf{A}]^{-1} = \begin{bmatrix} 1/(s-2) & (4s-6)/\{s^2(s-2)\} & -(2s+6)/\{s^2(s-2)\} \\ 0 & (s-1)/s^2 & -1/s^2 \\ 0 & 1/s^2 & (s+1)/s^2 \end{bmatrix},$$

$$\text{and } \mathbf{F}(s) = \begin{bmatrix} 1/s \\ 1/(s^2+1) \\ 0 \end{bmatrix}.$$

Combining terms and substituting into (6.37) gives

$$\mathbf{Z}(s) = [s\mathbf{I} - \mathbf{A}]^{-1}[\mathbf{x}(0) + \mathbf{F}(s)] = \begin{bmatrix} \dfrac{s^4 + s^3 + s^2 + 5s - 6}{s^2(s-2)(s^2+1)} \\ \dfrac{s-1}{s^2(s^2+1)} \\ \dfrac{1}{s^2(s^2+1)} \end{bmatrix}.$$

Using partial fractions to simplify the expressions in s, writing $\mathbf{Z}(s) = [Z_1(s), Z_2(s), Z_3(s)]^{\mathrm{T}}$, and taking the inverse Laplace transform shows the solution of the initial-value problem is given by

$$x_1(t) = \mathcal{L}^{-1}\{Z_1(s)\} = -1 + 3t + \tfrac{2}{5}\cos t - \tfrac{16}{5}\sin t + \tfrac{8}{5}e^{2t},$$

$$x_2(t) = \mathcal{L}^{-1}\{Z_2(t)\} = 1 - t - \cos t + \sin t,$$

$$x_1(t) = \mathcal{L}^{-1}\{Z_3(s)\} = t - \sin t. \text{ for } t \geq 0.$$

Once again the eigenvalues and eigenvectors of $\mathbf{A}$ were not used in these calculations, though in this case they were

$$\lambda_1 = \lambda_2 = 0\,, \quad \mathbf{x}_{1,2} = \begin{bmatrix} 3 \\ -1 \\ 1 \end{bmatrix}, \quad \lambda_3 = 2\,,\ \mathbf{x}_2 = \begin{bmatrix} 1 \\ 0 \\ 0 \end{bmatrix}.$$

Notice that here $\mathbf{A}$ only has two linearly independent eigenvectors, so in this case diagonalization of $\mathbf{A}$ could *not* have been used to construct a general solution.

◇

Example 6.12. Solve the initial-value problem

$$\frac{dx_1}{dt} = 2x_1 + x_2 + t\,, \quad \frac{dx_2}{dt} = -x_1 + 2x_2 + 3\,, \quad x_1(0) = 2\,,\ x_2(0) = -1\,.$$

Solution. For this system

$$\mathbf{x}(t) = \begin{bmatrix} x_1(t) \\ x_2(t) \end{bmatrix}, \quad \mathbf{A} = \begin{bmatrix} 2 & 1 \\ -1 & 2 \end{bmatrix}, \quad \mathbf{f}(t) = \begin{bmatrix} t \\ 3 \end{bmatrix}, \text{ and } \mathbf{x}(0) = \begin{bmatrix} 2 \\ -1 \end{bmatrix}.$$

Here

$$s\mathbf{I} - \mathbf{A} = \begin{bmatrix} s-2 & -1 \\ 1 & s-2 \end{bmatrix}, \text{ so } [s\mathbf{I} - \mathbf{A}]^{-1} = \frac{1}{(s^2 - 4s + 5)} \begin{bmatrix} s-2 & 1 \\ -1 & s-2 \end{bmatrix},$$

$$\mathbf{F}(s) = \begin{bmatrix} (2s^2+1)/s^2 \\ -(s-3)/s \end{bmatrix}.$$

Substituting into (6.37) gives

$$\mathbf{Z}(s) = [s\mathbf{I} - \mathbf{A}]^{-1}[\mathbf{x}(0) + \mathbf{F}(s)] = \begin{bmatrix} \dfrac{2s^3 - 5s^2 + 4s - 2}{s^2(s^2 - 4s + 5)} \\ -\dfrac{(s^3 - 3s^2 + 6s + 1)}{s^2(s^2 - 4s + 5)} \end{bmatrix} .$$

Using partial fractions to simplify the expressions in s, writing $\mathbf{Z}(s) = [Z_1(s), Z_2(s)]^{\mathrm{T}}$, and taking the inverse Laplace transform, the solution is found to be

$$x_1(t) = \mathcal{L}^{-1}\{Z_1(s)\} = \tfrac{1}{25}e^{2t}(9\sin t + 38\cos t) + \tfrac{12}{25} - \tfrac{2}{5}t ,$$

$$x_2(t) = \mathcal{L}^{-1}\{Z_2(t)\} = \tfrac{1}{25}e^{2t}(9\cos t - 38\sin t) - \tfrac{34}{25} - \tfrac{1}{5}t , \quad t \geq 0 .$$

As before, the eigenvalues and eigenvectors of $\mathbf{A}$ were not used when finding $x_1(t)$ and $x_2(t)$, though in this case they were complex with

$$\lambda_1 = 2 + i , \ \mathbf{x}_1 = \begin{bmatrix} -i \\ 1 \end{bmatrix} , \ \lambda_2 = 2 - i , \ \mathbf{x}_2 = \begin{bmatrix} i \\ 1 \end{bmatrix} .$$

Diagonalization of $\mathbf{A}$ could have been used to solve this system, though it would have involved working with complex eigenvalues and eigenvectors.

◊

Finally, we show by example how the above method can be extended to solve an initial-value problem for a linear second-order system of matrix differential equations.

Example 6.13. Solve the initial-value problem

$$\frac{d^2x_1}{dt^2} = -4x_1 + x_2 + \cos 2t , \quad \frac{d^2x_2}{dt^2} = -4x_2 + x_1 ,$$
$$x_1(0) = 1 , \ x'_1(0) = 0 , \ x_2(0) = 0 , \ x'_2(0) = 0.$$

Solution. The argument proceeds as before, but this time making use of the Laplace transform of a second derivative $\mathcal{L}\{d^2x_i(t)/dt^2\} = s^2X_i(s) - sx_i(0) - x_i'(0)$ for $i = 1$, 2, involving the initial conditions for both the $x_i(0)$ and for the derivative$x_i'(0)$ for $i = 1, 2$. Consequently, after taking the Laplace transform of each equation, and using the initial conditions, the equations become

$$s^2X_1(s) - s = -4X_1(s) + X_2(s) + \frac{s}{s^2+4}$$

and

$$s^2X_2(s) = -4X_2(s) + X_1(s) = 0.$$

When expressed in matrix form these can be written as

$$\mathbf{A}(s)\mathbf{Z}(s) = \mathbf{B}(s),$$

with

$$\mathbf{A}(s) = \begin{bmatrix} s^2+4 & -1 \\ -1 & s^2+4 \end{bmatrix}, \quad \mathbf{B}(s) = \begin{bmatrix} (s^3+5s)/(s^2+4) \\ 0 \end{bmatrix} \text{ and } \mathbf{Z}(s) = \begin{bmatrix} X_1(s) \\ X_2(s) \end{bmatrix}.$$

After computing $\mathbf{A}(s)^{-1}$ it is found that

$$\mathbf{Z}(s) = \mathbf{A}(s)^{-1}\mathbf{B}(s) = \begin{bmatrix} \dfrac{s^2+4}{s^4+8s^2+15} & \dfrac{1}{s^4+8s^2+15} \\ \dfrac{1}{s^4+8s^2+15} & \dfrac{s^2+4}{s^4+8s^2+15} \end{bmatrix} \begin{bmatrix} \dfrac{s^3+5s}{s^4+4} \\ 0 \end{bmatrix}$$

$$= \begin{bmatrix} \dfrac{s}{s^2+3} \\ \dfrac{s}{s^2+3} - \dfrac{s}{s^2+4} \end{bmatrix}.$$

Again using partial fractions to simplify the expressions in s, and then taking the inverse transformation, shows the solution of the initial-value problem to be

$$x_1(t) = \mathcal{L}^{-1}\{s/(s^2+3)\} = \cos\sqrt{3}t,$$

$$x_2(t) = \mathcal{L}^{-1}\{s/(s^2+3) - s/(s^2+4)\} = x_2(t) = \mathcal{L}^{-1}\{s/(s^2+3) - s/(s^2+4)\}$$

$$= \cos\sqrt{3}t - \cos 2t, \quad t \geq 0.$$

◇

It is important to understand that although a high-order equation in the dependent variable $y(t)$ can be reduced to a set of first-order equations by introducing derivatives of $y(t)$ as new dependent variables, such a reduction is *not* unique. Nevertheless, in whatever way a linear change of variables is used in a single linear higher-order equation for $y(t)$ to reduce it to a linear first-order system of equations, the solution of an initial-value problem for $y(t)$ will remain the same.

Suppose, for example, it is required to solve the third-order initial-value problem

$$\frac{d^3y}{dt^3} + 2\frac{d^2y}{dt^2} - \frac{dy}{dt} - 2y = 1 + t.$$

subject to the homogeneous initial conditions $y(0) = 0$, $y'(0) = 0$, $y''(0) = 0$. By introducing the new functions $u(t) = dy/dt$, and $v(t) = du/dt$, so that $v(t) = d^2y/dt^2$ and $dv/dt = d^3y/dt^3$, the third-order equation for $y(t)$ is replaced by the equivalent first-order system

$$\frac{dv}{dt} + 2v - u - 2y = 1 + t, \quad \frac{du}{dt} = v \text{ and } \frac{dy}{dt} = u,$$

with the initial conditions $y(0) = 0$, $u(0) = 0$, $v(0) = 0$. The solution of this system will give $y(t)$, $u(t) = dy/dt$ and $v(t) = d^2y/dt^2$ as functions of t. Consequently, if the Laplace transform method is used to solve the system in matrix form, and only $y(t)$ is required, it would only be necessary for the Laplace transform $Y(s) = \mathcal{L}^{-1}\{y(t)\}$ to be inverted.

The solution of this system is easily found to be $y(t) = -\frac{1}{4} - \frac{1}{2}t + \frac{1}{3}e^{-t} - \frac{1}{12}e^{-2t}$, though its derivation is left as an exercise. To show that although the reduction of the equation to a system is not unique, the solution is unchanged, we could introduced the different variables $u(t) = dy/dt$ and $v(t) = 2du/dt$, when the system would have become

$$\frac{1}{2}\frac{dv}{dt} + v - u - 2y = 1 + t, \frac{du}{dt} = \frac{1}{2}v \text{ and } \frac{dy}{dt} = u,$$

with the homogeneous initial conditions $u(0) = 0$, $v(0) = 0$, $y(0) = 0$. The functions $u(t)$ and $v(t)$ will now differ from the ones found previously, though the solution $y(t)$ will remain unchanged at $y(t) = -\frac{1}{4} - \frac{1}{2}t + \frac{1}{3}e^{-t} - \frac{1}{12}e^{-2t}$. Here again, the details of this solution are left as an exercise.

◇

6.6 The Matrix Exponential and Differential Equations

This section provides a brief introduction to the matrix exponential $e^{t\mathbf{A}}$, where $\mathbf{A}$ is an $n \times n$ constant coefficient matrix and t is a scalar variable. The matrix exponential generalizes in a natural way the solution of the scalar differential equation $dx/dt = ax$ to the solution of the homogeneous first-order system $d\mathbf{x}/dt = \mathbf{A}\mathbf{x}$, which in turn leads to the solution of the nonhomogeneous matrix differential equation $d\mathbf{x}/dt = \mathbf{A}\mathbf{x} + \mathbf{f}(t)$.

There are many different ways of finding $e^{t\mathbf{A}}$, though only the simplest will be described here once the matrix exponential has been defined in the classical algebraic manner. Some of the ways in which $e^{t\mathbf{A}}$ can be computed will then be described, and the results will be applied to both homogeneous and nonhomogeneous linear systems of matrix differential equations.

The idea of a matrix exponential originates from the definition of the ordinary exponential function defined as the infinite series

$$e^{at} = 1 + \frac{at}{1!} + \frac{a^2t^2}{2!} + \frac{a^3t^3}{3!} + \cdots = \sum_{n=0}^{\infty} \frac{a^n t^n}{n!}, \tag{6.39}$$

which is absolutely convergent for all real at. This suggests that if $\mathbf{A}$ is an $n \times n$ constant matrix, and the convention $\mathbf{A^0} = \mathbf{I}$ is adopted, it is natural to try to define the **matrix exponential** $e^{t\mathbf{A}}$ as

$$e^{t\mathbf{A}} = \sum_{n=0}^{\infty} \frac{t^n}{n!}\mathbf{A}^r = \mathbf{I} + t\mathbf{A} + \frac{1}{2!}t^2\mathbf{A}^2 + \frac{1}{3!}t^3\mathbf{A}^3 + \cdots \quad . \tag{6.40}$$

The expression on the right is an infinite sum of $n \times n$ matrices, so for this to make sense it must be interpreted as summing corresponding elements of matrices, in which case each element of $e^{t\mathbf{A}}$ will become an infinite series in the variable t. Furthermore, if the result is to be applied to a differential equation, these series must be absolutely convergent and capable of being differentiated term by term with respect to t.

We will start by proving that the infinite series forming the elements of $e^{t\mathbf{A}}$ are absolutely convergent, and to do this we will make use of the norm $\|\mathbf{A}\|_M$ introduced in Section 3.1. If each of the powers of $t\mathbf{A}$ occurring in (6.40) is replaced by its norm the result will be the ordinary power series in t

$$e^{t\mathbf{A}} = 1 + t\|\mathbf{A}\|_M + \frac{1}{2!}t^2\|\mathbf{A}\|_M^2 + \frac{1}{3!}t^3\|\mathbf{A}\|_M^3 + \cdots \quad .$$

The power series on the right is simply $e^{t\|\mathbf{A}\|_M}$, which is absolutely convergent for all t. The absolute convergence of this series involving the norm $\|\mathbf{A}\|_M$ implies the absolute convergent of all of the power series that form the elements of $e^{t\mathbf{A}}$, so the absolute convergence of the expression on the right of (6.40) has been established. We mention in passing that any norm of matrix $\mathbf{A}$ could have been used in the above argument, but the norm $\|\mathbf{A}\|_M$ is the simplest.

The next step is to discover how, when given a matrix $\mathbf{A}$, the matrix exponential $e^{t\mathbf{A}}$ can be computed. We begin by considering a special type of matrix A, and although it is a very special case it is still a useful one. Some $n \times n$ matrices $\mathbf{A}$ have the property that integral powers of $\mathbf{A}$ up to $n - 1$ all yield nonzero matrices, whereas $\mathbf{A}^n = \mathbf{0}$ is the null matrix, and thereafter all higher powers of $\mathbf{A}$ are null matrices. A matrix with this property is called a *nilpotent* matrix, and the number n is called the *nilpotent index* of matrix $\mathbf{A}$.

When a matrix like this is substituted into (6.40), only the terms up to $\mathbf{A}^{n-1}$ will be retained, causing the infinite series in t in each of the elements of the matrix to degenerate into finite polynomials in t of order less than or equal to $n - 1$. A typical case now follows.

Example 6.14. Show that matrix $\mathbf{A}$ is nilpotent, find its nilpotent index, and find $e^{t\mathbf{A}}$ if

$$\mathbf{A} = \begin{bmatrix} 0 & 2 & 1 & 1 \\ 0 & 0 & 1 & 2 \\ 0 & 0 & 0 & 3 \\ 0 & 0 & 0 & 0 \end{bmatrix}.$$

Solution.

$$\mathbf{A}^2 = \begin{bmatrix} 0 & 0 & 2 & 1 \\ 0 & 0 & 0 & 1 \\ 0 & 0 & 0 & 0 \\ 0 & 0 & 0 & 0 \end{bmatrix}, \quad \mathbf{A}^3 = \begin{bmatrix} 0 & 0 & 0 & 6 \\ 0 & 0 & 0 & 0 \\ 0 & 0 & 0 & 0 \\ 0 & 0 & 0 & 0 \end{bmatrix}, \quad \mathbf{A}^4 = \begin{bmatrix} 0 & 0 & 0 & 0 \\ 0 & 0 & 0 & 0 \\ 0 & 0 & 0 & 0 \\ 0 & 0 & 0 & 0 \end{bmatrix}.$$

This shows $\mathbf{A}$ is nilpotent, with nilpotent index 4.

Substituting these powers of $\mathbf{A}$ into (6.40) and combining terms we find that

$$e^{t\mathbf{A}} = \begin{bmatrix} 1 & 2t & t^2 - t & t + \frac{1}{2}t^2 + t^3 \\ 0 & 1 & t & 2t + \frac{3}{2}t^2 \\ 0 & 0 & 1 & 3t \\ 0 & 0 & 0 & 1 \end{bmatrix}.$$

◇

Before finding $e^{t\mathbf{A}}$ for more general matrices $\mathbf{A}$, let us first find $e^{t\mathbf{A}}$ when $\mathbf{A}$ is the diagonal matrix $\mathbf{A} = \operatorname{diag}\{\lambda_1, \lambda_2, \ldots, \lambda_n\}$. We have $t\mathbf{A} = \operatorname{diag}\{\lambda_1 t,\ \lambda_2 t,\ \ldots,\ \lambda_n t\}$, after which a simple calculation shows that

$$(t\mathbf{A})^r = \begin{bmatrix} \lambda_1^r t^r & 0 & \cdots & 0 \\ 0 & \lambda_2^r t^r & \cdots & 0 \\ \vdots & \vdots & \vdots & \vdots \\ 0 & 0 & \cdots & \lambda_n^r t^r \end{bmatrix}.$$

Substituting this result into (6.40) gives

$$e^{t\mathbf{A}} = \begin{bmatrix} 1 & 0 & \cdots & 0 \\ 0 & 1 & \cdots & 0 \\ \vdots & \vdots & \vdots & \vdots \\ 0 & 0 & \cdots & 1 \end{bmatrix} + \frac{1}{1!}\begin{bmatrix} \lambda_1 t & 0 & \cdots & 0 \\ 0 & \lambda_2 t & \cdots & 0 \\ \vdots & \vdots & \vdots & \vdots \\ 0 & 0 & \cdots & \lambda_n t \end{bmatrix}$$
$$+ \frac{1}{2!}\begin{bmatrix} \lambda_1^2 t^2 & 0 & \cdots & 0 \\ 0 & \lambda_2^2 t^2 & \cdots & 0 \\ \vdots & \vdots & \vdots & \vdots \\ 0 & 0 & \cdots & \lambda_n^2 t^2 \end{bmatrix} + \cdots,$$

and summing the matrices on the right we find that

$$e^{t\mathbf{A}} = \begin{bmatrix} 1 + \lambda_1 t + \frac{1}{2!}\lambda_1^2 t^2 + \cdots & 0 & \cdots & 0 \\ 0 & 1 + \lambda_2 t + \frac{1}{2!}\lambda_2^2 t^2 + \cdots & \cdots & 0 \\ \vdots & \vdots & \vdots & \vdots \\ 0 & 0 & \cdots & 1 + \lambda_n t + \frac{1}{2!}\lambda_n^2 t^2 + \cdots \end{bmatrix}.$$

In the limit, as the number of terms tends to infinity, so the ith entry on the leading diagonal of $e^{t\mathbf{A}}$ becomes $e^{\lambda_i t}$. This has established the important result that will be needed later that when $\mathbf{A} = \operatorname{diag}\ \{\lambda_1, \lambda_2, \ldots, \lambda_n\}$, the matrix exponential

$$e^{t\mathbf{A}} = \begin{bmatrix} e^{\lambda_1 t} & 0 & \cdots & 0 \\ 0 & e^{\lambda_2 t} & \cdots & 0 \\ \vdots & \vdots & \vdots & \vdots \\ 0 & 0 & \cdots & e^{\lambda_n t} \end{bmatrix}. \tag{6.41}$$

The matrices **A** in the first two examples that follow have structures that are sufficiently simple for powers of **A** to be calculated in a straightforward manner. In each case the matrix exponential $e^{t\mathbf{A}}$ is found by direct substitution into (6.40), followed by recognizing that the Maclaurin series in t that form the elements of $e^{t\mathbf{A}}$ are series expansions of familiar functions. Unfortunately, this method cannot be used with more general matrices **A**, because then the series comprising the elements of the matrix become too complicated to be recognized as series expansions of familiar functions.

Example 6.15. Find $e^{t\mathbf{A}}$ given that $\mathbf{A} = \begin{bmatrix} 0 & a \\ -a & 0 \end{bmatrix}$, where a is real.

Solution. In this case substitution into (6.40) is simplified because $(t\mathbf{A})^n$ takes on one of two different forms, depending whether n is even or odd. Routine calculation shows that

$$t\mathbf{A} = \begin{bmatrix} 0 & at \\ -at & 0 \end{bmatrix}, \ (t\mathbf{A})^2 = \begin{bmatrix} -a^2t^2 & 0 \\ 0 & -a^2t^2 \end{bmatrix}, \ (t\mathbf{A})^3 = \begin{bmatrix} 0 & -a^3t^3 \\ a^3t^3 & 0 \end{bmatrix},$$

$$(t\mathbf{A})^4 = \begin{bmatrix} a^4t^4 & 0 \\ 0 & a^4t^4 \end{bmatrix}, \ (t\mathbf{A})^5 = \begin{bmatrix} 0 & a^5t^5 \\ -a^5t^5 & 0 \end{bmatrix},$$

after which this pattern is repeated, so that

$$(t\mathbf{A})^0 = \mathbf{I}, \ (t\mathbf{A})^1 = ta\begin{bmatrix} 0 & 1 \\ -1 & 0 \end{bmatrix}, \ (t\mathbf{A})^2 = -a^2t^2\mathbf{I},$$

$$(t\mathbf{A})^3 = -a^3t^3\begin{bmatrix} 0 & 1 \\ -1 & 0 \end{bmatrix},$$

$$(t\mathbf{A})^4 = a^4t^4\mathbf{I}, \ \ldots .$$

Substituting these results into (6.40) and collecting terms gives

$$e^{t\mathbf{A}} = \begin{bmatrix} \sum\limits_{n=0}^{\infty}(-1)^n \frac{(at)^{2n}}{(2n)!} & \sum\limits_{n=1}^{\infty}(-1)^n \frac{(at)^{2n+1}}{(2n+1)!} \\ -\sum\limits_{n=1}^{\infty}(-1)^n \frac{(at)^{2n+1}}{(2n+1)!} & \sum\limits_{n=0}^{\infty}(-1)^n \frac{(at)^{2n}}{(2n)!} \end{bmatrix} = \begin{bmatrix} \cos at & \sin at \\ -\sin at & \cos at \end{bmatrix}.$$

Notice this same form of argument generalizes and shows that if

$$\mathbf{A} = \begin{bmatrix} 0 & a & 0 & 0 \\ -a & 0 & 0 & 0 \\ 0 & 0 & 0 & b \\ 0 & 0 & -b & 0 \end{bmatrix}, \text{ then } e^{t\mathbf{A}} = \begin{bmatrix} \cos at & \sin at & 0 & 0 \\ -\sin at & \cos at & 0 & 0 \\ 0 & 0 & \cos bt & \sin bt \\ 0 & 0 & -\sin bt & \cos bt \end{bmatrix}.$$

This result extends immediately in an obvious way when $\mathbf{A}$ is a larger diagonal block matrix of similar form.

◇

Example 6.16. Find $e^{t\mathbf{A}}$ given that $\mathbf{A} = \begin{bmatrix} a & b \\ 0 & a \end{bmatrix}$ where a and b are real numbers.

Solution.

$$t\mathbf{A} = \begin{bmatrix} ta & tb \\ 0 & tb \end{bmatrix}, \quad (t\mathbf{A})^2 = \begin{bmatrix} t^2a^2 & 2tab \\ 0 & t^2a^2 \end{bmatrix}, \quad (t\mathbf{A})^3 = \begin{bmatrix} t^3a^3 & 3t^2a^2b \\ 0 & t^3a^3 \end{bmatrix},$$

$$(t\mathbf{A})^4 = \begin{bmatrix} t^4a^4 & 4t^3a^3b \\ 0 & t^4a^4 \end{bmatrix},$$

and in general

$$(t\mathbf{A})^n = \begin{bmatrix} t^na^n & nt^{n-1}a^{n-1}b \\ 0 & t^na^n \end{bmatrix}.$$

Substitution into (6.40) gives

$$e^{t\mathbf{A}} = \begin{bmatrix} \sum\limits_{n=0}^{\infty} (at)^n/n! & tb\sum\limits_{n=0}^{\infty} (at)^n/n! \\ 0 & \sum\limits_{n=0}^{\infty} (at)^n/n! \end{bmatrix} = \begin{bmatrix} e^{at} & tbe^{at} \\ 0 & e^{at} \end{bmatrix}.$$

Each series that forms an element of $e^{t\mathbf{A}}$ defines an exponential function, so when required these functions can be differentiated with respect to t as many times as required.

◇

So far the matrix exponential $e^{t\mathbf{A}}$ has been computed for matrices $\mathbf{A}$ which have a convenient structure. This may, for example, be when matrices are nilpotent, leading to exponential matrices with polynomial elements, or when the matrices have a structure that allows the elements generated by (6.40) as Maclaurin series to be sufficiently simple for them to be identified as exponential or trigonometric functions. This leaves open the question of how $e^{t\mathbf{A}}$ can be computed for a matrix like

$$\mathbf{A} = \begin{bmatrix} 1 & -2 \\ 1 & 4 \end{bmatrix}.$$

In this case, when $\mathbf{A}$ is substituted into (6.40), the series that are generated to form the elements of $e^{t\mathbf{A}}$ are not recognizable elementary functions (try it). This problem can be solved if **A** is diagonalizable, though to show how diagonalization can be used it is first necessary to establish the following useful result concerning exponential matrices.

Let **M** and **D** be a nonsingular $n \times n$ matrices, and let us find the matrix exponential $e^{\mathbf{M}(t\mathbf{D})\mathbf{M}^{-1}}$ by substituting $\mathbf{M}(t\mathbf{D})\mathbf{M}^{-1}$ into (6.40). The result of the substitution is

$$\begin{aligned} e^{\mathbf{M}(t\mathbf{D})\mathbf{M}^{-1}} &= \mathbf{I} + \mathbf{M}(t\mathbf{D})\mathbf{M}^{-1} + \frac{1}{2!}\left(\mathbf{M}(t\mathbf{D})\mathbf{M}^{-1}\right)^2 + \frac{1}{3!}\left(\mathbf{M}(t\mathbf{D})\mathbf{M}^{-1}\right)^3 + \cdots \\ &= \sum_{n=0}^{\infty}\frac{1}{n!}\left(\mathbf{M}(t\mathbf{D})\mathbf{M}^{-1}\right)^n. \end{aligned}$$

The general term in this series is $\left(\mathbf{M}(t\mathbf{D})\mathbf{M}^{-1}\right)^n/n!$, so expanding it we have

$$\frac{1}{n!}\left(\mathbf{M}(t\mathbf{D})\mathbf{M}^{-1}\right)^n = \frac{1}{n!}\left[\underbrace{\left(\mathbf{M}(t\mathbf{D})\mathbf{M}^{-1}\right)\left(\mathbf{M}(t\mathbf{D})\mathbf{M}^{-1}\right)\left(\mathbf{M}(t\mathbf{D})\mathbf{M}^{-1}\right)\ldots\left(\mathbf{M}(t\mathbf{D})\mathbf{M}^{-1}\right)}_{n\text{ times}}\right].$$

Removing the brackets, and using the fact that $\mathbf{M}^{-1}\mathbf{M} = \mathbf{I}$, reduces this result to

$$\frac{1}{n!}\left(\mathbf{M}(t\mathbf{D})\mathbf{M}^{-1}\right)^n = \frac{1}{n!}\mathbf{M}(t\mathbf{D})^n\mathbf{M}^{-1},$$

causing the expression $e^{\mathbf{M}(t\mathbf{D})\mathbf{M}^{-1}}$ to simplify to the useful result

$$e^{\mathbf{M}(t\mathbf{D})\mathbf{M}^{-1}} = \mathbf{M}e^{t\mathbf{D}}\mathbf{M}^{-1}. \tag{6.42}$$

Now consider an $n \times n$ diagonalizable matrix $t\mathbf{A} = \left[ta_{ij}\right]$. Its eigenvalues $\lambda_i t$ with $i = 1, 2, \ldots, n$ are the roots of the characteristic determinant

$$|t\mathbf{A} - \lambda\mathbf{I}| = 0,$$

where the eigenvectors $\mathbf{x}_i$ are the solutions of the n equations

$$[t\mathbf{A} - \lambda_i t\mathbf{I}]\mathbf{x}_i = 0.$$

The initial assumption that **A** is diagonalizable ensures there are n linearly independent eigenvectors $\mathbf{x}_1, \mathbf{x}_2, \ldots, \mathbf{x}_n$. The variable t enters linearly into each element of matrix $t\mathbf{A} - \lambda_i t\mathbf{I}$, so each eigenvector $\mathbf{x}_i$ will be scaled by t. We have seen that when an eigenvector is scaled, it always remains an eigenvector, so the scale factor can be chosen arbitrarily. Consequently, for convenience when considering the eigenvectors of $t\mathbf{A}$ we can set $t = 1$, and then for the eigenvectors of $t\mathbf{A}$ we can use the eigenvectors $\mathbf{x}_i$ of **A**.

We are now ready to use the above results to find $e^{t\mathbf{A}}$ for a matrix $\mathbf{A}$ that is diagonalizable. If the eigenvectors of $t\mathbf{A}$ are $\lambda_i t$, and the corresponding eigenvectors of $t\mathbf{A}$ are $\mathbf{x}_i$, we know that if $\mathbf{P}$ is the matrix of eigenvectors of $t\mathbf{A}$, arranged say in the order $\mathbf{x}_1, \mathbf{x}_2, \ldots, \mathbf{x}_n$, and $\mathbf{D}$ is the diagonal matrix $t\mathbf{D} = \mathrm{diag}\{\lambda_1 t, \lambda_2 t, \ldots, \lambda_n t\}$ with its elements $\lambda_i t$ arranged in the same order as the eigenvalues in $\mathbf{P}$, then

$$t\mathbf{A} = \mathbf{P}(t\mathbf{D})\mathbf{P}^{-1}.$$

Using (6.42) we then find that

$$e^{\mathbf{P}(t\mathbf{D})\mathbf{P}^{-1}} = \mathbf{P}e^{t\mathbf{D}}\mathbf{P}^{-1}, \tag{6.43}$$

but $t\mathbf{D} = \mathrm{diag}\{\lambda_1 t,\ \lambda_2 t,\ \ldots,\ \lambda_n t\}$, so from (6.41) we arrive at the important result

$$e^{t\mathbf{A}} = e^{\mathbf{P}(t\mathbf{D})\mathbf{P}^{-1}} = \mathbf{P}\begin{bmatrix} e^{\lambda_1 t} & 0 & \cdots & 0 \\ 0 & e^{\lambda_2 t} & \cdots & 0 \\ \vdots & \vdots & \cdots & 0 \\ 0 & 0 & 0 & e^{\lambda_n t} \end{bmatrix}\mathbf{P}^{-1}. \tag{6.44}$$

Example 6.17. Find $e^{t\mathbf{A}}$, given that $\mathbf{A} = \begin{bmatrix} 1 & -2 \\ 1 & 4 \end{bmatrix}$.

Solution. Matrix $\mathbf{A}$ is diagonalizable, because its eigenvalues and corresponding eigenvectors are, respectively,

$$\lambda_1 = 2 \text{ with } \mathbf{x}_1 = [-2,\ 1]^{\mathrm{T}}, \text{ and } \lambda_2 = 3 \text{ with } \mathbf{x}_2 = [-1,\ 1]^{\mathrm{T}}.$$

Thus

$$\mathbf{P} = \begin{bmatrix} -2 & -1 \\ 1 & 1 \end{bmatrix},\ \mathbf{P}^{-1} = \begin{bmatrix} -1 & -1 \\ 1 & 2 \end{bmatrix} \text{ and } \mathbf{D} = \begin{bmatrix} 2 & 0 \\ 0 & 3 \end{bmatrix}.$$

Substituting into (6.44) we find that

$$e^{t\mathbf{A}} = \begin{bmatrix} -2 & -1 \\ 1 & 1 \end{bmatrix}\begin{bmatrix} e^{2t} & 0 \\ 0 & e^{3t} \end{bmatrix}\begin{bmatrix} -1 & -1 \\ 1 & 2 \end{bmatrix} = \begin{bmatrix} 2e^{2t} - e^{-3t} & -2e^{3t} + 2e^{2t} \\ e^{3t} - e^{2t} & -e^{2t} + 2e^{3t} \end{bmatrix}.$$

Although direct substitution of $t\mathbf{A}$ into (6.40) would produce Maclaurin series as the elements in $e^{t\mathbf{A}}$, it is unlikely these series would be recognized as the functions that occur in the elements of the matrix on the right – hence the need for the approach that has just been described.

◇

Example 6.18. Find $e^{t\mathbf{A}}$, given that $\mathbf{A} = \begin{bmatrix} 1 & -2 \\ 1 & 1 \end{bmatrix}$.

Solution. Matrix $\mathbf{A}$ is diagonalizable, because its eigenvalues and corresponding eigenvectors are, respectively,

$$\lambda_1 = 1 + i\sqrt{2} \text{ with } \mathbf{x}_1 = [i\sqrt{2},\ 1]^{\mathrm{T}}, \text{ and } \lambda_2 = 1 - i\sqrt{2} \text{ with } \mathbf{x}_2 = [-i\sqrt{2},\ 1]^{\mathrm{T}}.$$

Thus

$$\mathbf{P} = \begin{bmatrix} i\sqrt{2} & -i\sqrt{2} \\ 1 & 1 \end{bmatrix}, \quad \mathbf{P}^{-1} = \begin{bmatrix} -\frac{1}{4}i\sqrt{2} & \frac{1}{2} \\ \frac{1}{4}i\sqrt{2} & \frac{1}{2} \end{bmatrix} \text{ and } t\mathbf{D} = \begin{bmatrix} e^{\lambda_1 t} & 0 \\ 0 & e^{\lambda_2 t} \end{bmatrix},$$

Thus

$$e^{t\mathbf{A}} = \mathbf{P}(t\mathbf{D})\mathbf{P}^{-1} = \begin{bmatrix} \frac{1}{2}(e^{\lambda_1 t} + e^{\lambda_2 t}) & \frac{1}{\sqrt{2}}i(e^{\lambda_1 t} - e^{\lambda_2 t}) \\ -\frac{1}{2\sqrt{2}}i(e^{\lambda_1 t} - e^{\lambda_2 t}) & \frac{1}{2}(e^{\lambda_1 t} + e^{\lambda_2 t}) \end{bmatrix},$$

and after simplification this becomes

$$e^{t\mathbf{A}} = \begin{bmatrix} e^t\cos(t\sqrt{2}) & -\sqrt{2}e^t\sin(t\sqrt{2}) \\ \frac{1}{\sqrt{2}}e^t\sin(t\sqrt{2}) & e^t\cos(t\sqrt{2}) \end{bmatrix}.$$

◇

When a and b are real numbers we have the familiar result

$$e^{at}e^{bt} = e^{(a+b)t},$$

so it is necessary to discover if this property of exponential functions remains true when the numbers a and b are replaced by real $n \times n$ matrices $\mathbf{A}$ and $\mathbf{B}$. The first step when answering this question involves examining the relationship between $e^{\mathbf{A}}$ and $e^{\mathbf{B}}$ where $\mathbf{A}$ and $\mathbf{B}$ are diagonal matrices.

Example 6.19. Find $e^{\mathbf{A}}$ and $e^{\mathbf{B}}$ for the matrices

$$\mathbf{A} = \begin{bmatrix} 1 & 0 & 0 \\ 0 & -2 & 0 \\ 0 & 0 & 4 \end{bmatrix} \text{ and } \mathbf{B} = \begin{bmatrix} 2 & 0 & 0 \\ 0 & 3 & 0 \\ 0 & 0 & -1 \end{bmatrix},$$

and examine the relationship between $e^{\mathbf{A}}$, $e^{\mathbf{B}}$ and $e^{\mathbf{A}+\mathbf{B}}$.

Solution. Notice first that two diagonal matrices that are compatible for multiplication always commute, so $\mathbf{AB} = \mathbf{BA}$, where here

$$\mathbf{AB} = \mathbf{BA} = \begin{bmatrix} 2 & 0 & 0 \\ 0 & -6 & 0 \\ 0 & 0 & -4 \end{bmatrix} \text{ and } \mathbf{A}+\mathbf{B} = \begin{bmatrix} 3 & 0 & 0 \\ 0 & 1 & 0 \\ 0 & 0 & 3 \end{bmatrix}.$$

so using (6.41) we see that

$$e^{t\mathbf{A}} = \begin{bmatrix} e & 0 & 0 \\ 0 & e^{-2t} & 0 \\ 0 & 0 & e^{4t} \end{bmatrix}, \quad e^{t\mathbf{B}} = \begin{bmatrix} e^{2t} & 0 & 0 \\ 0 & e^{3t} & 0 \\ 0 & 0 & e^{-t} \end{bmatrix}, \quad e^{t\mathbf{A}+t\mathbf{B}} = \begin{bmatrix} e^{3t} & 0 & 0 \\ 0 & e^{t} & 0 \\ 0 & 0 & e^{3t} \end{bmatrix},$$

after which matrix multiplication confirms that

$$e^{t\mathbf{A}}e^{t\mathbf{B}} = \begin{bmatrix} e^{3t} & 0 & 0 \\ 0 & e^{t} & 0 \\ 0 & 0 & e^{3t} \end{bmatrix} = e^{t\mathbf{A}+t\mathbf{B}}.$$

◇

The result of Example 6.18 would seem to suggest that when $\mathbf{A}$ and $\mathbf{B}$ are general $n \times n$ matrices, the rule for a product of ordinary exponential functions extends to the product of matrix exponentials, allowing us to write $e^{t\mathbf{A}}e^{t\mathbf{B}} = e^{t(\mathbf{A}+\mathbf{B})}$. In fact this assumption is *not true*, and $e^{t\mathbf{A}}e^{t\mathbf{B}} = e^{t(\mathbf{A}+\mathbf{B})}$ if, and only if, $\mathbf{A}$ and $\mathbf{B}$ *commute*, which was the case in Example 6.19, because the product of two $n \times n$ diagonal matrices is always commutative.

Theorem 6.1 *The condition that* $e^{\mathbf{A}}e^{\mathbf{B}} = e^{\mathbf{A}+\mathbf{B}}$. *Let* $\mathbf{A}$ *and* $\mathbf{B}$ *be* n n *matrices. Then the results* $e^{\mathbf{A}}e^{\mathbf{B}} = e^{\mathbf{A}+\mathbf{B}}$ *and* $e^{t\mathbf{A}}e^{t\mathbf{B}} = e^{t(\mathbf{A}+\mathbf{B})}$ *are true if, and only if, the product of the matrices* $\mathbf{A}$ *and* $\mathbf{B}$ *is commutative.*

Proof. Let $\sum_{i=0}^{\infty} R_i$and $\sum_{i=0}^{\infty} S_i$ be two absolutely convergent series, with the respective sums R and S. Expanding the product $RS = (R_0 + R_1 + R_2 + \cdots) \times (S_0 + S_1 + S_2 + \cdots)$ and arranging the result as follows gives

$$RS = \begin{cases} R_0S_0 + R_0S_1 + R_0S_2 + R_0S_3 + \cdots \\ \qquad\quad R_1S_0 + R_1S_1 + R_1S_2 + \cdots \\ \qquad\quad R_2S_0 + R_2S_1 + \cdots \\ \qquad\qquad\quad R_3S_0 + \cdots \\ \qquad\qquad\qquad \vdots \end{cases} \tag{6.45}$$

Summing the columns of (6.45) and grouping the results we find that

$$(R_0S_0) + (R_0S_1 + R_1S_0) + (R_0S_2 + R_1S_1 + R_2S_0) + (R_0S_3 + R_1S_2 + R_2S_1 + R_3S_0) + \cdots .$$

Proceeding in this way we arrive at the result

$$RS = \sum_{m=0}^{\infty} \left(\sum_{n=0}^{m} R_n S_{m-n} \right). \tag{6.46}$$

This formal manipulation of infinite series is justified, because it is shown in calculus texts that (6.46) is true for the product of two absolutely convergent series.

To relate this result to the product $e^{\mathbf{A}}e^{\mathbf{B}}$, let $e^{\mathbf{A}}$ and $e^{\mathbf{B}}$ be absolutely convergent series, and set $R_m = \mathbf{A}^m/m!$ and $S_n = \mathbf{B}^n/n!$, then (6.46) becomes

$$e^{\mathbf{A}}e^{\mathbf{B}} = \sum_{m=0}^{\infty}\left(\sum_{n=0}^{m}\frac{\mathbf{A}^n\mathbf{B}^{m-n}}{n!(m-n)!}\right).$$

Provided matrices $\mathbf{A}$ and $\mathbf{B}$ commute, the ordinary binomial expansion can be used to determine $(\mathbf{A}+\mathbf{B})^n$, because then $\mathbf{A}^{n-r}\mathbf{B}^r = \mathbf{B}^r\mathbf{A}^{n-r}$, in which case

$$\begin{aligned} e^{(\mathbf{A}+\mathbf{B})} &= \sum_{m=0}^{\infty}\frac{1}{m!}(\mathbf{A}+\mathbf{B})^m = \sum_{m=0}^{\infty}\frac{1}{m!}\left(\sum_{n=0}^{m}\frac{m!}{n!(m-n)!}\mathbf{A}^n\mathbf{B}^{m-n}\right) \\ &= \sum_{m=0}^{\infty}\left(\sum_{n=0}^{m}\frac{\mathbf{A}^n\mathbf{B}^{m-n}}{n!(m-n)!}\right) = e^{\mathbf{A}}e^{\mathbf{B}}, \end{aligned}$$

and the proof of the first result is complete, because $e^{(\mathbf{A}+\mathbf{B})} = e^{(\mathbf{B}+\mathbf{A})}$ so $e^{\mathbf{A}}\,e^{\mathbf{B}} = e^{\mathbf{B}}e^{\mathbf{A}}$. The proof of the second statement follows by replacing $\mathbf{A}$ by $t\mathbf{A}$ and $\mathbf{B}$ by $t\mathbf{B}$. ◊

When required, the matrix exponential $e^{-t\mathbf{A}}$ follows from the expression for $e^{t\mathbf{A}}$ by reversing the sign of t. As $\mathbf{A}$ commutes with itself, an important consequence of Theorem 6.1 is obtained by considering the product $e^{t\mathbf{A}}e^{-t\mathbf{A}}$, which becomes

$$e^{t\mathbf{A}}e^{-t\mathbf{A}} = e^{t\mathbf{0}} = \mathbf{I}, \tag{6.47}$$

because from (6.40) it follows that $e^{t\mathbf{0}} = \mathbf{I}$. This confirms that $e^{-t\mathbf{A}}$ is the inverse of $e^{t\mathbf{A}}$, so it is permissible to write

$$\left(e^{t\mathbf{A}}\right)^{-1} = e^{-t\mathbf{A}}, \tag{6.48}$$

while a similar argument shows that $e^{-t\mathbf{A}}e^{t\mathbf{A}} = \mathbf{I}$.

Because $\mathbf{A}$ commutes with itself, it follows at once that

$$e^{(t+\tau)\mathbf{A}} = e^{t\mathbf{A}}e^{\tau\mathbf{A}}, \tag{6.49}$$

The series produced by (6.40) as the elements of $e^{t\mathbf{A}}$ are all absolutely convergent and have infinite radii of convergence, so the series of matrices in (6.40) may be differentiated term by term, to give

$$\frac{de^{t\mathbf{A}}}{dt} = \frac{d}{dt}\sum_{r=0}^{\infty}\frac{t^r}{r!}\mathbf{A}^r = \mathbf{A} + t\mathbf{A}^2 + \frac{1}{2!}t^2\mathbf{A}^3 + \frac{1}{3!}t^3\mathbf{A}^4 + \cdots, \tag{6.50}$$

showing that

$$\frac{de^{t\mathbf{A}}}{dt} = \mathbf{A}e^{t\mathbf{A}}. \tag{6.51}$$

Removing the matrix factor $\mathbf{A}$, first from the left, and then from the right, in (6.50), shows that $\mathbf{A}e^{t\mathbf{A}} = e^{t\mathbf{A}}\mathbf{A}$, and we have established the fundamental result that

$$\frac{de^{t\mathbf{A}}}{dt} = \mathbf{A}e^{t\mathbf{A}} = e^{t\mathbf{A}}\mathbf{A}. \tag{6.52}$$

This result demonstrates that $e^{t\mathbf{A}}$ is a solution of the homogeneous differential equation $d\mathbf{x}/dt = \mathbf{A}\mathbf{x}$, so any linear combination of the columns of $e^{t\mathbf{A}}$ must be a solution vector of the homogeneous linear matrix differential equation $d\mathbf{x}/dt = \mathbf{A}\mathbf{x}$. Consequently $e^{t\mathbf{A}}$ is a *fundamental solution matrix* for the differential equation $d\mathbf{x}/dt = \mathbf{A}\mathbf{x}$, and setting $\mathbf{c} = [C_1, C_2, \ldots, C_n]^{\mathrm{T}}$, with the C_i arbitrary constants, allows the general solution of the differential equation to be written

$$\mathbf{x}(t) = e^{t\mathbf{A}}\mathbf{c}. \tag{6.53}$$

This result forms the statement of the following Theorem.

Theorem 6.2 *The General Solution of a Linear Homogeneous System. Let* $\mathbf{A}$ *be a diagonalizable* $n \times n$ *constant matrix. Then* $e^{t\mathbf{A}}$ *is a fundamental solution matrix for the homogeneous matrix differential equation*

$$\frac{d\mathbf{x}}{dt} = \mathbf{A}\mathbf{x},$$

and if an initial condition $\mathbf{x}(t_0) = \mathbf{c}$ *is imposed at time* $\mathrm{t} = \mathrm{t}_0$ *the unique solution of this initial value problem is* $\mathbf{x}(t) = e^{t\mathbf{A}}\mathbf{c}$ *for* $t \geq t_0$.

Proof. This main part of this theorem has already been proved, leaving only the justification of the assertion that the solution of the initial-value problem is unique. Uniqueness is easily established by assuming, if possible, that the initial-value problem has two *different* solutions $\mathbf{x}$ and $\mathbf{y}$ that satisfy the same initial condition, then

$$\frac{d\mathbf{x}}{dt} = \mathbf{A}\mathbf{x} \text{ and } \frac{d\mathbf{y}}{dt} = \mathbf{A}\mathbf{y}, \text{ where } \mathbf{x}(t_0) = \mathbf{y}(t_0).$$

Subtracting the second equation from the first one and setting $\mathbf{x} - \mathbf{y} = \mathbf{u}$ we find that

$$\frac{d\mathbf{u}}{dt} = \mathbf{A}\mathbf{u}, \text{ subject to the initial condition } \mathbf{u}(t_0) = \mathbf{x}(t_0) - \mathbf{y}(t_0) = \mathbf{0}.$$

The only solution of this initial-value problem given by the main result of the theorem is $\mathbf{u}(t) \equiv 0$ for $t > t_0$, so $\mathbf{x}(t) \equiv \mathbf{y}(t)$ for $t > t_0$, and the uniqueness is proved. ◇

Example 6.20. Use the matrix exponential to solve the initial-value problem

$$\dot{x}_1 = 3x_1 + x_2 - x_3, \quad \dot{x}_2 = 3x_1 + x_2 - 3x_3, \quad \dot{x} = 2x_1 - 2x_2,$$

subject to the initial conditions $x_1(0) = 1,\ x_2(0) = 0,\ x_3(0) = -1$.

Solution. The matrix of coefficients **A**, the solution vector **x**, the eigenvalues and eigenvectors of **A** are, respectively,

$$\mathbf{A} = \begin{bmatrix} 3 & 1 & -1 \\ 3 & 1 & -3 \\ 2 & -2 & 0 \end{bmatrix}, \quad \mathbf{x} = \begin{bmatrix} x_1 \\ x_2 \\ x_3 \end{bmatrix}, \quad \lambda_1 = 2,\ \mathbf{x}_1 = \begin{bmatrix} 1 \\ 0 \\ 1 \end{bmatrix}, \quad \lambda_2 = -2,$$
$$\mathbf{x}_2 = \begin{bmatrix} 0 \\ 1 \\ 1 \end{bmatrix}, \quad \lambda_3 = 4, \quad \mathbf{x}_3 = \begin{bmatrix} 1 \\ 1 \\ 0 \end{bmatrix}.$$

The matrix of eigenvectors **P**, its inverse $\mathbf{P}^{-1}$, the matrix $t\mathbf{D} = \mathrm{diag}\{e^{2t},\ e^{-2t},\ e^{4t}\}$ and the initial condition vector **c** are

$$\mathbf{P} = \begin{bmatrix} 1 & 0 & 1 \\ 0 & 1 & 1 \\ 1 & 1 & 0 \end{bmatrix}, \quad \mathbf{P}^{-1} = \begin{bmatrix} \frac{1}{2} & -\frac{1}{2} & \frac{1}{2} \\ -\frac{1}{2} & \frac{1}{2} & \frac{1}{2} \\ \frac{1}{2} & \frac{1}{2} & -\frac{1}{2} \end{bmatrix}, \quad \mathbf{D} = \begin{bmatrix} e^{2t} & 0 & 0 \\ 0 & e^{-2t} & 0 \\ 0 & 0 & e^{4t} \end{bmatrix}, \quad \mathbf{c} = \begin{bmatrix} 1 \\ 0 \\ -1 \end{bmatrix}.$$

So, from (6.44) we find that

$$e^{t\mathbf{A}} = \mathbf{PDP}^{-1} = \begin{bmatrix} \frac{1}{2}(e^{2t} + e^{4t}) & \frac{1}{2}(-e^{2t} + e^{4t}) & \frac{1}{2}(e^{2t} - e^{4t}) \\ \frac{1}{2}(-e^{-2t} + e^{4t}) & \frac{1}{2}(e^{-2t} + e^{4t}) & \frac{1}{2}(e^{-2t} - e^{4t}) \\ \frac{1}{2}(e^{2t} - e^{-2t}) & \frac{1}{2}(-\frac{1}{2}e^{2t} + e^{-2t}) & \frac{1}{2}(e^{2t} + e^{-2t}) \end{bmatrix}.$$

From Theorem 6.1 the solution vector $\mathbf{x} = e^{t\mathbf{A}}\mathbf{c}$ becomes

$$\mathbf{x} = \left[e^{4t}, -e^{-2t} + e^{4t}, -e^{-2t}\right]^{\mathrm{T}},$$

and so

$$x_1(t) = e^{4t}, \quad x_2(t) = -e^{-2t} + e^{4t}, \quad x_3(t) = -e^{-2t}.$$

◇

It is a straightforward matter to generalize the result of Theorem 6.1 to non homogeneous systems of the form

$$\frac{d\mathbf{x}}{dt} = \mathbf{Ax} + \mathbf{f}(t), \tag{6.54}$$

where $\mathbf{f}(t)$ is the column vector $\mathbf{f}(t) = [f_1(t), f_2(t), \ldots, f_n(t)]^{\mathrm{T}}$ whose elements are integrable functions of t.

The approach used will be the analogue of the way nonhomogeneous linear first-order scalar equations are solved by means of an integrating factor, though here the analog of the integrating factor will be $e^{-t\mathbf{A}}$. A review of the use of an integrating factor when solving a linear first-order differential equation will be found in Appendix 1 at the end of this chapter. Rearranging the terms in (6.54) and pre-multiplying the result by $e^{-t\mathbf{A}}$ gives

$$e^{-t\mathbf{A}}(d\mathbf{x}/dt - \mathbf{A}\mathbf{x}) = e^{-t\mathbf{A}}\mathbf{f}(t). \tag{6.55}$$

To simplify this result notice that

$$\frac{d}{dt}\left(e^{-t\mathbf{A}}\mathbf{x}\right) = -\mathbf{A}e^{-t\mathbf{A}}\mathbf{x} + e^{-t\mathbf{A}}\frac{d\mathbf{x}}{dt},$$

but $\mathbf{A}e^{-t\mathbf{A}} = e^{-t\mathbf{A}}\mathbf{A}$, so

$$-\mathbf{A}e^{-t\mathbf{A}}\mathbf{x} + e^{-t\mathbf{A}}\frac{d\mathbf{x}}{dt} = -e^{-t\mathbf{A}}\mathbf{A}\mathbf{x} + e^{-t\mathbf{A}}\frac{d\mathbf{x}}{dt},$$

allowing (6.55) to be written

$$\frac{d}{dt}\left(e^{-t\mathbf{A}}\mathbf{x}\right) = e^{-t\mathbf{A}}\mathbf{f}(t).$$

Integrating this result with respect to t and pre-multiplying the result by $e^{t\mathbf{A}}$ shows the solution vector to be

$$\mathbf{x}(t) = e^{t\mathbf{A}}\mathbf{c} + e^{t\mathbf{A}}\int e^{-t\mathbf{A}}\mathbf{f}(t)dt. \tag{6.56}$$

where $\mathbf{c}$ is an arbitrary n element column vector that contains the arbitrary integration constants.

On occasions it is convenient to take $e^{t\mathbf{A}}$ under the integral sign in (6.56). Then, to avoid confusion with the variable t, the variable of integration must be changed from t to τ, when (6.56) becomes

$$\mathbf{x}(t) = e^{t\mathbf{A}}\mathbf{c} + \int e^{(t-\tau)\mathbf{A}}\mathbf{f}(\tau)d\tau. \tag{6.57}$$

The solution of an initial-value problem at time $t = t_0$ follows from either (6.56) or (6.57) by matching the arbitrary constants in $\mathbf{c}$ to suit the initial conditions $\mathbf{x}(t_0)$. The following theorem has been proved.

Theorem 6.3 *The Solution of Nonhomogeneous Linear First-Order Equations. The general solution of the matrix differential equation*

$$\frac{d\mathbf{x}}{dt} = \mathbf{A}\mathbf{x} + \mathbf{f}(t)$$

is

$$\mathbf{x}(t) = e^{t\mathbf{A}}\mathbf{c} + e^{t\mathbf{A}} \int e^{-t\mathbf{A}}\mathbf{f}(t)dt$$

or equivalently

$$\mathbf{x}(t) = e^{t\mathbf{A}}\mathbf{c} + \int e^{(t-\tau)\mathbf{A}}\mathbf{f}(\tau)d\tau.$$

◇

Example 6.21. Use the matrix exponential to solve the initial-value problem

$$\dot{x}_1 = 2x_1 + 2x_2 + t, \;\; \dot{x}_2 = x_1 + 3x_2 - 1, \;\; x_1(0) = 2, \; x_2(0) = -1.$$

Solution. In this solution, because the intermediate calculations are straightforward, only the key result will be given. We are required to solve the nonhomogeneous differential equation $d\mathbf{x}/dt = \mathbf{A}\mathbf{x} + \mathbf{f}(t)$, where

$$\mathbf{A} = \begin{bmatrix} 2 & 2 \\ 1 & 3 \end{bmatrix}, \mathbf{x} = \begin{bmatrix} x_1 \\ x_2 \end{bmatrix}, \; \lambda_1 = 1, \; \mathbf{x}_1 = \begin{bmatrix} -2 \\ 1 \end{bmatrix}, \; \lambda_2 = 4, \; \mathbf{x}_2 = \begin{bmatrix} 1 \\ 1 \end{bmatrix},$$
$$\mathbf{x}(0) = \begin{bmatrix} 2 \\ -1 \end{bmatrix}, \; \mathbf{f}(t) = \begin{bmatrix} t \\ -1 \end{bmatrix}.$$

A routine calculation shows that

$$\mathbf{P} = \begin{bmatrix} -2 & 1 \\ 1 & 1 \end{bmatrix}, \; \mathbf{P}^{-1} = \begin{bmatrix} -\frac{1}{3} & \frac{1}{3} \\ \frac{1}{3} & \frac{2}{3} \end{bmatrix}, \; t\mathbf{D} = \begin{bmatrix} e^t & 0 \\ 0 & e^{4t} \end{bmatrix}, \text{ so}$$

$$e^{t\mathbf{A}} = \mathbf{P}(t\mathbf{D})\mathbf{P}^{-1} = \begin{bmatrix} \frac{2}{3}e^t + \frac{1}{3}e^{4t} & -\frac{2}{3}e^t + \frac{2}{3}e^{4t} \\ -\frac{1}{3}e^t + \frac{1}{3}e^{4t} & \frac{1}{3}e^t + \frac{2}{3}e^{4t} \end{bmatrix},$$

from which $e^{-t\mathbf{A}}$ follows by changing the sign of t. Routine integration gives

$$\int e^{-t\mathbf{A}}\mathbf{f}(t)dt = \begin{bmatrix} -\frac{2}{3}t\,e^{-t} - \frac{4}{3}e^{-t} - \frac{1}{12}t\,e^{-4t} + \frac{7}{48}e^{-4t} \\ \frac{1}{3}te^{-t} + \frac{2}{3}e^{-t} - \frac{1}{12}te^{-4t} + \frac{7}{48}e^{-4t} \end{bmatrix}.$$

Once the result $\mathbf{x}(t) = e^{t\mathbf{A}}\mathbf{c} + e^{t\mathbf{A}} \int e^{-t\mathbf{A}}\mathbf{f}(t)dt$ has been evaluated, and the arbitrary constants C_1 and C_2 in the vector $\mathbf{c} = [C_1, C_2]^{\mathrm{T}}$ have been matched to the initial conditions $x_1(0) = 2$ and $x_2(0) = -1$, the solution vector $\mathbf{x}(t)$ is found to be

$$\begin{bmatrix} x_1(t) \\ x_2(t) \end{bmatrix} = \begin{bmatrix} -\frac{19}{16} - \frac{3}{4}t + \frac{10}{3}e^t - \frac{7}{48}e^{4t} \\ \frac{13}{16} + \frac{1}{4}t - \frac{5}{3}e^t - \frac{7}{48}e^{4t} \end{bmatrix} \quad \text{for } t \geq 0,$$

so

$$x_1(t) = -\tfrac{19}{16} - \tfrac{3}{4}t + \tfrac{10}{3}e^t - \tfrac{7}{48}e^{4t} \text{ and } x_2(t) = \tfrac{13}{16} + \tfrac{1}{4}t - \tfrac{5}{3}e^t - \tfrac{7}{48}e^{4t} \text{ for } t \geq 0.$$

◇

It has been shown how $e^{t\mathbf{A}}$ can be computed when an $n \times n$ matrix $\mathbf{A}$ is diagonalizable, but not how it can be computed when this is not the case.

Various methods exist for finding $e^{t\mathbf{A}}$ when $\mathbf{A}$ is an arbitrary $n \times n$ matrix, but the method described here depends for its success on using the Laplace transform to interpret the meaning of $e^{t\mathbf{A}}$. Although, in principle, this method is applicable for any $n \times n$ matrix, because of the algebraic manipulation involved it is really only practical when $n \leq 4$.

Consider the homogeneous constant coefficient differential equation

$$\frac{d\mathbf{x}}{dt} = \mathbf{A}\mathbf{x} \text{ subject to the initial condition } \mathbf{x}(0) = \mathbf{c} \text{ when } t = 0, \tag{6.58}$$

where $\mathbf{A}$ is a constant $n \times n$ matrix and $\mathbf{c} = [c_1, c_2, \ldots, c_n]^{\mathrm{T}}$ is a constant n element column vector. Defining the Laplace transform $\mathbf{V}(s)$ of a vector

$$\mathbf{v}(t) = [v_1(t), v_2(t), \ldots, v_n(t)]^{\mathrm{T}} \text{ as}$$

$$\mathbf{V}(s) = \mathcal{L}\{\mathbf{v}(t)\} = [\mathcal{L}\{v_1(t), \mathcal{L}\{v_2(t), \ldots, \mathcal{L}\{v_n(t)]^{\mathrm{T}}, \tag{6.59}$$

taking the Laplace transform of (6.58) using $\mathcal{L}\{dv_i/dt\} = sV_i(s) - v_i(0)$, and setting $\mathcal{L}\{\mathbf{x}(t)\} = \mathbf{X}(s)$ gives $s\mathbf{X}(s) - \mathbf{c} = \mathbf{A}\mathbf{X}(s)$, and so $[s\mathbf{I} - \mathbf{A}]\mathbf{X}(s) = \mathbf{c}$. After pre-multiplying this result by the inverse of $[s\mathbf{I} - \mathbf{A}]$ it becomes

$$\mathbf{X}(s) = [s\mathbf{I} - \mathbf{A}]^{-1}\mathbf{c}. \tag{6.60}$$

Taking the inverse Laplace transform gives the solution of the initial-value problem as

$$\mathbf{x}(t) = \mathcal{L}^{-1}\{[s\mathbf{I} - \mathbf{A}]^{-1}\}\mathbf{c}, \tag{6.61}$$

where if $\mathbf{H}(s) = [h_{ij}(s)]$, then $\mathcal{L}^{-1}\{\mathbf{H}(s)\} = [\mathcal{L}^{-1}\{h_{ij}(s)\}]$.

A comparison of (6.53) and (6.61) establishes the following representation of $e^{t\mathbf{A}}$ in terms of an inverse Laplace transform.

6.6.1 *Finding the Matrix Exponential Using the Laplace Transform*

$$e^{t\mathbf{A}} = \mathcal{L}^{-1}\{[s\mathbf{I} - \mathbf{A}]^{-1}\}. \tag{6.62}$$

Example 6.22. Find the matrix exponential (the fundamental solution matrix) for the system of equations

$$\dot{x}_1 = x_1 + x_2, \quad \dot{x}_2 = x_2,$$

and hence find its general solution.

Solution. Writing this homogeneous system of equations in the form $d\mathbf{x}/dt = \mathbf{A}\mathbf{x}$, we see the matrices $\mathbf{A}$ and $\mathbf{x}$, and the eigenvalues and eigenvector, are

$$\mathbf{A} = \begin{bmatrix} 1 & 1 \\ 0 & 1 \end{bmatrix}, \quad \mathbf{x} = \begin{bmatrix} x_1 \\ x_2 \end{bmatrix}, \quad \lambda = 1 \text{ (twice), and the single eigenvector } \mathbf{x}_1 = \begin{bmatrix} 1 \\ 1 \end{bmatrix}.$$

Consequently matrix $\mathbf{A}$ cannot be diagonalized. Because of this, the matrix exponential $e^{t\mathbf{A}}$ will be found from (6.62). We have

$$[s\mathbf{I} - \mathbf{A}] = \begin{bmatrix} s-1 & -1 \\ 0 & s-1 \end{bmatrix}, \quad \text{so } [s\mathbf{I} - \mathbf{A}]^{-1} = \begin{bmatrix} 1/(s-1) & 1\Big/(s-1)^2 \\ 0 & 1/(s-1) \end{bmatrix}.$$

Taking the inverse Laplace transform gives

$$e^{t\mathbf{A}} = \mathcal{L}^{-1}\left\{[s\mathbf{I} - \mathbf{A}]^{-1}\right\} = \begin{bmatrix} e^t & te^t \\ 0 & e^t \end{bmatrix},$$

so denoting the constant arbitrary integration vector by $\mathbf{c} = [C_1, C_2]^{\mathrm{T}}$, the general solution vector

$$\mathbf{x}(t) = \begin{bmatrix} x_1(t) \\ x_2(t) \end{bmatrix} = \begin{bmatrix} e^t & te^t \\ 0 & e^t \end{bmatrix} \begin{bmatrix} C_1 \\ C_2 \end{bmatrix},$$

so the general solution is

$$x_1(t) = C_1 e^t + C_2 te^t \text{ and } x_2(t) = C_2 e^t.$$

◇

Example 6.23. Use (6.62) to find $e^{t\mathbf{A}}$ when $\mathbf{A} = \begin{bmatrix} a & b \\ 0 & a \end{bmatrix}$, and show by example the result is the same as the one found in Example 6.15 by direct substitution of $\mathbf{A}$ into (6.40).

Solution. From (6.62) we have

$$\mathbf{sI}-\mathbf{A}=\begin{bmatrix} s-a & -b \\ 0 & s-a \end{bmatrix}, \text{ so } [\mathbf{sI}-\mathbf{A}]^{-1}=\begin{bmatrix} 1/(s-a) & b/(s-a)^2 \\ 0 & 1/(s-a) \end{bmatrix}.$$

Using the table of Laplace transform pairs in Appendix 2 we find that

$$\mathcal{L}^{-1}\left\{[\mathbf{sI}-\mathbf{A}]^{-1}\right\}=\begin{bmatrix} e^{at} & bte^{at} \\ 0 & e^{at} \end{bmatrix},$$

which is precisely the result found in Example 6.15. In this case the result was found more simply by using (6.62) than by the direct method used in the example. ♢

Appendix 1: The Solution of a Linear First-Order Differential Equation

The most general linear first-order differential equation has the form

$$\frac{dy}{dx}+p(x)y=q(x).$$

The *integrating factor* for this equation is

$$\mu(x)=\exp\left(\int p(x)dx\right),$$

where no arbitrary constant is to be added when $\int p(x)dx$ is evaluated. The general solution of the general linear first-order equation is then given by

$$y(x)=\frac{1}{\mu(x)}\left[C+\int q(x)\mu(x)dx\right],$$

where C is the arbitrary integration constant introduced when $\int q(x)\mu(x)dx$ is evaluated.

When $p(x)=a$ (a constant), as is the case with linear constant coefficient first-order differential equations, the integrating factor simplifies to $\mu(x)=\exp(ax)$, and then the general solution becomes

$$y(x) = \exp(-ax)\left[C + \int \exp(ax)q(x)dx\right].$$

$$y(x) = \exp(-ax)\left[C + \int \exp(ax)q(x)dx\right].$$

Appendix 2: A Summary of the Laplace Transform and a Short Table of Laplace Transform Pairs

The Laplace transform $\mathcal{L}\{y(t)\} = Y(s)$ of the function $y(t)$, is defined as

$$Y(s) = \int_0^\infty e^{-st}y(t)dt,$$

for those functions $y(t)$ such that the improper integral on the right exists.

Linearity of the Laplace transform

If a and b are constants and $f(t)$ and $g(t)$ have the respective Laplace transforms $F(s)$ and $G(s)$, then

$$\mathcal{L}\{af(t) + bg(t)\} = aF(s) + bG(s).$$

The Laplace transform of derivatives

$$\mathcal{L}\{dy/dt\} = sY(s) + y(0),$$

$$\mathcal{L}\{d^2y/dt^2\} = s^2Y(s) - sy(0) - y'(0),$$

$$\mathcal{L}\{d^ny/dt^n\} = s^nY(s) - s^{n-1}y(0) - s^{n-2}y'(0) - s^{n-3}y''(0) - \cdots - y^{(n-1)}(0).$$

The first shift theorem

$$\mathcal{L}\{e^{-at}f(t)\} = F(s+a).$$

The second shift theorem

$$\mathcal{L}\{H(t-a)f(t-a)\} = e^{-as}F(s),$$

where $H(t-a) = \begin{cases} 0, & t < a \\ 1, & t \geq a \end{cases}$ is the Heaviside unit step function

The Laplace convolution theorem

$$\mathcal{L}\left\{\int_0^t f(\tau)g(t-\tau)d\tau\right\} = F(s)G(s),$$

where $\mathcal{L}\{f(t)\} = F(s)$ and $\mathcal{L}\{g(t)\} = G(s)$

Table of useful Laplace transform pairs

$f(t)$	$F(s)$
k	k/s
t	$1/s^2$
t^n (n a positive integer)	$n!/s^{n+1}$
e^{at}	$1/(s-a)$
$t^n e^{at}$ (n a positive integer)	$n!/(s-a)^{n+1}$
$\sin at$	$a/(s^2+a^2)$
$\cos at$	$s/(s^2+a^2)$
$t \sin at$	$2as/(s^2+a^2)^2$
$t \cos at$	$(s^2-a^2)/(s^2+a^2)^2$
$e^{at}\sin bt$	$b/\left[(s-a)^2+b^2\right]$
$e^{at}\cos bt$	$(s-a)/\left[(s-a)^2+b^2\right]$
$\sinh at$	$a/(s^2-a^2)$
$\cosh at$	$s/(s^2-a^2)$
$H(t-a)$ $(a \geq 0)$	e^{-as}/s (the Heaviside unit step function)
$\delta(t-a)$ $(a \geq 0)$	e^{-as} (the Dirac delta function)

Exercises

1. Construct any two matrices F(t) and G(t) conformable for multiplication. Compute $d[\mathbf{G}(t)\mathbf{H}(t)]/dt$ directly, and by adding the matrix products $\frac{d\mathbf{G}(t)}{dt}\mathbf{H}(t)$ and $\mathbf{G}(t)\frac{d\mathbf{H}(t)}{dt}$, verify that

$$\frac{d}{dt}[\mathbf{G}(t)\mathbf{H}(t)] = \frac{d\mathbf{G}(t)}{dt}\mathbf{H}(t) + \mathbf{G}(t)\frac{d\mathbf{H}(t)}{dt}.$$

2. Construct a 3×3 matrix $\mathbf{G}(t)$ of your own choice. Find $\mathbf{G}^{-1}(t)$ and differentiate it to find $d\mathbf{G}^{-1}(t)/dt$. Use the result to verify that

$$\frac{d\mathbf{G}^{-1}(t)}{dt} = -\mathbf{G}^{-1}(t)\frac{d\mathbf{G}(t)}{dt}\mathbf{G}^{-1}(t).$$

3. If $\mathbf{G}(t)$ and $\mathbf{H}(t)$ are any two nonsingular $n \times n$ matrices, find an expression for

$$\frac{d}{dt}[\mathbf{G}(t)\mathbf{H}(t)]]^{-1}.$$

4. Construct a nonsingular 2×2 constant matrix $\mathbf{A}$ and a nonsingular 2×2 matrix $\mathbf{F}(t)$ with differentiable elements of your own choice. Setting $\mathbf{G}(t) = \mathbf{AF}(t)$, find $d\mathbf{G}^{-1}(t)/dt$ by differentiation of $\mathbf{AG}^{-1}(t)$, and also by using result (6.7).

5. Set

$$\mathbf{A} = \begin{bmatrix} 2 & -4 \\ 1 & 3 \end{bmatrix}, \ \mathbf{H}(t) = \begin{bmatrix} \sin t & \cos t \\ -\cos t & \sin t \end{bmatrix} \text{ and } \mathbf{G}(t) = \mathbf{AH}(t) .$$

Find $d\mathbf{G}^{-1}(t)/dt$ directly, and also by using the result $\mathbf{G}^{-1}(t) = \mathbf{H}^{-1}(t)\mathbf{A}^{-1}$.

6. If matrices $\mathbf{A}(t) = [a_{ij}(t)]$ and $\mathbf{B}(t) = [b_{ij}(t)]$ are conformable for addition, prove that $\int (\alpha\mathbf{A}(t) + \beta\mathbf{B}(t))dt = \alpha \int \mathbf{A}(t)dt + \beta \int \mathbf{B}(t)dt$, for any scalars α and β.
7. Prove that if $\mathbf{A}(t)$ and $\mathbf{B}(t)$ are conformable for multiplication, and each matrix is differentiable, then the matrix analogue of integration by parts is

$$\int \mathbf{A}\frac{d\mathbf{B}}{dt}dt = \mathbf{AB} - \int \frac{d\mathbf{A}}{dt}\mathbf{B}dt .$$

In Exercises 8 through 13 find the general solution vector $\mathbf{x}(t)$ with elements $x_i(t)$ of the homogeneous equation $d\mathbf{x}/dt = \mathbf{Ax}$ using the given matrix $\mathbf{A}$, and hence find the solution that satisfies the given initial conditions.

8.

$$\mathbf{A} = \begin{bmatrix} 0 & 0 & -1 \\ 1 & -2 & 1 \\ -1 & 0 & 0 \end{bmatrix}, \quad x_1(0) = 2 , \ x_2(0) = 2 , \ x_3(0) = -1 .$$

9.

$$\mathbf{A} = \begin{bmatrix} -1 & 0 & 0 \\ 2 & -1 & 2 \\ -4 & 0 & 3 \end{bmatrix}, \quad x_1(0) = 2 , \ x_2(0) = 1 , \ x_3(0) = -1 .$$

10.

$$\mathbf{A} = \begin{bmatrix} -1 & -4 \\ 1 & -1 \end{bmatrix}, \quad x_1(0) = 0 , \ x_2(0) = 1 .$$

11.

$$\mathbf{A} = \begin{bmatrix} 2 & 4 \\ -1 & 2 \end{bmatrix}, \quad x_1(0) = -2 , \ x_2(0) = 1 .$$

12.

$$\mathbf{A} = \begin{bmatrix} 0 & 0 & -2 \\ -1 & -1 & -1 \\ 1 & 0 & 3 \end{bmatrix}, \quad x_1(0) = 1 , \ x_2(0) = -1 , \ x_3(0) = 2 .$$

13.

$$\mathbf{A} = \begin{bmatrix} 2 & -3 & 3 \\ -1 & 2 & -1 \\ -1 & 3 & -2 \end{bmatrix}, \quad x_1(0) = 1\,, \quad x_2(0) = -1\,, \quad x_3(0) = 0\,.$$

In Exercises 14 through 19 find the general solution $\mathbf{x}(t)$ with elements $x_i(t)$ of the nonhomogeneous equation $d\mathbf{x}/dt = \mathbf{A}\mathbf{x} + \mathbf{f}(t)$ using the given matrices $\mathbf{A}$ and $\mathbf{f}(t)$, and hence find the solution satisfying the stated initial conditions.

14.

$$\mathbf{A} = \begin{bmatrix} -1 & 1 \\ -2 & 1 \end{bmatrix}, \quad \mathbf{f}(t) = \begin{bmatrix} t^2 \\ -2e^{-t} \end{bmatrix}, \quad \text{(a) } x(0) = 1,\ y(0) = -1,$$
$$\text{(b) } x(1) = 0,\ y(1) = 0\,.$$

15.

$$\mathbf{A} = \begin{bmatrix} -1 & 1 \\ 3 & 1 \end{bmatrix}, \quad \mathbf{f}(t) = \begin{bmatrix} 3t \\ -\sin t \end{bmatrix}, \quad x(0) = -1\,, \ y(0) = 0\,.$$

16.

$$\mathbf{A} = \begin{bmatrix} 2 & -2 \\ 1 & -1 \end{bmatrix}, \quad \mathbf{f}(t) = \begin{bmatrix} 2-t \\ -4 \end{bmatrix}, \quad \text{(a) } x(0) = -2\,, \ y(0) = 1\,,$$
$$\text{(b) } x(1) = 1\,, \ y(1) = 0\,.$$

17.

$$\mathbf{A} = \begin{bmatrix} -1 & 3 \\ 3 & -1 \end{bmatrix}, \quad \mathbf{f}(t) = \begin{bmatrix} -e^{-2t} \\ 2\cos t \end{bmatrix}, \quad x(0) = 1\,, \ y(0) = -1\,.$$

18.

$$\mathbf{A} = \begin{bmatrix} 1 & 1 & 0 \\ 0 & 1 & 1 \\ 0 & 1 & 1 \end{bmatrix}, \quad \mathbf{f}(t) = \begin{bmatrix} -2t \\ -3 \\ 3\sin 3t \end{bmatrix}, \quad x(0) = 0\,, \ y(0) = -2\,,$$
$$z(0) = -1\,.$$

19.

$$\mathbf{A} = \begin{bmatrix} 1 & -1 & 0 \\ 0 & 1 & 1 \\ 0 & -1 & 1 \end{bmatrix}, \quad \mathbf{f}(t) = \begin{bmatrix} 2 \\ -\sin t \\ 1 + e^{-2t} \end{bmatrix}, \quad x(0) = 1\,, \ y(0) = 2\,,$$
$$z(0) = -1\,.$$

In Exercises 20 and 21, transform the system into the standard form $d\mathbf{x}/dt = \mathbf{A}\mathbf{x} + \mathbf{f}(t)$. Find the general solution, and the solution subject to the given initial conditions.

20.

$$\frac{dx_1}{dt} + 2\frac{dx_2}{dt} = 7x_1 + x_2 - 5 + 4t\,,\ \ 2\frac{dx_1}{dt} + \frac{dx_2}{dt} = 8x_1 - x_2 - 1 + 2t\ ;$$

$$x_1(0) = -2\ ,\ x_2(0) = 1\ .$$

21.

$$\frac{dx_1}{dt} - 2\frac{dx_2}{dt} = -2x_1 + x_2 + 2 - 6t^2, -\frac{dx_1}{dt} + \frac{dx_2}{dt} = x_1 - x_2 - 2 + 3t^2\ ;$$

$$x(0) = 2,\ y(0) = -1.$$

In Exercises 22 through 32, use the Laplace transform with matrix methods to solve the given initial-value problem.

22.

$$\frac{dx_1}{dt} = -x_1 + x_2 + te^{-t}\ ,\ \ \frac{dx_2}{dt} = x_1 - x_2 + 4\ ,\ \ x_1(0) = 0\ ,\ x_2(0) = -2\ .$$

23.

$$\frac{dx_1}{dt} = 2x_1 + x_2 - \sin t\ ,\ \ \frac{dx_2}{dt} = 2x_1 + x_2 + 2\cos t\ ,$$
$$x_1(0) = 1\ ,\ x_2(0) = -1\ .$$

24.

$$\frac{dx_1}{dt} = x_2 + \cos t\ ,\ \ \frac{dx_2}{dt} = x_1 + 3t\ ,\ \ x_1(0) = 1\ ,\ x_2(0) = 0\ .$$

25.

$$\frac{dx_1}{dt} = x_1 + x_2 + e^{-t}\ ,\ \ \frac{dx_2}{dt} = 6x_1 - 2t\ ,\ \ x_1(0) = 0\ ,\ x_2(0) = 0\ .$$

26.

$$\frac{dx_1}{dt} = x_3 + 2\ ,\ \ \frac{dx_2}{dt} = x_3\ ,\ \ \frac{dx_3}{dt} = x_1 + 2\cos 2t\ \ x_1(0) = 0\ ,$$
$$x_2(0) = 1\ ,\ x_3(0) = -1\ .$$

27.

$$\frac{dx_1}{dt} = x_1 + x_3 + 3\cos t\ , \ \frac{dx_2}{dt} = -x_3 + 1\ , \ \frac{dx_3}{dt} = x_2 - \sin t,$$
$$x_1(0) = 1\ , \ x_2(0) = 0\ , \ x_3(0) = 0\ .$$

28.

$$\frac{dx_1}{dt} = 2x_1 + 2x_3 + 3\ , \ \frac{dx_2}{dt} = 2x_3 + 1\ , \ \frac{dx_3}{dt} = 2x_2 - t,$$
$$x_1(0) = 1\ , \ x_2(0) = 0\ , \ x_3(0) = -1\ .$$

29.

$$\frac{dx_1}{dt} = x_1 - x_3 + 2t\ , \ \frac{dx_2}{dt} = -x_3 + t\ , \ \frac{dx_3}{dt} = x_2 - \sin t,\ x_1(0) = 0\ ,$$
$$x_2(0) = 1\ , \ x_3(0) = 0\ .$$

30.

$$\frac{d^2x_1}{dt^2} = x_2 + t\ , \ \frac{d^2x_2}{dt^2} = x_1 + \sin t\ , \ x_1(0) = 1\ , \ x_1{}'(0) = 0\ ,$$
$$x_2(0) = 0\ , \ x_2{}'(0) = -1\ .$$

31.

$$\frac{d^2x_1}{dt^2} + 3\frac{dx_1}{dt} + 7x_1 + x_2 = 3\ , \ \frac{dx_2}{dt} = 5x_1 + 1\ , \ x_1(0) = 1\ ,$$
$$x_1{}'(0) = 0\ , \ x_2(0) = 2\ .$$

32. The initial-value problem

$$\frac{d^3y}{dt^3} + 2\frac{d^2y}{dt^2} - \frac{dy}{dt} - 2y = 1 + \sin\ t,\ \ y(0) = 0,\ y'(0) = 0,\ y''(0) = 0$$

has the solution

$$y(t) = \tfrac{1}{4}e^{-t} + \tfrac{1}{4}e^{t} - \tfrac{1}{10}e^{-2t} - \tfrac{1}{2} + \tfrac{1}{10}(\cos t - 2\sin t)\ \text{ for } t \geq 0.$$

Find this solution $y(t)$ by converting the equation into a system: (a) as in the text using the higher-order derivatives as the new dependent variables, and (b) by introducing the new dependent variables $y(t)$, $u(t)$, $v(t)$ with $u(t) = 2dy/dt$, and $v(t) = 4\,du/dt$. The fact that the solutions will be identical will confirm that a linear change of variables used when reducing the third-order differential equation to a system of first-order equations will not alter the solution $y(t)$.

33. Show matrix $\mathbf{A}$ is nilpotent and find its nilpotent index given that

$$\mathbf{A} = \begin{bmatrix} 0 & 3 & 1 & 2 \\ 0 & 0 & 1 & 3 \\ 0 & 0 & 0 & 2 \\ 0 & 0 & 0 & 0 \end{bmatrix}.$$

Use the fact that $\mathbf{A}$ is nilpotent to find $e^{t\mathbf{A}}$.

34. Given that

$$\mathbf{A} = \begin{bmatrix} -1 & 0 & 0 & 0 \\ 0 & -2 & 0 & 0 \\ 0 & 0 & 1 & 0 \\ 0 & 0 & 0 & 2 \end{bmatrix} \text{ and } \mathbf{B} = \begin{bmatrix} 1 & 0 & 0 & 0 \\ 0 & 2 & 0 & 0 \\ 0 & 0 & -1 & 0 \\ 0 & 0 & 0 & 1 \end{bmatrix}$$

find $e^{\mathbf{A}}$, $e^{\mathbf{B}}$ and show that in this case $e^{\mathbf{A}+\mathbf{B}} = e^{\mathbf{A}}e^{\mathbf{B}}$. Why is this so?

35. Find $e^{t\mathbf{A}}$ from (6.40), given that $\mathbf{A} = \begin{bmatrix} 0 & a \\ a & 0 \end{bmatrix}$.

36. Find $e^{t\mathbf{A}}$, given that $\mathbf{A} = \begin{bmatrix} 1 & 2 \\ 2 & 1 \end{bmatrix}$.

37. Find $e^{t\mathbf{A}}$, given that $\mathbf{A} = \begin{bmatrix} 1 & -4 \\ 1 & 1 \end{bmatrix}$.

38. Use the matrix exponential to solve the initial-value problem for the system $dx_1(t)/dt = x_2(t),\ dx_2(t)/dt = -x_1(t)$ if $x_1(0) = 1$ and $x_2(0) = -1$.

39. Use the matrix exponential to solve the initial-value problem for the system

$$dx_1(t)/dt = x_2(t) + t,\quad dx_2(t)/dt = x_1(t) + 1 \text{ if } x_1(0) = -1 \text{ and } x_2(0) = 1.$$

40. Use the Laplace transform method to find (a) $e^{t\mathbf{A}}$ given that $\mathbf{A} = \begin{bmatrix} 1 & 1 \\ -1 & 1 \end{bmatrix}$, and (b) $e^{t\mathbf{A}}$ given that $\mathbf{A} = \begin{bmatrix} 2 & 1 \\ 1 & 2 \end{bmatrix}$.

41. Find $e^{t\mathbf{A}}$, given that $\mathbf{A} = \begin{bmatrix} 1 & 1 & 0 \\ -1 & 1 & 1 \\ 0 & 2 & 1 \end{bmatrix}$.

42. Use the Laplace transform method to find $e^{t\mathbf{A}}$ given that $\mathbf{A} = \begin{bmatrix} 1 & 2 & 0 \\ 2 & 1 & 0 \\ 1 & 0 & 0 \end{bmatrix}$.

Hence find the general solution of the system $d\mathbf{x}/dt = \mathbf{A}\mathbf{x}$ if $\mathbf{x} = [x_1, x_2, x_3]^{\mathrm{T}}$.

Chapter 7
An Introduction to Vector Spaces

7.1 A Generalization of Vectors

The theory of matrices developed in Chapters 1–6, and the algebra of vectors in three dimensions used throughout calculus and physics, hereafter called *space vectors*, both belong to the part of mathematics called *linear algebra*. Each is a particular example of a linear algebra, with matrices being the more general of the two. At first sight the algebra of matrices and of space vectors appear be very different, but this is due to the use of different notations when describing vectors themselves, and the operations of vector addition and the scaling of vectors by a real number λ. General space vectors $\mathbf{r} = a\mathbf{i} + b\mathbf{j} + c\mathbf{k}$ are constructed by the scaling and addition of the unit vectors $\mathbf{i}$, $\mathbf{j}$ and $\mathbf{k}$ that are parallel to the orthogonal x, y and z-axes, and thereafter the algebra of space vectors is developed in terms of these unit vectors. However, vector $\mathbf{r}$ with its components a, b and c can equally well be defined as a three element row or column matrix, after which the linear operations of the scaling and addition of matrix vectors can be developed using the rules of matrix algebra.

The purpose of this chapter is to show how the algebra of space vectors can be regarded as a special case of matrix algebra. Then, by using the properties of matrix algebra as a model, the formal definition of a linear vector space will be developed.

Although this approach may appear to be somewhat abstract, it is this very abstraction that enables the notion of a linear vector space to find wide-ranging applications throughout mathematics, engineering and physics.

Some of the most familiar examples of space vectors occur in engineering and physics, where a vector is considered to be an entity with a magnitude, a line of action and a direction along that line in which the vector acts. Typical examples of such vectors are a force, a velocity, a momentum, an angular velocity, a magnetic field and a heat flow vector, all of which are represented by directed line segments. However, the quantities to be introduced in this chapter, also called *vectors*, generalize the familiar idea of the space vectors in three-dimensional space.

A. Jeffrey, *Matrix Operations for Engineers and Scientists*,
DOI 10.1007/978-90-481-9274-8_7, © Springer Science+Business Media B.V. 2010

In elementary calculus the vectors **u** and **v** are added using an algebraic operation denoted by the symbol $+$ to produce a unique vector sum written $\mathbf{u} + \mathbf{v}$, which is again a vector. Also, a vector **u** can be multiplied by a real number λ (scaled) to produce a unique vector $\lambda\mathbf{u}$, which is again another vector, where the juxtaposition of λ and **u** indicates the scaling of vector **u** by λ to yield the vector $\lambda\mathbf{u}$.

When more general vectors are involved, the algebraic operations of vector addition and scaling usually needs to be defined in ways that differ from those used with space vectors. The two algebraic operations of addition and scaling are used with all vectors, and they are called *binary operations*. The term *binary operation* is used to describe these operations because the addition of *two* vectors **u** and **v** produces a sum $\mathbf{w} = \mathbf{u} + \mathbf{v}$ that is also a vector, while the *two* quantities comprising a number λ and a vector **u** can be combined to produce a scaled quantity $\lambda\mathbf{u}$ which is again a vector. In the case of space vectors, *scaling* a vector by the scale factor λ means changing the "length" of the line segment that represents a vector by a factor λ, where the sense in which the vector acts is reversed when the scale factor λ is negative. The set of all real numbers λ used to scale vectors is usually denoted by the symbol $\mathbb{R}$, where $\mathbb{R}$ is said to describe the *field* over which the numbers λ are defined. This *field* contains the sum, difference, product and quotient of any two real numbers, where only division by zero is excluded. We preface what is to follow by summarizing some of the familiar ideas that will be generalized.

The set of all real numbers $\mathbb{R}$ can be displayed as points on a straight line, where each point represents a unique real number. This straight line forms a one-dimensional space that will denoted by R^1, or simply by R, where the superscript 1 indicates that the line represents a one-dimensional space, and such a line will be called an *axis*. On an axis a real number x is identified with a point at a distance proportional to x from a point on the axis called the *origin* O, which in turn corresponds to the number 0. By convention, a point x on R^1 will be taken to lie on one side of the origin when x is positive, and on the other side when x is negative. In the Euclidean geometry of three-dimensional space R^3, where the superscript 3 indicates the number of dimensions, it is convenient to work with three mutually *orthogonal* (perpendicular) axes that all pass through a common origin O. In this space a vector is represented by a straight line segment drawn from its *base*, located at the origin O of the system of axes, to its *tip* located at a given point P in space. The line segment from the base to the tip of a vector is the vector's *line of action*, and the *magnitude* of the vector is proportional to the length of this line segment. The *sense* of the vector is taken to be the *direction* along the line segment. In R^3 the line segment representing a vector usually has an arrowhead added to it to indicate the sense of the vector.

For convenience, the three axes in R^3 are taken to be orthogonal, and they are then called the ***x*, *y*** and ***z*-axes**. It is a standard convention when working with orthogonal axes in three space dimensions to orient the axes in such a way that they form a *right-handed set*. Here, a right-handed set of axes is one in which, given the x and y-axes, the positive sense along the z-axis is the direction in which a right-handed screw aligned with the z-axis will advance when rotated around the z-axis from the x-axis to the y-axis. In terms of these axes, a space vector with its base at the origin and its tip at the point P with coordinates (x_1, y_1, z_1) can also be represented in matrix form as an *ordered number triple* $[x_1, y_1, z_1]$, when matrix row notation has been used to identify the tip of the vector. The numbers x_1, y_1 and z_1, called the *components* of the vector, represent the points on the respective axes formed by the perpendicular projections of the tip P of the vector onto the corresponding x, y and z-axes, as shown in Fig. 7.1. The ordering of the components in the number triple is important, because the first component is the x-coordinate, the second is the y-coordinate and the third is the z-coordinate of the tip of the vector. So changing the order of the elements in a number triple changes the vector that is represented.

Unlike the notation used in vector analysis, in the notation used here a vector $\mathbf{r}$ in R^3 will be written as the matrix row vector $\mathbf{r} = [x_1, y_1, z_1]$. The length of the line segment representing a vector, measured from its base to its tip determines the "*strength*" of the vector, and it is a *nonnegative* scalar quantity called the *magnitude* of the vector. When working with space vectors in R^3, the magnitude of vector $\mathbf{r}$, that is the length of its line segment, is usually denoted by $|\mathbf{r}|$. However when these ideas are generalized to an n-dimensional space, hereafter denoted by R^n, it is customary to use a different notation, and to represent the magnitude of a vector $\mathbf{r}$ by $\|\mathbf{r}\|$, and to call it the *norm* of vector $\mathbf{r}$. From now on, for consistency with the notation of vector spaces to be introduced later, the symbol $|\mathbf{r}|$ will be dropped, and in its place $\|\mathbf{r}\|$ will be used to signify the norm all vectors $\mathbf{r}$, including the space vectors in R^3.

When the axes in R^3 are mutually orthogonal, the *norm* of a vector in R^3 is determined by successive applications of Pythagoras' theorem, as can be seen from Fig. 7.1. We have

$$\|\mathbf{r}\| = \left(OQ^2 + QP^2\right)^{1/2}, \tag{7.1}$$

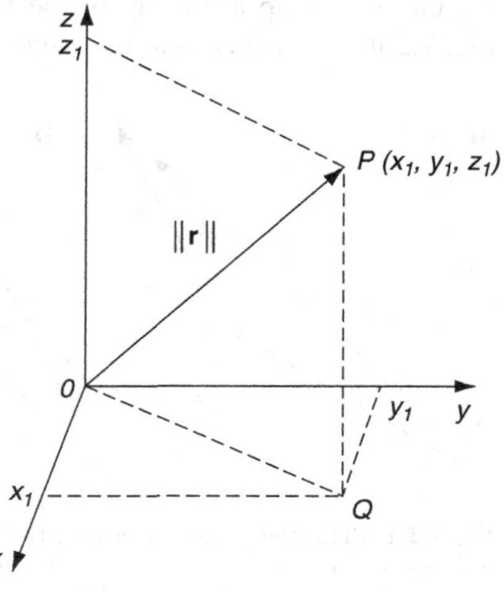

Fig. 7.1 A space vector $\mathbf{r}$ in a right-handed orthogonal system of axes in R^3

but $QP = z_1$, and $OQ^2 = x_1^2 + y_1^2$, so this becomes

$$\|\mathbf{r}\| = \left(x_1^2 + y_1^2 + z_1^2\right)^{1/2}, \tag{7.2}$$

where the positive square root is taken because, by convention, the norm of a vector is a nonnegative quantity.

When space vectors in R^3 are confined to a plane they become two-dimensional vectors, and then they are said to belong to the space R^2. We will see later that because the space R^2 is a special case of the space R^3, and it has strictly analogous algebraic properties, it is called a *subspace* of R^3.

When generalizing the concept of a vector to n dimensions, the notation $O\{x, y, z\}$ used for axes in three dimensions cannot be extended alphabetically, so instead the n axes will be denoted respectively by $x_1, x_2, \ldots, x_n$, when the system of axes will become $O\{x_1, x_2, \ldots, x_n\}$.

The *equality* of two space vectors $\mathbf{u}$ and $\mathbf{v}$, written $\mathbf{u} = \mathbf{v}$, is only possible if $\mathbf{u}$ and $\mathbf{v}$ have the same number of components, and corresponding components are equal. So in R^3, if $\mathbf{u} = [u_1, u_2, u_3]$ and $\mathbf{v} = [v_1, v_2, v_3]$, writing $\mathbf{u} = \mathbf{v}$ means that $u_1 = v_1$, $u_2 = v_2$ and $u_3 = v_3$. It is also necessary to define the vector $\mathbf{0}$ called the *null vector*, also known as the *zero vector*, as a vector in which each component is zero, so in R^3 the null vector $\mathbf{0} = [0, 0, 0]$. The null vector has neither magnitude nor direction.

The sum $\mathbf{w}$ of the space vectors $\mathbf{u}$ and $\mathbf{v}$ has for its respective components the sum of the corresponding components of vectors $\mathbf{u} = [u_1, u_2, u_3]$ and $\mathbf{v} = [v_1, v_2, v_3]$, so using the matrix row vector notation

$$\mathrm{w} = \mathbf{u} + \mathbf{v} = \left[u_1 + v_1,\; u_2 + v_2,\; u_3 + v_3\right]. \tag{7.3}$$

In geometrical terms, the addition of space vectors is performed by translating (sliding) vector $\mathbf{v}$ parallel to itself, without change of scale, until its base coincides with the tip of vector $\mathbf{u}$, when vector $\mathbf{w} = \mathbf{u} + \mathbf{v}$ is the vector with the base of $\mathbf{u}$ as its origin, and its tip at the tip of the repositioned vector $\mathbf{v}$. From the geometry in Fig. 7.2 it is it is clear that the same result follows by translating vector $\mathbf{u}$ until its

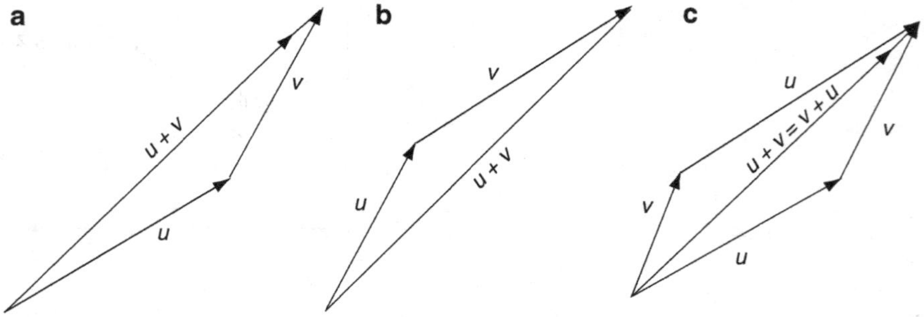

Fig. 7.2 (**a**) The triangle rule $\mathbf{w} = \mathbf{u} + \mathbf{v}$ (**b**) The triangle rule $\mathbf{w} = \mathbf{v} + \mathbf{u}$ (**c**) The parallelogram rule $\mathbf{w} = \mathbf{u} + \mathbf{v} = \mathbf{v} + \mathbf{u}$

base coincides with the tip of vector **v**, with the result that $\mathbf{w} = \mathbf{v} + \mathbf{u}$, which also follows directly from (7.3). So the addition of vectors is *commutative* because $\mathbf{u} + \mathbf{v} = \mathbf{v} + \mathbf{u}$. The geometrical description of the addition of space vectors **u** and **v** to form the third side **w** of the triangles in Figs. 7.2a and b is called the *triangle rule for vector addition*, while the diagonal of the parallelogram in Fig. 7.2c to form $\mathbf{w} = \mathbf{u} + \mathbf{v} = \mathbf{v} + \mathbf{u}$ is called the *parallelogram rule* for vector addition.

When a vector **u** is scaled by the real number λ, with the result written $\lambda\mathbf{u}$, each component of **u** is multiplied by λ, so in matrix row vector notation

$$\lambda\mathbf{u} = [\lambda u_1, \lambda u_2, \lambda u_3]. \tag{7.4}$$

Geometrically, the scaling of vector **u** by a real number λ amounts to leaving the line of action of the vector unchanged, multiplying the norm $\|\mathbf{u}\|$ of vector **u** by $|\lambda|$, and keeping the sense of the vector unchanged if $\lambda > 0$ but reversing it if $\lambda < 0$.

The abstract mathematical concept of a vector is far more general than that of a space vector in R^3. For example, a vector may be taken to be any one of a set of n element column or row matrices with real elements, or any one of a set of $m \times n$ matrices with real elements on which the usual operations of matrix addition and scaling may be performed. More generally still, other quite different mathematical objects may also be regarded as vectors as, for example, a member of a set of functions on which the mathematical operations defined as addition and scaling may be performed.

Despite the many different forms that may be taken by vectors, for the applications to space vectors in R^3 that are to follow, attention will be confined to matrix row and column vectors with real elements, when the operation of addition denoted by + will be the operation of the addition of matrices, while scaling by a real number λ will be interpreted as the scaling of matrices by the number λ. When two or three element row or column matrix vectors arise they may be interpreted, respectively, as representing the coordinates of geometrical vectors in R^2 and R^3. More generally still, n element row or column matrix vectors will be interpreted as vectors in an n-dimensional Euclidean space R^n in which n mutually orthogonal (to be defined later) axes are defined. There, by analogy with (7.2), the norm $\|\mathbf{r}\|$ of the vector $\mathbf{r} = [\alpha_1, \alpha_2, \ldots, \alpha_n]$ is defined by the expression

$$\|\mathbf{r}\| = \left(\alpha_1^2 + \alpha_2^2 + \cdots + \alpha_n^2\right)^{1/2}, \tag{7.5}$$

and called the *Euclidean norm* of the vector (see the digression on norms in Section 3.1).

7.2 Vector Spaces and a Basis for a Vector Space

We now use the familiar properties of space vectors to formulate the *axioms* that define a *general vector space*.

Definition 7.1. *The Axioms for a Real Vector Space*

A set of objects V with typical elements **u**, **v** *and* **w** *is said to form a* real vector space, *with* **u**, **v** *and* **w** *called* vectors *in the space, if they satisfy the following set of axioms, where* λ *and* μ *are arbitrary real numbers in the field* $\mathbb{R}$, *and the operations of the addition of vectors, and multiplication of a vector by a scalar are defined in an appropriate manner.*

1. If **u** *and* **v** *belong to V, then* $\mathbf{u} + \mathbf{v}$ *also belongs to V (*closure *of vectors in* V *under addition).*
2. $\mathbf{u} + \mathbf{v} = \mathbf{v} + \mathbf{u}$ *(addition of vectors in V is* commutative*).*
3. $\mathbf{u} + (\mathbf{v} + \mathbf{w}) = (\mathbf{u} + \mathbf{v}) + \mathbf{w}$ *(addition of vectors in V is* distributive*).*
4. For every **u** *in V there exists a unique vector* **0** *in V such that* $\mathbf{0} + \mathbf{u} = \mathbf{u} + \mathbf{0}$ *(there exists a unique* zero element *in V, denoted here by* **0***).*
5. Associated with each vector **u** *in V there exists a vector* $-\mathbf{u}$, *also in V, such that* $\mathbf{u} + (-\mathbf{u}) = (-\mathbf{u}) + \mathbf{u} = \mathbf{0}$ *(associated with each vector* **u** *in V there is a* negative *vector* $-\mathbf{u}$ *in V, also called the* additive inverse *of* **u** *).*
6. If vector **u** *belongs to V, so also does the vector* $\lambda\mathbf{u}$, *for any real scalar* λ *(scaling of a vector* **u** *in V by* λ *produces another vector* $\lambda\mathbf{u}$ *also in V).*
7. $\lambda(\mathbf{u} + \mathbf{v}) = \lambda\mathbf{u} + \lambda\mathbf{v}$ (scalar multiplication is distributive *over vector addition).*
8. $(\lambda + \mu)\mathbf{u} = \lambda\mathbf{u} + \mu\mathbf{u}$ (scalar multiplication of a vector is distributive*).*
9. $\lambda(\mu\mathbf{u}) = (\lambda\mu)\mathbf{u}$ (scalar multiplication is homogeneous*).*
10. $1\mathbf{u} = \mathbf{u}$ *for each vector* **u** *in V (scalar multiplication of any vector* **u** *by unity leaves* **u** *unchanged).*

It is left as a routine exercise to show space vectors in the Euclidean spaces R^2 and R^3 form vector spaces.

The concept of a real vector space is far-reaching, and there are many quite different types of vector spaces, of which some are given in the example that follows, while the subsequent example shows that not every set of vectors forms a vector space.

Example 7.1. The following are examples of vector spaces.

(a) The set V comprising all $m \times n$ matrices with real entries subject to the usual rules for matrix addition and multiplication by a scalar form a vector space. This is so because, clearly, the rules of matrix algebra satisfy the axioms of Definition 7.1.

(b) The set V of real-valued functions $\mathbf{f} = f(x)$ and $\mathbf{g} = g(x)$ defined for $-\infty < x < \infty$, with their sum and multiplication by a scalar defined in the usual way, form a vector space. To show these functions form a real vector space we start from the two obvious results $(\mathbf{f} + \mathbf{g})(x) = f(x) + g(x)$ and $(\lambda\mathbf{f})(x) = \lambda f(x)$, and then proceed to check that the axioms of Definition 7.1.1 are satisfied, while using the fact that **0**, the zero vector, is considered to be the function of x that is identically zero for all x. The details are left as an exercise.

(c) The set V of all real differentiable functions form a vector space. This follows because the sum of two differentiable functions is a differentiable function and a differentiable function scaled by a real number λ is also a differentiable function, after which the other axioms of Definition 7.1.1 are easily seen to be satisfied.

◇

Example 7.2. To see that not every set of vectors forms a vector space, it is only necessary to consider the set of all points *inside* a unit sphere V centred on the origin in R^3, subject to the usual rules for vector addition and multiplication by a scalar. Let vector **u** be any vector drawn from the origin with its tip at the point (x, y, z) inside the unit sphere V. Then vector **u** can be written $\mathbf{u} = [x, y, z]$, and if **u** is to lie inside the unit sphere it is necessary that $x^2 + y^2 + z^2 < 1$. Then, although the vector $\mathbf{u} = \left[\frac{1}{2}, 0, 0\right]$ lies inside V , the vector $3\mathbf{u} = 3 \times \left[\frac{1}{2}, 0, 0\right] = \left[\frac{3}{2}, 0, 0\right]$ lies outside V, showing that axiom 6 of Definition 7.1 is *not* satisfied, so the vectors **u** in V do not form a vector space. This is sufficient to establish that V is *not* a vector space, though this conclusion also follows by considering the vectors $\mathbf{u} = \left[\frac{1}{2},\ 0,\ 0\right]$ and $\mathbf{v} = \left[\frac{3}{2},\ 0,\ 0\right]$, both of which lie inside V, though their sum $\mathbf{u} + \mathbf{v} = [2,\ 0,\ 0]$ lies outside V, showing that axiom 1 of Definition 7.1 is also *not* satisfied. Notice that to prove a set of vectors does not form a vector space it is only necessary to show that any one of the axioms defining a vector space is not satisfied, and thereafter it is unnecessary to check which, if any, of the other axioms also fail to be satisfied.

◇

Following from Definition 7.1, we now formulate the definition of a *subspace* of a real vector space V.

Definition 7.2. *The Subspace of a Real Vector Space V*

Let V be a real vector space on which are defined the operations of vector addition and the scaling of a vector by a real number. Then a subset W of vectors V, that itself forms a vector space with the same operations of vector addition and scaling as the ones in V, is called a subspace *of V.*

For example, if V is the set of all geometrical vectors in the three-dimensional Euclidean space R^3, a subspace W of V comprises the set of all geometrical vectors in a plane in R^2. Let V be the set of all $m \times n$ matrices with real entries subject to the rules for matrix addition and multiplication by a scalar. Then a subspace W of V is the set of all $m \times n$ matrices with real entries in which each of the m elements in their first columns is equal to 0. To see why this set of vectors forms a subspace, notice that when the vectors are added or scaled by a real number λ, the result will again be a matrix of the same form in which each element in the first column is equal to zero. Conversely, consider the case where W is the set of all $m \times n$ matrices with real elements, in which each of the m elements in their first columns is equal to 1. Then, although every matrix in W belongs to V, the set of matrices in W does *not* form subspace of V, because when such matrices are added, their first columns no longer contain elements equal to 1.

A simple but important consequence of the definition of a subspace is that every vector space V has at least two subspaces, comprising the space V, which is a subspace of itself, and the *null* or *zero* subspace comprising the single vector $\mathbf{0}$.

Of fundamental importance in vector spaces is the concept of a *linear combination* of vectors. A vector $\mathbf{w}$ is said to be a *linear combination* of the m vectors $\mathbf{v}_1, \mathbf{v}_2, \ldots, \mathbf{v}_m$ belonging to a vector space V if it can be written

$$\mathbf{w} = c_1 v_1 + c_2 v_2 + \cdots + c_m \mathbf{v}_m, \tag{7.6}$$

where $c_1, c_2, \ldots, c_k$ are real numbers (scalars), not all of which are zero.

Let a set of vectors $Q = \{\mathbf{v}_1, \mathbf{v}_2, \ldots, \mathbf{v}_m\}$ belong to a vector space V, and consider the vector equation

$$c_1\mathbf{v}_1 + c_2\mathbf{v}_2 + \cdots + c_m\mathbf{v}_m = 0. \tag{7.7}$$

The set of vectors Q will be said to be *linearly independent* if (7.7) is only true when

$$c_1 = c_2 = \cdots = c_m = 0. \tag{7.8}$$

If, however, (7.7) is true when not all of the numbers $c_1, c_2, \ldots, c_m$ vanish, then the set of vectors in Q will be said to be *linearly dependent*.

For example the vectors $\mathbf{u} = [1, 2, 3]$, $\mathbf{v} = [0, 1, 2]$ and $\mathbf{w} = [1, 3, 5]$ are linearly dependent, because $\mathbf{u} + \mathbf{v} - \mathbf{w} = \mathbf{0}$, showing that in the notation of (7.7) it follows that $c_1 = 1$, $c_2 = 1$ and $c_3 = -1$. However, the vectors $\mathbf{u} = [2, 0, 0]$, $\mathbf{v} = [0, 3, 0]$ and $\mathbf{w} = [0, 0, 5]$ are linearly independent, because $c_1\mathbf{u} + c_2\mathbf{v} + c_3\mathbf{w} = \mathbf{0}$ if and only if $c_1 = c_2 = c_3 = 0$.

Let $Q = \{\mathbf{v}_1, \mathbf{v}_2, \ldots, \mathbf{v}_m\}$ be a set of m vectors belonging to a vector space V. If there are vectors in V that cannot be expressed in the form the linear combination of vectors in Q given by (7.6), these vectors must belong to a subspace W of V. When this occurs the set of vectors $Q = \{\mathbf{v}_1, \mathbf{v}_2, \ldots, \mathbf{v}_m\}$ is said to *span* the subspace W.

If, instead, *every* vector in V can be expressed as a linear combination of the m linearly independent vectors in (7.6), the set of vectors Q is said to *span* the finite-dimensional vector space V, which then has the *dimension m*. It is convenient to denote the dimension of V by dim(V) so, for example, $\dim(R^3) = 3$.

When vectors $\mathbf{v}_1, \mathbf{v}_2, \ldots, \mathbf{v}_m$ in (7.6) belonging to a vector space V are linearly independent and span the vector space V, this set of m vectors is said to form a *basis* for vector space V , which is then called an ***m*-dimensional** vector space. Clearly although a basis for a vector space V spans V, a basis is not unique. This follows because any other set of linearly independent vectors $\mathbf{w}_1, \mathbf{w}_2, \ldots, \mathbf{w}_n$, each of which is formed by linear combinations of the linearly independent vectors $\mathbf{v}_1, \mathbf{v}_2, \ldots, \mathbf{v}_n$ forming a basis for V, will itself form a different though equivalent basis for V. This simple result is often used to advantage when choosing a basis that is computationally convenient in a given situation.

A vector space V whose dimension is a finite number is called a *finite-dimensional vector space*, and only such spaces will be considered here. There are also *infinite-dimensional vector spaces*, a familiar example of which involves the Fourier series expansion of a function $f(x)$ over some interval $a \le x \le b$. In such an expansion, each of the Fourier coefficients can be regarded as a coordinate of the function $f(x)$ in an infinite-dimensional vector space, where the vectors in the space are the sine and cosine functions of multiple angles, while the Fourier coefficients are the coordinates. The analysis of an infinite-dimensional vector space is more complicated than that of a finite dimensional vector space, so this topic will only be mentioned here.

A familiar example of a basis for a finite-dimensional vector space is provided by the three unit vectors **i**, **j** and **k** used with elementary space vectors in the calculus. These can represent *every* vector **r** in the space R^3 by writing $\mathbf{r} = c_1\mathbf{i} + c_2\mathbf{j} + c_3\mathbf{k}$. Here, the numbers c_1, c_2 and c_3 are the components of vector **r**, while (c_1, c_2, c_3) specifies the coordinates of the tip of vector **r** with its base at the origin, so in the vector notation used here $\mathbf{r} = [c_1, c_2, c_3]$. This is, of course, not the only basis for R^3 that can be used, because any set of three noncoplanar vectors will serve equally well, though such a basis may not always be as convenient to use as the orthogonal system **i**, **j** and **k**.

By way of example, let us consider the real n-dimensional Euclidean vector space R^n represented by all n element vectors **w** with real elements that can be written

$$\mathbf{w} = c_1\mathbf{v}_1 + c_2\mathbf{v}_2 + \cdots + c_n\mathbf{v}_n, \tag{7.9}$$

where the vectors $\mathbf{v}_1, \mathbf{v}_2, \ldots, \mathbf{v}_n$ are linearly independent. A simple and convenient basis for this space is provided by the set of n vectors $Q = \{\mathbf{e}_1, \mathbf{e}_2, \ldots, \mathbf{e}_n\}$, where the n element vectors $\mathbf{e}_r$ with $r = 1, 2, \ldots, n$ have the form

$$\mathbf{e}_1 = [1, 0, 0, \ldots, 0],\ \mathbf{e}_2 = [0, 1, 0, \ldots, 0], \ldots,\ \mathbf{e}_n = [0, 0, 0, \ldots, 1], \tag{7.10}$$

in which $\mathbf{e}_r$ is the vector in which all elements are zero, with the exception of the rth element which is 1.

The vectors $\mathbf{e}_1, \mathbf{e}_2, \ldots, \mathbf{e}_n$, in this order, are said to form a *standard ordered basis* for R^n. Clearly these vectors satisfy the linear independence condition (7.7) and (7.8) with $m = n$, because only when $c_1 = c_2 = \cdots = c_n = 0$ will the linear combination $c_1\mathbf{e}_1 + c_2\mathbf{e}_2 + \cdots + c_n\mathbf{e}_n = \mathbf{0}$. This linear independence also follows from the determinant test for linear independence given in Theorem 2.3, because if **A** is a matrix with vectors $\mathbf{e}_1, \mathbf{e}_2, \ldots, \mathbf{e}_n$ as its rows (columns), we see that $\mathbf{A} = \mathbf{I}$, and $\det \mathbf{A} = 1 \neq 0$.

A different, but equivalent, basis for R^n is provided by the set of n vectors $S = \{\mathbf{v}_1, \mathbf{v}_2, \ldots, \mathbf{v}_n\}$, each with n components, where

$$\mathbf{v}_1 = [1,\ 0,\ 0, \ldots, 0], \mathbf{v}_2 = [1,\ 1,\ 0,\ \ldots,\ 0], \mathbf{v}_3 = [1,\ 1,\ 1, \ldots,\ 0], \ldots,$$

$$\mathbf{v}_n = [1, 1, 1, \ldots, 1], \tag{7.11}$$

where in vector $\mathbf{v}_r$, with $r = 1, 2, \ldots, n$, the first r elements are 1, and the remainder are zeros.

The linear independence of these vectors follows from the determinant test given in Theorem 2.3, because a matrix $\mathbf{A}$ with these vectors as its rows is a lower triangular matrix with 1's on its leading diagonal, so det $\mathbf{A} = 1 \neq 0$.

7.3 Changing Basis Vectors

On occasions it is necessary to represent a vector expressed in terms of one set of basis vectors in terms of a different set of basis vectors, and the way this can be done is illustrated in the next example.

Example 7.3. In terms of the standard ordered basis $Q = \{\mathbf{e}_1, \mathbf{e}_2, \ldots, \mathbf{e}_n\}$, a vector $\mathbf{r} = [1, 3, 4, 2]$. Find the form of this vector in terms of the basis $S = \{\mathbf{v}_1, \mathbf{v}_2, \mathbf{v}_3, \mathbf{v}_4\}$, where the vectors $\mathbf{v}_i$ for $i = 1, 2, 3, 4$ in R^4 are given in (7.11) for $n = 4$.

Solution. Some notation is necessary, so let $\mathbf{r_e} = \mathbf{r} = [1,\ 3,\ 4,\ 2]$ be vector $\mathbf{r}$ expressed in terms of the standard ordered basis $Q = \{\mathbf{e}_1, \mathbf{e}_2, \mathbf{e}_3, \mathbf{e}_4\}$ associated with (7.10), and let $\mathbf{r_v}$ be the vector $\mathbf{r}$ expressed in terms of the new basis involving the vectors $\mathbf{v}_i$. Then to find the new representation we must set

$$\begin{aligned}[1,3,4,2]_Q &= \alpha\mathbf{v}_1 + \beta\mathbf{v}_2 + \gamma\mathbf{v}_3 + \delta\mathbf{v}_4 \\ &= \alpha[1,0,0,0] + \beta[1,1,0,0] + \gamma[1,1,1,0] + \delta[1,1,1,1],\end{aligned}$$

and find the constants α, β, γ, δ, to obtain the representation $\mathbf{r_v} = [\alpha, \beta, \gamma, \delta]_S$. In applications, both $\mathbf{r_e}$ and $\mathbf{r_v}$ may be considered to be either matrix row or column vectors.

Equating corresponding components on each side of this equation gives

$$1 = \alpha + \beta + \gamma + \delta,\ \ 3 = \beta + \gamma + \delta,\ 4 = \gamma + \delta,\ 2 = \delta,$$

with the solution set $\alpha = -2$, $\beta = -1$, $\gamma = 2$, $\delta = 2$ so $\mathbf{r_v} = -2\mathbf{v}_1 - \mathbf{v}_2 + 2\mathbf{v}_3 + 2\mathbf{v}_4 = [-2, -1,\ 2,\ 2]_S$.

7.4 Row and Column Rank

It was stated without proof in Chapter 4, that if $\mathbf{A}$ is an arbitrary $m \times n$ matrix with real elements, then row rank($\mathbf{A}$) $=$ column rank($\mathbf{A}$). This result is sufficiently important for it to be formulated as a theorem, and then proved.

Theorem 7.1. *The Equivalence of Row and Column Ranks*

If $\mathbf{A}$ *is an* $m \times n$ *matrix, then*

$$\textit{row rank}\ (\mathbf{A}) = \textit{column rank}\ (\mathbf{A}).$$

Proof. The proof of this result involves straightforward use of the basis vectors for a space. Consider an arbitrary $m \times n$ matrix **A** with real elements a_{ij} with $i = 1, 2, \ldots,$ m and $j = 1, 2, \ldots, n$. Then each of its m rows forms an n element row vector belonging to R^n, while each of its n columns forms an m element column vector belonging to R^m.

If **A** is the zero matrix **0**, its rows and columns all contain the zeros, in which case the result is certainly true, because then row rank(**A**) = column rank(**A**) = 0. Now let us suppose row rank (**A**) = r with $0 < r \leq m$. It follows from this that r matrix row vectors $\mathbf{b}_j = [b_{j1}, b_{j2}, \ldots, b_{jn}]$, with $j = 1, 2, \ldots, m$, can be found that form a basis for the *row space* of **A**, namely the space to which all of the row vectors in **A** belong. Then each row $\mathbf{a}_i = [a_{i1}, a_{i2}, \ldots, a_{in}]$ of **A** can be expressed as the following linear combination of the basis vectors

$$\mathbf{a}_i = \lambda_{i1}\mathbf{b}_1 + \lambda_{i2}\mathbf{b}_2 + \cdots + \lambda_{ir}\mathbf{b}_r \; i = 1, 2, \ldots, m,$$

where the λ_{ij} are constants.

Equality of vectors occurs when their corresponding elements are equal, so the expression for $\mathbf{a}_i$ implies that

$$a_{ik} = \lambda_{i1}b_{1j} + \lambda_{i2}b_{2j} + \cdots + \lambda_{ir}b_{rj},$$

for

$$1 \leq i \leq m, \; 1 \leq j \leq n.$$

Consequently, the jth column vector of matrix **A** can be written in the form

$$\begin{bmatrix} a_{1j} \\ a_{2j} \\ \vdots \\ a_{mj} \end{bmatrix} = \begin{bmatrix} b_{1j}\lambda_{11} \\ b_{1j}\lambda_{21} \\ \vdots \\ b_{1j}\lambda_{m1} \end{bmatrix} + \begin{bmatrix} b_{2j}\lambda_{12} \\ b_{2j}\lambda_{22} \\ \vdots \\ b_{2j}\lambda_{m2} \end{bmatrix} + \cdots + \begin{bmatrix} b_{rj}\lambda_{1r} \\ b_{rj}\lambda_{2r} \\ \vdots \\ b_{rj}\lambda_{mr} \end{bmatrix}.$$

Denoting the jth column vector of **A** by $\mathbf{c}_j = [a_{1j}, a_{2j}, \ldots, a_{mj}]^{\mathrm{T}}$, and defining the column vector $\mathbf{l}_s = [\lambda_{1s}, \lambda_{2s}, \cdots, \lambda_{ms}]^{\mathrm{T}}$, for $s = 1, 2, \ldots, r$, allows the last result to be written in the simpler form

$$\mathbf{c}_j = b_{1j}\mathbf{l}_1 + b_{2j}\mathbf{l}_2 + \cdots + b_{rj}\mathbf{l}_r, \text{ for } j = 1, 2, \ldots, n.$$

This shows that each of the n column vectors of **A** can be expressed as a linear combination of the r column vectors $\mathbf{l}_1, \mathbf{l}_2, \ldots, \mathbf{l}_r$, that form a basis for the *column space* of A, namely the space to which all columns of **A** belong. So the dimension of the column space of **A**, although unknown at present, cannot exceed r. However, dim(row space of **A**) = r, so we have established that

$$\text{dim(column space of A)} \leq \text{dim (row space of A)} = r. \qquad \text{(I)}$$

If matrix **A** is now transposed, the result becomes an $n \times m$ matrix, in which the rows of **A** become the columns of $\mathbf{A}^{\mathrm{T}}$, while the columns of **A** become the rows of $\mathbf{A}^{\mathrm{T}}$. An application of the previous argument to $\mathbf{A}^{\mathrm{T}}$ then shows that

$$\dim\left(\text{column space of } \mathbf{A}^{\mathrm{T}}\right) \leq \dim\left(\text{row space of } \mathbf{A}^{\mathrm{T}}\right),$$

but the row space of $\mathbf{A}^{\mathrm{T}}$ = the column space of **A**, and the column space of $\mathbf{A}^{\mathrm{T}}$ = the row space of **A**, so this last result implies that

$$r = \dim(\text{row space of } \mathbf{A}) \leq \dim(\text{column space of } \mathbf{A}). \qquad \text{(II)}$$

The two inequalities (I) and (II) concerning the dimensions of spaces can only be true if

$$\dim(\text{row space of } \mathbf{A}) = \dim(\text{column space of } \mathbf{A}) = r,$$

from which it then follows that row rank(**A**) = column rank(**A**) = r, and the proof is complete.

7.5 The Inner Product

The *magnitude* of a vector in R^n has already been defined in (7.5), but now the *distance* between the tips of two vectors in the Euclidean space R^n must be defined, and both quantities related to the concept of an *inner product* of vectors.

Definition 7.3. *(The axioms of an inner product)*

Let $\mathbf{u}$, $\mathbf{v}$ *and* $\mathbf{w}$ *be any three vectors in a finite-dimensional vector space* V, *and let* k *be an arbitrary real number. Then the* inner product *of vectors* $\mathbf{u}$ *and* $\mathbf{v}$ *in the space* V, *written* $\langle \mathbf{u}, \mathbf{v} \rangle$, *is defined as a real number that satisfies the following axioms:*

P1. $\langle \mathbf{u}, \mathbf{v} \rangle = \langle \mathbf{v}, \mathbf{u} \rangle$ *(the inner product is* symmetric *with respect to* $\mathbf{u}$ *and* $\mathbf{v}$*)*
P2. $\langle \mathbf{u}+\mathbf{v}, \mathbf{w} \rangle = \langle \mathbf{u}, \mathbf{w} \rangle + \langle \mathbf{v}, \mathbf{w} \rangle$ *(the inner productive is* additive*)*
P3. $\langle k\mathbf{u}, \mathbf{v} \rangle = k\langle \mathbf{u}, \mathbf{v} \rangle$ *(the inner product is* homogeneous*)*
P4. $\langle \mathbf{u}, \mathbf{u} \rangle \geqslant 0$ with $\langle \mathbf{u}, \mathbf{u} \rangle = 0$ *if, and only if,* $\mathbf{u} = \mathbf{0}$ *(the inner product of* $\mathbf{u}$ *with itself is positive, and vanishes only when* $\mathbf{u} = \mathbf{0}$*)*

□

In the n-dimensional Euclidean space R^n an *inner product* of the vectors $\mathbf{u} = (u_1, u_2, \ldots, u_n)$ and $\mathbf{v} = (v_1, v_2, \ldots, v_n)$ is defined as the real scalar quantity

$$\langle \mathbf{u}, \mathbf{v} \rangle = u_1 v_1 + u_2 v_2 + \cdots + u_n v_n, \qquad (7.12)$$

where $\langle \mathbf{u}, \mathbf{v} \rangle$ may be positive, negative, or zero.

Notice that if **u** and **v** are treated as matrix row vectors, the operation of matrix multiplication combined with the matrix transpose operation allows the inner product $\langle \mathbf{u}, \mathbf{v} \rangle$ to be written

$$\langle \mathbf{u}, \mathbf{v} \rangle = [u_1, u_2, \ldots, u_n]\,[v_1, v_2, \ldots, v_n]^{\mathrm{T}}. \tag{7.13}$$

It is left as a routine exercise to show the definition of an inner product of matrix vectors **u** and **v** satisfies the axioms of Definition 7.3.

In terms of the inner product, two vectors **u** and **v**, neither of which is a zero vector, are said to be *orthogonal* (a generalization of the mutual perpendicularity of vectors in R^3) if

$$\langle \mathbf{u}, \mathbf{v} \rangle = 0, \tag{7.14}$$

where the justification for the term *orthogonal* will be given later.

In engineering and physics, when working with space vectors in R^2 and R^3, the inner product $\langle \mathbf{u}, \mathbf{v} \rangle$ is called the *dot product* and denoted either by $\mathbf{u} \cdot \mathbf{v}$, or sometimes by (**u**,**v**). For example, setting $\mathbf{u} = u_1\mathbf{i} + u_2\mathbf{j} + u_3\mathbf{k}$ and $\mathbf{v} = v_1\mathbf{i} + v_2\mathbf{j} + v_3\mathbf{k}$, where **i**, **j** and **k** are the usual mutually orthogonal unit vectors in the Euclidean space R^3, and using the familiar results from unit space vectors that

$$\mathbf{i}\cdot\mathbf{i} = \mathbf{j}\cdot\mathbf{j} = \mathbf{k}\cdot\mathbf{k} = 1,\ \mathbf{i}\cdot\mathbf{j} = \mathbf{j}\cdot\mathbf{i} = 0, \mathbf{i}\cdot\mathbf{k} = \mathbf{k}\cdot\mathbf{i} = 0, \mathbf{j}\cdot\mathbf{k} = \mathbf{k}\cdot\mathbf{j} = 0,$$

it follows that

$$\mathbf{u}\cdot\mathbf{v} = u_1v_1 + u_2v_2 + u_3v_3,$$

which should be compared with (7.12).

A set of basis vectors $\mathbf{v}_1, \mathbf{v}_2, \ldots, \mathbf{v}_n$ for the vector space R^n will be said to form an *orthogonal* set of vectors if

$$\langle \mathbf{u}_i, \mathbf{u}_j \rangle = \begin{cases} 0 \text{ for } i \neq j, \\ k_i \neq 0 \text{ for } i = j, \end{cases} \tag{7.15}$$

for some set of positive numbers k_i, with $i = 1, 2, \ldots, n$. Furthermore, the set of basis vectors will be said to form an *orthonormal set* of vectors if every number k_i in (7.15) is equal to unity. When expressed more concisely, a set of vectors $\mathbf{u}_1, \mathbf{u}_2, \ldots, \mathbf{u}_n$ will form an *orthonormal set* if

$$\langle \mathbf{u}_i, \mathbf{u}_j \rangle = \delta_{ij} \text{ for } i, j = 1, 2, \ldots, n, \tag{7.16}$$

where δ_{ij} is the *Kronecker delta* symbol defined as

$$\delta_{ij} = \begin{cases} 1, & i = j \\ 0, & i \neq j. \end{cases} \tag{7.17}$$

It follows directly from (7.16) that the vectors $\mathbf{e}_1, \mathbf{e}_2, \ldots, \mathbf{e}_n$ in the standard basis for R^n given in (7.10) form an *orthonormal set*, as do the unit space vectors $\mathbf{i}, \mathbf{j}$ and $\mathbf{k}$.

To define the *length* of a vector represented by an n element matrix row or column vector $\mathbf{u}$ with real elements in the space R^n, and the *distance* between the tips of two n element matrix row vectors $\mathbf{u}$ and $\mathbf{v}$, also with real elements, we proceed as follows.

As already seen in (7.5), the *norm* (*magnitude*) of the vector $\mathbf{u} = [u_1, u_2, \ldots, u_n]$ in the n-dimensional Euclidean space R^n is obtained by generalizing the magnitude of a three-dimensional Euclidean vector, by setting

$$\|\mathbf{u}\| = \left(u_1^2 + u_2^2 + \cdots + u_n^2\right)^{1/2} \geq 0. \tag{7.18}$$

It follows directly that

$$\langle \mathbf{u}, \mathbf{u} \rangle^{1/2} = \|\mathbf{u}\| = \left(u_1^2 + u_2^2 + \cdots + u_n^2\right)^{1/2} \geq 0, \tag{7.19}$$

where the *norm* of vector $\mathbf{u}$ is zero only when $\mathbf{u} = \mathbf{0}$.

An *orthogonal* set of vectors $\mathbf{u}_1, \mathbf{u}_2, \ldots, \mathbf{u}_n$ can be converted to an *orthonormal* set by dividing each vector $\mathbf{u}_i$ by its norm $\|\mathbf{u}_i\|$, because then the set of vectors

$$\mathbf{v}_i = \mathbf{u}_i / \|\mathbf{u}_i\| \text{ for } i = 1, 2, \ldots, n \tag{7.20}$$

satisfies condition (7.16).

The concept of *distance* plays an essential role in the geometry of R^n, as does *orthogonality*, so we now generalize this concept to the space R^n. Let vector $\mathbf{u} = [u_1, u_2, \ldots, u_n]$ and vector $\mathbf{v} = [v_1, v_2, \ldots, v_n]$. Then, by analogy with (7.2), the *distance* $d(\mathbf{u}, \mathbf{v})$ between the tips of the vectors $\mathbf{u}$ and $\mathbf{v}$, that is the distance between the points represented by of the vectors $\mathbf{u} = [u_1, u_2, \ldots, u_n]$ and $\mathbf{v} = [v_1, v_2, \ldots, v_n]$ in the space R^n, is defined as the nonnegative number

$$d(\mathbf{u}, \mathbf{v}) = \left((v_1 - u_1)^2 + (v_2 - u_2)^2 + \cdots + (v_n - u_n)^2\right)^{1/2} \tag{7.21}$$

When $n = 2$ or 3 result (7.21) reduces to the familiar expressions for the Euclidean distance between points in two or three space dimensions. A distance function like (7.21) in a vector space is called a *metric* for the space, so (7.21) is a metric for the space R^n. A vector space in which a metric is defined is called a *metric space*, so the Euclidean space R^n with the metric (7.21) is an example of a metric space. A metric for a vector space may take many different forms, though all metrics must satisfy the following conditions that are based on the familiar properties of distance in the two and three-dimensional Euclidean spaces R^2 and R^3.

Definition 7.4 *Properties of Length and of a Metric. Let* $\mathbf{u}$ *and* $\mathbf{v}$ *be vectors in a space U, and let k be an arbitrary real number. Then the* norm *(*length *or* magnitude*) of a vector must satisfy the following conditions:*

N1. $\|\mathbf{u}\| \geqslant 0$ *(a norm is a nonnegative scalar).*
N2. $\|\mathbf{u}\| = 0$ *if, and only if,* $\mathbf{u} = \mathbf{0}$ *(a vector has a zero norm only if the vector is* $\mathbf{0}$*).*
N3. $\|k\mathbf{u}\| = |k|\|\mathbf{u}\|$ *(when a vector* $\mathbf{u}$ *is scaled by a number k the norm of* $k\mathbf{u}$ *is scaled by the number* $|k|$*).*
N4. $\|\mathbf{u} + \mathbf{v}\| \leqslant \|\mathbf{u}\| + \|\mathbf{v}\|$ *(the* triangle inequality *for norms).*

The distance $d(\mathbf{u}, \mathbf{v})$ *between the vectors* $\mathbf{u} = [u_1, u_2, \ldots, u_n]$ *and* $\mathbf{v} = [v_1, v_2, \ldots, v_n]$ *must satisfy the following conditions:*

D1. $d(\mathbf{u}, \mathbf{v})$ 0 *(the distance between two points must be nonnegative).*
D2. $d(\mathbf{u}, \mathbf{v}) = 0$ *if and only if* $\mathbf{u} = \mathbf{v}$ *(the distance between two points is zero only when the points are coincident).*
D3. $d(\mathbf{u}, \mathbf{v}) = d(\mathbf{v}, \mathbf{u})$ *(The distance from a point* P *to a point* Q *equals the distance from point* Q *to point* P*).*
D4. $d(\mathbf{u}, \mathbf{v})$ $d(\mathbf{u}, \mathbf{w}) + d(\mathbf{w}, \mathbf{v})$ *(the* triangle inequality *for distances).*

□

The name *triangle inequality* used in N4 and D4 is derived from the familiar Euclidean result that the length of the hypotenuse of a triangle is less than or equal to the sum of the lengths of the other two sides of the triangle (see Fig. 7.2b). In R^2 and R^3, equality in D4 is only possible when the triangle degenerates in such a way that all three of its vertices A, B and C lie on a straight line, with vertex B between vertices A and C.

When $\mathbf{u}$ and $\mathbf{v}$ are vectors in R^n, the verification of conditions N1 to N3 and D1 to D3 is straightforward, so these results will be omitted. However, showing that conditions N4 and D4 are satisfied by vectors in a vector space is a little harder.

We prove only condition D4, because the proof of condition N4 proceeds along similar lines. The starting point involves proving the *Cauchy–Schwarz* inequality for the real n element matrix row vectors $\mathbf{u} = [u_1, u_2, \ldots, u_n]$ and $\mathbf{v} = [v_1, v_2, \ldots, v_n]$

.

Theorem 7.2 *The Cauchy–Schwarz Inequality. If* $\mathbf{u} = [u_1, u_2, \ldots, u_n]$ *and* $\mathbf{v} = [v_1, v_2, \ldots, v_n]$ *are real vectors, then*

$$|\langle \mathbf{u}, \mathbf{v} \rangle| \leqslant \|\mathbf{u}\|\|\mathbf{v}\|.$$

Proof. The proof of this inequality starts from the fact that a sum of the squares of real numbers is nonnegative, so as the elements of $\mathbf{u}$ and $\mathbf{v}$ are real,

$$\sum_{k=1}^{n} (\lambda u_k + v_k)^2 \geqslant 0,$$

for all real. Expanding the expression on the left and grouping terms gives

$$A\lambda^2 + 2B\lambda + C \geqslant 0,$$

where

$$A = \sum_{k=1}^{n} u_k^2, \quad B = \sum_{k=1}^{n} u_k v_k \text{ and } C = \sum_{k=1}^{n} v_k^2.$$

If $A > 0$, setting $= B/A$ reduces the original inequality to B^2 AC, where $B^2 = \langle \mathbf{u}, \mathbf{v} \rangle^2$, $A = \|\mathbf{u}\|^2$ and $C = \|\mathbf{v}\|^2$. As $\langle \mathbf{u}, \mathbf{v} \rangle$ may be negative, we will first replace $\langle \mathbf{u}, \mathbf{v} \rangle$ by $|\langle \mathbf{u}, \mathbf{v} \rangle|$ in the inequality, which is permissible because $\langle \mathbf{u}, \mathbf{v} \rangle^2 = |\langle \mathbf{u}, \mathbf{v} \rangle|^2$. The positive square root of each side of the inequality can be taken, yielding the Cauchy–Schwarz inequality. If $A = 0$, then $u_1 = u_2 = \cdots = u_n$ and the result is trivial. The Cauchy–Schwarz inequality shows the inequality sign can only be replaced by an equality sign when **u** is proportional to **v**, in which case $\mathbf{u} = k\mathbf{v}$ for some real k, so the Cauchy–Schwarz inequality is proved for all **u** and **v**.

To establish result N4 for the Euclidean metric in R^n we begin with the result

$$\begin{aligned} \|\mathbf{u} + \mathbf{v}\|^2 &= \langle \mathbf{u} + \mathbf{v}, \mathbf{u} + \mathbf{v} \rangle \\ &= \|\mathbf{u}\|^2 + 2\langle \mathbf{u}, \mathbf{v} \rangle + \|\mathbf{v}\|^2. \end{aligned}$$

The inner product $\langle \mathbf{u}, \mathbf{v} \rangle$ may be negative, so $\langle \mathbf{u}, \mathbf{v} \rangle \le |\langle \mathbf{u}, \mathbf{v} \rangle|$, and after using the Cauchy–Schwarz inequality we find that

$$\|\mathbf{u} + \mathbf{v}\|^2 \le \|\mathbf{u}\|^2 + 2\|\mathbf{u}\|\|\mathbf{v}\| + \|\mathbf{v}\|^2 = (\|\mathbf{u}\| + \|\mathbf{v}\|)^2.$$

Taking the positive square root of each side of this inequality then gives the triangle inequality, and the result is established.

7.6 The Angle Between Vectors and Orthogonal Projections

When working with geometrical vectors in the Euclidean spaces R^2 and R^3, it is a standard result that the angle θ between vectors **u** and **v** is given in terms of the scalar (dot) product,

$$\cos\theta = \frac{\mathbf{u} \cdot \mathbf{v}}{|\mathbf{u}||\mathbf{v}|}, \text{ for } 0 \le \theta \le \pi. \tag{7.22}$$

From the Cauchy–Schwarz inequality, because $\langle \mathbf{u}, \mathbf{v} \rangle$ may be positive or negative, it follows that in R^n the analogue of result (7.23) is

$$-1 \le \frac{\langle \mathbf{u}, \mathbf{v} \rangle}{\|\mathbf{u}\|\|\mathbf{v}\|} < 1, \tag{7.23}$$

where equality only occurs when $\mathbf{u} = k\mathbf{v}$, with k real. So, by analogy with the spaces R^2 and R^3, in the Euclidean space R^n an angle θ between vectors $\mathbf{u}$ and $\mathbf{v}$ can be defined by using the result

$$\cos\theta = \frac{\langle \mathbf{u}, \mathbf{v}\rangle}{\|\mathbf{u}\|\|\mathbf{v}\|}, \text{ for } 0 \le \theta \le \pi. \tag{7.24}$$

This result provides the justification for saying that vectors $\mathbf{u}$ and $\mathbf{v}$ in R^n are *orthogonal* when $\langle \mathbf{u} \cdot \mathbf{v}\rangle = 0$, because this occurs when $\cos\theta = 0$, so that $\theta = \frac{1}{2}\pi$.

Example 7.4. Find the angle between $\mathbf{u} = [1, -1, 2, 3]$ and $\mathbf{v} = [2, 0, -1, 1]$.

Solution.

$$\langle \mathbf{u}, \mathbf{v}\rangle = 3, \ \|\mathbf{u}\| = \sqrt{15}, \ \|\mathbf{v}\| = \sqrt{6},$$

hence

$$\cos\theta = 3/(\sqrt{15}\sqrt{6}) = 0.3162, \text{ so } \theta = 71.56^{\circ}.$$

◇

It is useful to relate (7.24) to the concept of the orthogonal projection of a vector $\mathbf{u}$ in the direction of a vector $\mathbf{v}$. This is best understood by first considering the two-dimensional case involving geometrical vectors, because the concept generalizes immediately to the space R^n. Figure 7.3 shows two arbitrary vectors $\mathbf{u}$ and $\mathbf{v}$, each with its base at the origin O, where the tip of vector $\mathbf{u}$ is at P, the line PQ is perpendicular to vector $\mathbf{v}$, and θ is the included angle between the vectors $\mathbf{u}$ and $\mathbf{v}$. Then the length OQ is the *orthogonal projection* of $\mathbf{u}$ in the direction of vector $\mathbf{v}$, which will be denoted by $\text{proj}_{\mathbf{v}}\ \mathbf{u}$ is $OQ = OP\cos\theta$. When working with space vectors and using the vector dot product notation, we can write

$$\text{proj}_{\mathbf{v}}\mathbf{u} = \cos\theta|\mathbf{u}| = \frac{\mathbf{u}\cdot\mathbf{v}}{|\mathbf{u}||\mathbf{v}|}|\mathbf{u}| = \frac{\mathbf{u}\cdot\mathbf{v}}{|\mathbf{v}|}. \tag{7.25}$$

Generalizing this notation to the space R^n, using inner product notation, this becomes

$$\text{proj}_{\mathbf{v}}\mathbf{u} = \frac{\langle \mathbf{u}, \mathbf{v}\rangle}{\|\mathbf{v}\|}. \tag{7.26}$$

It is important to understand that, in general, $\text{proj}_{\mathbf{v}}\ \mathbf{u} \neq \text{proj}_{\mathbf{u}}\ \mathbf{v}$.

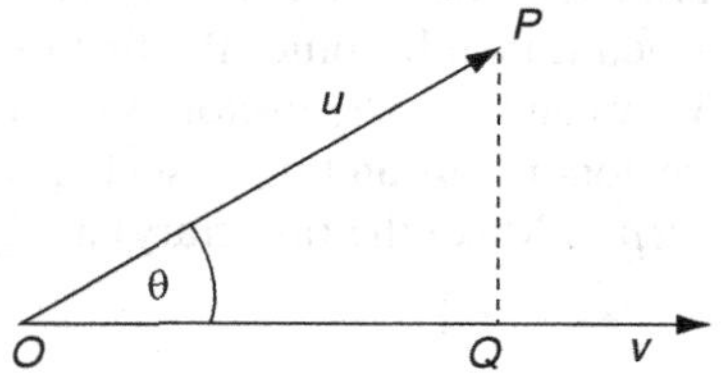

Fig. 7.3 The orthogonal projection of $\mathbf{u}$ in the direction of $\mathbf{v}$

Denoting the vector in the direction of OQ by $\mathbf{w}$, we see that

$$\mathbf{w} = (\mathrm{proj}_{\mathbf{v}}\mathbf{u})\frac{\mathbf{v}}{\|\mathbf{v}\|} = \frac{\langle \mathbf{u}, \mathbf{v}\rangle}{\langle \mathbf{v}, \mathbf{v}\rangle}\mathbf{v}, \tag{7.27}$$

where use has been made of the fact that $\mathbf{v}/\|\mathbf{v}\|$ is a vector of unit length in the direction of $\mathbf{v}$ and $\langle \mathbf{v}, \mathbf{v}\rangle = \|\mathbf{v}\|^2$.

The ease with which space vectors can be manipulated when expressed in component form using the triad of orthogonal unit vectors $\mathbf{i}$, $\mathbf{j}$ and $\mathbf{k}$, is shared by vectors in the Euclidean space R^n when they are expressed in terms of an orthonormal set of basis vectors. This raises the question of how, given an arbitrary set of n basis vectors in the vector space R^n, the set can be replaced by an equivalent orthonormal set of basis vectors.

7.7 Gram–Schmidt Orthogonalization

The method of construction we now describe is called the *Gram–Schmidt orthogonalization process*, and for simplicity the method will first be developed using vectors in the space R^3, before being generalized to vectors in the space R^n.

Letting $\mathbf{u}_1$, $\mathbf{u}_2$ and $\mathbf{u}_3$ be any three linearly independent vectors in the space R^3, we now show how they may be used to construct an equivalent set of orthogonal basis vectors $\mathbf{v}_1$, $\mathbf{v}_2$ and $\mathbf{v}_3$. Once this has been done, if required, an equivalent orthonormal set of vectors $\mathbf{w}_1$, $\mathbf{w}_2$ and $\mathbf{w}_3$ follows directly by dividing each vector by its norm, leading to the results

$$\mathbf{w}_1 = \mathbf{v}_1/\|\mathbf{v}_1\|,\ \mathbf{w}_2 = \mathbf{v}_2/\|\mathbf{v}_2\| \text{ and } \mathbf{w}_3 = \mathbf{v}_3/\|\mathbf{v}_3\|. \tag{7.28}$$

7.7.1 *The Gram–Schmidt Orthogonalization Process in R^3*

The purpose of this process is to take any three linearly independent vectors $\mathbf{u}_1$, $\mathbf{u}_2$ and $\mathbf{u}_3$ that form a basis for R^3, and to use them to construct an equivalent set of three orthogonal basis vectors $\mathbf{v}_1$, $\mathbf{v}_2$ and $\mathbf{v}_3$ in R^3.

The method of construction is straightforward, and it starts by making an arbitrary choice for $\mathbf{v}_1$ by setting it equal to one of the vectors $\mathbf{u}_1$, $\mathbf{u}_2$ and $\mathbf{u}_3$, when it then becomes the first of the three orthogonal linearly independent vectors $\mathbf{v}_1$, $\mathbf{v}_2$ and $\mathbf{v}_3$. The vectors $\mathbf{v}_2$ and $\mathbf{v}_3$ are then constructed as linear combinations of vectors $\mathbf{u}_1$, $\mathbf{u}_2$ and $\mathbf{u}_3$ in such a way that $\mathbf{v}_1$, $\mathbf{v}_2$ and $\mathbf{v}_3$ are mutually orthogonal.
Step 1. Make the (arbitrary) assignment

$$\mathbf{v}_1 = \mathbf{u}_1. \tag{7.29}$$

Step 2. Set $\mathbf{v}_2 = k_{12}\mathbf{v}_1 + \mathbf{u}_2$, and form the inner product of $\mathbf{v}_1$ and $\mathbf{v}_2$ to obtain $\langle \mathbf{v}_1, \mathbf{v}_2 \rangle = k_{12}\langle \mathbf{v}_1, \mathbf{v}_1 \rangle + \langle \mathbf{v}_1, \mathbf{u}_2 \rangle$. However, if $\mathbf{v}_1$ and $\mathbf{v}_2$ are to be orthogonal we must have $\langle \mathbf{v}_1, \mathbf{v}_2 \rangle = 0$, so $k_{12} = -\langle \mathbf{v}_1, \mathbf{u}_2 \rangle / \langle \mathbf{v}_1, \mathbf{v}_1 \rangle$, from which it follows that

$$\mathbf{v}_2 = \mathbf{u}_2 - \frac{\langle \mathbf{v}_1, \mathbf{u}_2 \rangle}{\langle \mathbf{v}_1, \mathbf{v}_1 \rangle}\mathbf{v}_1. \tag{7.30}$$

Step 3. Set $\mathbf{v}_3 = k_{13}\mathbf{v}_1 + k_{23}\mathbf{v}_2 + \mathbf{u}_3$ and form the inner product of $\mathbf{v}_3$ with $\mathbf{v}_1$. The orthogonality of $\mathbf{v}_1$, $\mathbf{v}_2$ and $\mathbf{v}_3$ means that $\langle \mathbf{v}_1, \mathbf{v}_2 \rangle = 0$ and $\langle \mathbf{v}_1, \mathbf{v}_3 \rangle = 0$, from which we see that $k_{13} = -\langle \mathbf{v}_1, \mathbf{u}_3 \rangle / \langle \mathbf{v}_1, \mathbf{v}_1 \rangle$.

Similarly, forming the inner product of $\mathbf{v}_3$ with $\mathbf{v}_2$ shows that $k_{23} = -\langle \mathbf{v}_2, \mathbf{u}_3 \rangle / \langle \mathbf{v}_2, \mathbf{v}_2 \rangle$. Finally, substituting for k_{13} and k_{23} in the expression for $\mathbf{v}_3$ we find that

$$\mathbf{v}_3 = \mathbf{u}_3 - \frac{\langle \mathbf{v}_1, \mathbf{u}_3 \rangle}{\langle \mathbf{v}_1, \mathbf{v}_1 \rangle}\mathbf{v}_1 - \frac{\langle \mathbf{v}_2, \mathbf{u}_3 \rangle}{\langle \mathbf{v}_2, \mathbf{v}_2 \rangle}\mathbf{v}_2. \tag{7.31}$$

The set of vectors $\mathbf{v}_1$, $\mathbf{v}_2$ and $\mathbf{v}_3$ in (7.29), (7.30) and (7.31), constructed from the arbitrary set of linearly independent vectors $\mathbf{u}_1$, $\mathbf{u}_2$ and $\mathbf{u}_3$, then form an orthogonal set of vectors. If an orthonormal set of vectors $\mathbf{w}_1$, $\mathbf{w}_2$ and $\mathbf{w}_3$ is required, these follow by using (7.28).

Notice that the orthogonal vectors $\mathbf{v}_1$, $\mathbf{v}_2$ and $\mathbf{v}_3$ found in this way will depend on the choice of vector used to form $\mathbf{u}_1$. Also, as any three linearly independent vectors formed by linear combinations of $\mathbf{u}_1$, $\mathbf{u}_2$ and $\mathbf{u}_3$ also forms a basis for the space R^3, it follows directly that there is no unique set of orthogonal basis vectors for R^3.

7.7.2 *The Extension of the Gram–Schmidt Orthogonalization Process to R^n*

An examination of the pattern of results (7.29) to (7.31) shows how this method of construction can be extended to the case where an orthogonal basis of n vectors $\mathbf{v}_1$, $\mathbf{v}_2$, ... , $\mathbf{v}_n$ is to be constructed from an arbitrary set of n linearly independent vectors $\mathbf{u}_1$, $\mathbf{u}_2$, ... , $\mathbf{u}_n$. Setting $\mathbf{v}_1 = \mathbf{u}_1$, and

$$\mathbf{v}_r = k_{1,r}\mathbf{v}_1 + k_{2,r}\mathbf{v}_2 + \cdots + k_{r-1,r}\mathbf{v}_{r-1} + \mathbf{u}_r, \text{ for } r = 2, 3, \ldots, n, \tag{7.32}$$

using the orthogonality of vector $\mathbf{v}_r$ with respect to the vectors $\mathbf{v}_i$, and forming the appropriate inner products, it is easily shown that the coefficients $k_{i,r}$ are given by

$$k_{i,r} = -\frac{\langle \mathbf{v}_i, \mathbf{u}_r \rangle}{\langle \mathbf{v}_i, \mathbf{v}_i \rangle} \text{ for } i = 1, 2, \ldots, r-1. \tag{7.33}$$

It is instructive to examine the geometrical interpretation of $\mathbf{v}_2$ and $\mathbf{v}_3$ in (7.30) and (7.31). Recalling the definition of the angle between two vectors given in

(7.24), it can be seen from (7.30) that $\mathbf{v}_2$ is obtained from $\mathbf{u}_2$ by subtracting from $\mathbf{u}_2$ a vector in the direction of $\mathbf{v}_1$ of magnitude equal to $\text{proj}_{\mathbf{v}_2}\mathbf{u}_2$, with a corresponding interpretation for $\mathbf{v}_3$ in (7.31).

If when using this construction only $m < n$ of the n vectors $\mathbf{u}_1, \mathbf{u}_2, \ldots, \mathbf{u}_n$ are linearly independent, the m vectors will span a subspace R^m of R^n of dimension m, with the result that the Gram–Schmidt orthogonalization process will only yield m orthogonal vectors that will together form a basis for the subspace R^m. (See Exercise 20 in Exercise Set 7.)

Example 7.5. Show the vectors $\mathbf{u}_1 = [1, -1, -1]$, $\mathbf{u}_2 = [1, -1, 1]$ and $\mathbf{u}_3 = [1, 1, -1]$ are linearly independent. Use the Gram–Schmidt orthogonalization process with these vectors to construct an orthogonal system, and hence an equivalent orthonormal system.

Solution. The vectors $\mathbf{u}_1$ to $\mathbf{u}_3$ are linearly independent because when they are arranged to form the first three rows of a third-order determinant

$$\det\mathbf{A} = \begin{vmatrix} 1 & -1 & -1 \\ 1 & -1 & 1 \\ 1 & 1 & -1 \end{vmatrix}$$

we find that $\det\mathbf{A} = 2\ 0$. Remember that the vectors will be linearly dependent if $\det\mathbf{A} = 0$.

From Step 1 we have

$$\mathbf{v}_1 = \mathbf{u}_1 = [1, -1, -1].$$

Omitting the details of the calculations involved, from Step 2 it turns out that

$$\mathbf{v}_2 = [1, -1, 1] - \tfrac{1}{3}[1, -1, -1] = \left[\tfrac{2}{3}, -\tfrac{2}{3}, \tfrac{4}{3}\right],$$

while Step 3 shows that

$$\mathbf{v}_3 = [1, 1, -1] - (-\tfrac{1}{3})[1, -1, -1] - (-\tfrac{1}{2})\left[\tfrac{2}{3}, -\tfrac{2}{3}, \tfrac{4}{3}\right] = [1, 1, 0].$$

Thus the three orthogonal vectors obtained from the Gram–Schmidt orthogonalization process are

$$\mathbf{v}_1 = [1, -1, -1], \mathbf{v}_2 = \left[\tfrac{2}{3}, -\tfrac{2}{3}, \tfrac{4}{3}\right] \text{ and } \mathbf{v}_3 = [1, 1, 0].$$

When these vectors are normalized using (7.20), the equivalent orthonormal system is found to be

$$\mathbf{w}_1 = \left[\frac{1}{\sqrt{3}}, -\frac{1}{\sqrt{3}}, -\frac{1}{\sqrt{3}}\right], \quad \mathbf{w}_2 = \left[\frac{1}{\sqrt{6}}, -\frac{1}{\sqrt{6}}, \frac{2}{\sqrt{6}}\right] \text{ and } \mathbf{w}_3 = \left[\frac{1}{\sqrt{2}}, \frac{1}{\sqrt{2}}, 0\right].$$

◇

Example 7.6. Use the orthonormal vectors $\mathbf{w}_1$, $\mathbf{w}_2$ and $\mathbf{w}_3$ in Example 7.1.5 to find the vector in R^3 represented by $\mathbf{z} = -\sqrt{3}\mathbf{w}_1 + \sqrt{6}\mathbf{w}_2 + \sqrt{2}\mathbf{w}_3$. What is the angle between $\mathbf{w}_1$ and $\mathbf{z}$?

Solution. Scaling the vectors and adding corresponding components gives $\mathbf{z} = [-1, 1, 1] + [1, -1, 2] + [1, 1, 0] = [1, 1, 3]$.

From (7.23)

$$\cos\theta = \frac{\langle \mathbf{w}_1, \mathbf{z}\rangle}{\|\mathbf{w}_1\|\|\mathbf{z}\|} = \frac{\left\langle \left(\frac{1}{\sqrt{3}}, -\frac{1}{\sqrt{3}}, -\frac{1}{\sqrt{3}}\right), (1, 1, 3)\right\rangle}{\|\mathbf{w}_1\|\|\mathbf{z}\|} = \frac{(-\sqrt{3})}{\sqrt{11}}.$$

As the numerator is negative, must lie in the second quadrant, so 121.5°.

◇

7.8 Projections

Now the Gram–Schmidt orthogonalization procedure is available, we are in a position to use it when developing the final topic in this chapter, which is how to project a vector in the space R^n onto a subspace. To understand the significance of such a projection, and why it is useful, it is only necessary to consider a practical application involving the architectural plans of a building, all of which are two-dimensional representations of a three-dimensional object. The plans all show the outline of the building when projected onto a plane perpendicular to the line of sight, corresponding to the building being viewed from different directions. Each of these diagrams (a *projection*) simplifies the task of understanding the detailed design of a building that exists in R^3, by considering those of its details that are shown when the building in R^3 is projected onto different planes in R^2, all of which are subspaces of R^3. This form of approach is also useful when applied to general mathematical results in R^n, whose meaning can be better understood by considering projections of the results in R^n onto different subspaces.

Consider the very simple situation in Fig. 7.4, where the perpendicular projection of the line OP in R^3 onto the (x_1, x_2)-plane is the line OQ that lies in a subspace R^2 of R^3. In this diagram the (x_1, x_2, x_3)-axes are the standard ordered orthogonal right-handed reference system with the associated unit vectors $\mathbf{e}_1$, $\mathbf{e}_2$ and $\mathbf{e}_3$ introduced in (7.10). In terms of this reference system, let OP to be the vector $\mathbf{p} = [a, b, c]$. Then vector $\mathbf{a} = \underline{OR} = [a, 0, 0] = a\mathbf{e}_1$ is a vector in the direction $\mathbf{e}_1$ with magnitude equal to the perpendicular projection of line OP in the direction $\mathbf{e}_1$, while $\mathbf{s} = \underline{OS} = b\mathbf{e}_2 = [0, b, 0]$ is the vector in the direction $\mathbf{e}_2$ with magnitude equal to the perpendicular projection of line OP in the direction $\mathbf{e}_2$. Vector addition now shows that vector $\mathbf{q}$ in the direction of the perpendicular projection OQ of OP with magnitude OQ is given by $\mathbf{q} = \underline{OR} + \underline{OS} = \mathbf{a} + \mathbf{s} = a\mathbf{e}_1 + b\mathbf{e}_2 = [a, 0, 0] + [0, b, 0]$ $= [a, b, 0]$.

Fig. 7.4 The perpendicular projection of OP in R^3 onto a subspace R^2 to form OQ

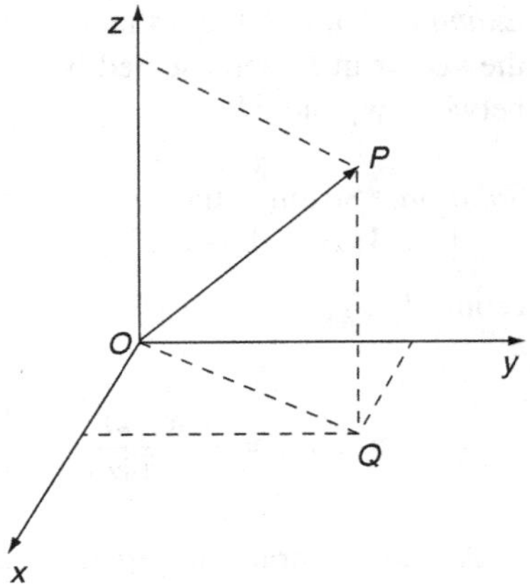

Now let us generalize the notation of a projection and its associated vector to take account of projections and their associated vectors from a space U onto a subspace W. To achieve this, we will use the notation

$$\mathrm{proj}_{V\to W(\mathbf{u})}(\mathbf{p}) = \frac{\langle \mathbf{p}, \mathbf{u} \rangle}{\|\mathbf{u}\|} \tag{7.34}$$

to represent the projection of a vector $\mathbf{p}$ in a vector space V onto a subspace W in the direction of a vector $\mathbf{u}$ in W. Naturally, as W is a subspace of V, it follows that vector $\mathbf{u}$ must also belong to V. By analogy, the vector $\mathbf{q}$ in W in the direction $\mathbf{u}$ with magnitude $\mathrm{proj}_{V\to W(\mathbf{u})}(\mathbf{r})$ is

$$\mathbf{q} = \mathrm{proj}_{V\to W(\mathbf{u})}(\mathbf{p}) \frac{\mathbf{u}}{\|\mathbf{u}\|} = \frac{\langle \mathbf{p}, \mathbf{u} \rangle}{\langle \mathbf{u}, \mathbf{u} \rangle}\mathbf{u}. \tag{7.35}$$

Let V be an n-dimensional vector space, and let the subspace W be m-dimensional, with an orthonormal basis for V provided by the set of vectors $V = \{\mathbf{v}_1, \mathbf{v}_2, \ldots, \mathbf{v}_m, \mathbf{w}_1, \mathbf{w}_2, \ldots, \mathbf{w}_{n\text{-}m}\}$, where the vectors $\mathbf{w}_1, \mathbf{w}_2, \ldots, \mathbf{w}_{n\text{–}m}$ form an orthonormal basis for the subspace W. Then, by analogy with the situation considered previously in R^2, if $\mathbf{q}$ is the vector in the direction of the projection of $\mathbf{p}$ onto W, we can write

$$\mathbf{q} = \mathrm{proj}_{V\to W(\mathbf{w}_1)}(\mathbf{p})\mathbf{w}_1 + \mathrm{proj}_{V\to W(\mathbf{w}_2)}(\mathbf{p})\mathbf{w}_2 + \cdots + \mathrm{proj}_{V\to W(\mathbf{w}_{n-m})}(\mathbf{p})\mathbf{w}_{n-m},$$

and so

$$\mathbf{q} = \langle \mathbf{p}, \mathbf{w}_1 \rangle \mathbf{w}_1 + \langle \mathbf{p}, \mathbf{w}_2 \rangle \mathbf{w}_2 + \cdots + \langle \mathbf{p}, \mathbf{w}_{n-m} \rangle \mathbf{w}_{n-m}. \tag{7.36}$$

Example 7.7. Find the vector $\mathbf{q}$ when $\mathbf{p} = [3, 1, 3]$ is projected onto the subspace W with the basis $W = \{\mathbf{b}_1, \mathbf{b}_2)$, where $\mathbf{b}_1 = [1,1,0]$ and $\mathbf{b}_2 = [1, 0, 1]$, and hence find $\|\mathbf{q}\|$.

Solution. The two vectors in W form a suitable basis for W because they are linearly independent, though they are not orthonormal. In general, the required projected vector $\mathbf{q}$ can be found by applying result (7.35) to each of the basis vectors $\mathbf{b}_1$ and $\mathbf{b}_2$, and then forming the vector sum of the results. However, inspection of vectors $\mathbf{p}$ and $\mathbf{b}_2$ shows that $\langle \mathbf{p}, \mathbf{b}_2 \rangle = 0$, so $\mathbf{p}$ is orthogonal to $\mathbf{b}_2$. Consequently, direct use cannot be made of the basis vector $\mathbf{b}_2$ in the projection process because it makes no contribution. This difficulty is easily overcome by using the Gram–Schmidt orthogonalization procedure to obtain two orthonormal vectors that are equivalent to the basis vectors. Starting with $\mathbf{v}_1 = \mathbf{b}_1$, the procedure yields the equivalent orthonormal basis vectors $\mathbf{w}_1 = \left[\frac{1}{\sqrt{2}}, -\frac{1}{\sqrt{2}}, 0\right]$ and $\mathbf{w}_2 = \left[\frac{1}{\sqrt{6}}, \frac{1}{\sqrt{6}}, -\frac{2}{\sqrt{6}}\right]$ for the subspace W, neither of which is orthogonal to $\mathbf{p}$. Using these vectors with $\mathbf{p} = [3, \ 1, \ 3]$ shows that $\langle \mathbf{p}, \mathbf{w}_1 \rangle = \sqrt{2}$ and $\langle \mathbf{p}, \mathbf{w}_2 \rangle = -2/\sqrt{6}$. Thus $\mathbf{q} = \langle \mathbf{p}, \mathbf{w}_1 \rangle \mathbf{w}_1 + \langle \mathbf{p}, \mathbf{w}_2 \rangle \mathbf{w}_2$ becomes

$$\mathbf{q} = \sqrt{2}\left[\frac{1}{\sqrt{2}}, -\frac{1}{\sqrt{2}}, 0\right] + \left(-\frac{2}{\sqrt{6}}\right)\left[\frac{1}{\sqrt{6}}, \frac{1}{\sqrt{6}}, -\frac{2}{\sqrt{6}}\right]$$

or

$$\mathbf{q} = [1, -1, 0] + \left[-\tfrac{1}{3}, -\tfrac{1}{3}, \tfrac{2}{3}\right] = \left[\tfrac{2}{3}, -\tfrac{4}{3}, \tfrac{2}{3}\right].$$

Hence the required norm of the projected vector is $\|\mathbf{q}\| = \left(\left(\tfrac{2}{3}\right)^2 + \left(-\tfrac{4}{3}\right)^2 + \left(\tfrac{2}{3}\right)^2\right)^{1/2} = \sqrt{8/3}$.

The final example illustrates the use of the projection operation to determine the projection of a finite section of a space curve onto a plane.

Example 7.8. In terms of the standard ordered basis for R^3, the position vector on a space curve C has the parametric representation $\mathbf{p} = [\cos t, \ \sin t, \ t]$ for 0 t 2. Find (a) the parametric representation of the position vector of the projection of curve C onto the (x_1, x_2)-plane, and hence determine its shape, and (b) find the parametric representation of the position vector of the projection of curve C onto the (x_2, x_3)-plane, and hence determine its shape.

Solution. The purpose of this example is to illustrate the use of the projection operation when applied to a simple space curve though, as will be seen later, in this particular case the results can be found more simply by using purely geometrical arguments. Examination of the form of the position vector $\mathbf{p}$, coupled with some elementary coordinate geometry, shows the space curve C, expressed here in parametric form, is a uniform helix about the x_3-axis. The tip of the position vector $\mathbf{p}$ on the helix is always at a unit perpendicular distance from the x_3-axis, and the helix starts at the point (1, 0, 0) and finishes at the point (1, 0, 2), having advanced uniformly in the x_3 direction through a distance after making one complete revolution around the x_3-axis.

Elementary geometrical reasoning shows that in case (a) the projection of helix C onto the (x_1, x_2)-plane must be a unit circle centred on the origin. In case (b), geometrical reasoning suggests that the projection of helix C onto the (x_2, x_3)-plane must be a plane curve with period 2 that oscillates in the plane between $x_2 = \pm 1$, while advancing uniformly in the x_3 direction. Intuition suggests this projection of the helix onto the (x_2, x_3)-plane may be a sinusoid.

Let us now give an analytical justification for these geometrical deductions. (a) Reasoning as in (7.35), let space V be the space R^3 to which $\mathbf{p}$ belongs, and let the subspace W be the (x_1, x_2)-plane in R^2. Set $\mathbf{p} = [\cos t,\ \sin t,\ t]$, and let $\mathbf{u}_1$ be a unit vector in the direction of the x_1-axis, so $\mathbf{u}_1 = [1, 0, 0]$. Then from (7.35) the vector $\mathbf{q}_1$ in the direction of $\mathbf{u}_1$ with its magnitude equal to the projection of $\mathbf{p}$ in the direction $\mathbf{u}_1$ is given by

$$\mathbf{q}_1 = \text{proj}_{V\to W(\mathbf{u}_1)}(\mathbf{p}) = \frac{\langle \mathbf{p}, \mathbf{u}_1\rangle}{\langle \mathbf{u}_1, \mathbf{u}_1\rangle}\mathbf{u}_1 = \frac{\langle [\cos t, \sin t, t], [1,0,0]\rangle}{\langle [1,0,0], [1,0,0]\rangle}[1,0,0] = [\cos t, 0, 0].$$

Similarly, let $\mathbf{u}_2$ be a unit vector in the direction of the x_2-axis, so $\mathbf{u}_2 = [0, 1, 0]$. Then from (7.35) the vector $\mathbf{q}_2$ in the direction of $\mathbf{u}_2$ with its magnitude equal to the projection of $\mathbf{p}$ in the direction of $\mathbf{u}_2$ is given by

$$\mathbf{q}_2 = \text{proj}_{V\to W(\mathbf{u}_2)}(\mathbf{p}) = \frac{\langle \mathbf{p}, \mathbf{u}_2\rangle}{\langle \mathbf{u}_2, \mathbf{u}_2\rangle}\mathbf{u}_2 = \frac{\langle [\cos t, \sin t, t], [0,1,0]\rangle}{\langle [0,1,0], [0,1,0]\rangle}[0,1,0] = [0, \sin t, 0].$$

So the parametric representation of the position vector $\mathbf{q}$ of a point on the projection of the helix onto the (x_1, x_2)-plane is

$$\mathbf{q} = \mathbf{q}_1 + \mathbf{q}_2 = [\cos t,\ \sin t,\ 0],$$

for

$$0 \leqslant t \leqslant 2\pi.$$

This is, of course, the parametric representation of a unit circle in the (x_1, x_2)-plane centred on the origin, as already deduced from purely geometrical considerations. This same result follows more simply by observing that in terms of the parametric representation of the helix, the (x_1, x_2)-plane corresponds to $t = 0$, so that $\mathbf{p} = [\cos t, \sin t, 0]$.

(b) The unit vectors along the x_2 and x_3-axes are, respectively, $\mathbf{v}_2 = [0,1,0]$, and $\mathbf{v}_3 = [0, 0, 1]$, and as before we take space V to be R^3, but this time the subspace W to be the (x_2, x_3)-plane. Then

$$\mathbf{q}_2 = \text{proj}_{V\to W(\mathbf{v}_2)}(\mathbf{p}) = \frac{\langle \mathbf{p}, \mathbf{v}_2\rangle}{\langle \mathbf{v}_2, \mathbf{v}_2\rangle}\mathbf{v}_2 = \frac{\langle [\cos t, \sin t, t], [0,1,0]\rangle}{\langle [0,1,0], [0,1,0]\rangle} = [0, \sin t, 0]$$

and

$$\mathbf{q}_3 = \text{proj}_{V\to W(\mathbf{v}_3)}(\mathbf{p}) = \frac{\langle \mathbf{p}, \mathbf{v}_3\rangle}{\langle \mathbf{v}_3, \mathbf{v}_3\rangle}\mathbf{v}_3 = \frac{\langle [\cos t, \sin t, t], [0,0,1]\rangle}{\langle [0,0,1], [0,0,1]\rangle} = [0, 0, t].$$

Hence the parametric representation of the position vector $\mathbf{q}$ of a point on the projection of the helix onto the (x_2, x_3)-plane is

$$\mathbf{q} = \mathbf{q}_2 + \mathbf{q}_3 = [0,\ \sin t,\ t],$$

for

$$0 \leqslant t \leqslant 2\pi.$$

As already conjectured from geometrical considerations, this curve is a sinusoid in the (x_2, x_3)-plane, though expressed here in parametric form. In fact this result also follows more simply from the definition of $\mathbf{p}$, because ignoring the x_1 coordinate by setting it equal to zero, we see that $\mathbf{p} = [0, \sin t, t]$.

◇

The geometrical consequence of each of the projections in Example 7.8 was easy to deduce intuitively because of the simplicity of the space curve involved, and also because in each case the projection was onto a plane on which one of the coordinate variables was constant. For example, the plane $x_3 = 0$ corresponds to the (x_1, x_2)-plane.

The purely geometrical approach used there to arrive at the form of a projection would not have been so successful had the space curve C been projected onto a general plane Π passing through the origin. This would happen, for example, when a plane Π through the origin is oriented relative to the usual x_1, x_2 and x_3-axes, so that mutually orthogonal axes x_1' and x_2' in the plane are directed, respectively, along the unit vectors $\mathbf{u}_1 = [1/\sqrt{3}, -1/\sqrt{3}, -1/\sqrt{3}]$ and $\mathbf{u}_2 = [2/\sqrt{6},\ 1/\sqrt{6},\ 1/\sqrt{6}]$. Such a projection would be difficult to visualize intuitively, because the helix would be projected onto a skew plane. Nevertheless, in such a case the approach used in Example 7.8 would proceed exactly as before, and would give the result automatically, and without difficulty.

Suffice it to say that projections from R^3 onto R^3 or onto R^2 are often needed in many practical situations. This happens, for example, when using a PC monitor to make three-dimensional plots of the surface of mathematical functions, which are best understood by viewing from different directions, and also by rotating the image on the monitor screen. In practice this is accomplished by using various different forms of readily available specialist software that is based on the projection operation.

Clearly, the analytical approach illustrated above will work in the case of the much more general situation when a projection is from R^n to R^m, with $n \geq m$, though this more general situation will not be considered here.

7.9 Some Comments on Infinite-Dimensional Vector Spaces

Although the main concern of this chapter is with finite-dimensional vector spaces, before closing the chapter, and because of their importance in applications, something should be said about the way an inner product is defined in an infinite-dimensional vector space. Suppose, for example, that a space V is a class of bounded real-valued functions defined over an interval $a \leq x \leq b$ like a special set of real polynomials, or the set of trigonometric functions $\{\sin x, \sin 2x, \ldots, 1, \cos x, \cos 2x, \ldots,\}$, defined for $-\pi \leq x \leq \pi$. In cases such as these each function can be considered to be a vector, when defining an appropriate inner product it is necessary to do so in terms of an integral. If $\mathbf{u}_i(x)$ and $\mathbf{u}_j(x)$ are any two vectors (functions) belonging to an infinite set of functions $\{\mathbf{u}_i(x)\}$, $i = 1, 2, \ldots$, defined in a vector space V, an *inner product* defined over V takes the form

$$\langle \mathbf{u}_i, \mathbf{u}_j \rangle = \int_a^b \rho(x)\mathbf{u}_i(x)\mathbf{u}_j(x)dx, \ i,j = 1, 2, \ldots, \tag{7.37}$$

where $\rho(x)$ is a nonnegative function called a *weight function*, whose form depends on the nature of the functions in V, while $\rho(x) \geq 0$ must be such that the integral (7.37) exists. In some cases the weight function $\rho(x) \equiv 1$, but in the exercises at the end of this chapter other forms of weight function occur like $\rho(x) = 1/(1-x^2)^{1/2}$ when integral (7.37) is taken over the interval $-1 \leq x \leq 1$.

Just as vectors in finite-dimensional vector spaces are orthogonal if their inner product vanishes, so also are vectors in infinite-dimensional vector spaces. The vectors $\mathbf{u}_i(x)$ and $\mathbf{u}_j(x)$ with $i \neq j$ from a set $\{\mathbf{u}_i(x)\}$, are said to be *orthogonal* over the interval $a \leq x \leq b$ with respect to the weight function $\rho(x) \geq 0$ if

$$\langle \mathbf{u}_i, \mathbf{u}_j \rangle = \int_a^b \rho(x)\mathbf{u}_i(x)\mathbf{u}_j(x)dx = 0. \tag{7.38}$$

The *norm* $\|\mathbf{u}_i(x)\|$ of a vector $\mathbf{u}_i(x)$ in an infinite-dimensional vector space V is defined as

$$\|\mathbf{u}_i\|^2 = \int_a^b \rho(x)[\mathbf{u}_i(x)]^2 dx, \ i = 1, 2, \ldots. \tag{7.39}$$

If the set of vectors $\{\mathbf{u}_i(x)\}$ forms an orthogonal set, the *normalized* vectors $\{\tilde{\mathbf{u}}_i(x)\}$, defined as

$$\tilde{\mathbf{u}}_1(x) = \mathbf{u}_i(x)/\|\mathbf{u}_i\|, i = 1, 2, \ldots, \tag{7.40}$$

are said to form an *orthonormal set*, because then

$$\langle \mathbf{u}_i, \mathbf{u}_j \rangle = \begin{cases} 0, \ i \neq j, \\ 1, \ i = j. \end{cases} \tag{7.41}$$

To show the definition in (7.37) satisfies the conditions required of an inner product it is necessary to demonstrate that it satisfies conditions P1 to P4. We have:

(i) $\langle \mathbf{u}, \mathbf{v} \rangle = \int_a^b \rho(x)\mathbf{u}(x)\mathbf{v}(x)dx = \int_a^b \rho(x)\mathbf{v}(x)\mathbf{u}(x)dx = \langle \mathbf{v}, \mathbf{u} \rangle$, so condition P1 is satisfied.

(ii) $\langle \mathbf{u} + \mathbf{v}, \mathbf{w} \rangle = \int_a^b \rho(x)(\mathbf{u}(x) + \mathbf{v}(x))\mathbf{w}(x)dx = \int_a^b \rho(x)\mathbf{u}(x)\mathbf{w}(x)dx + \int_a^b \rho(x)\mathbf{v}(x)\mathbf{w}(x)dx = \langle \mathbf{u}, \mathbf{w} \rangle + \langle \mathbf{v}, \mathbf{w} \rangle$, where $\mathbf{w}$ is any vector in V, so condition P2 is satisfied.

(iii) $\langle \lambda\mathbf{u}, \mathbf{v} \rangle = \int_a^b \lambda\rho(x)\mathbf{u}(x)\mathbf{v}(x)dx = \lambda \int_a^b \rho(x)\mathbf{u}(x)\mathbf{v}(x)dx = \lambda\langle \mathbf{u}, \mathbf{v} \rangle$, where λ is any real number, so condition P3 is satisfied.

(iv) As $\rho(x)$ is nonnegative, $\langle \mathbf{u}, \mathbf{u} \rangle = \int_a^b \rho(x)[\mathbf{u}(x)]^2 dx > 0$ if $\mathbf{u}(x) \neq 0$, and $\langle \mathbf{u}, \mathbf{u} \rangle = 0$ if, and only if, $\mathbf{u}(x) \equiv 0$ showing that condition P4 is satisfied.

◆

The exercise set at the end of this section contains examples of inner products associated with sets of orthogonal functions that arise in various applications, perhaps most frequently when solving partial differential equations.

Exercises

1. Verify that geometrical vectors in R^3 satisfy the axioms of a vector space.
2. Let V be the set of all 3×3 matrices with real elements. Does the subset W of all such matrices with zeros on their leading diagonal form a subspace of V? Give reasons for your answer.
3. Let V be the set of all 4×4 matrices with real elements. Does the subset W of all such matrices in which the first element in the leading diagonal is 1, while all other elements on the leading diagonal are zero, form a subspace of V ? Give reasons for your answer.
4. Does the set of all $m \times n$ matrices with complex entries form a (complex) vector space if the scalars λ and μ in the axioms in Definition 7.1.1 are complex numbers ? Give reasons for your answer.
5. Show that the set of all cubic polynomials $a_0 + a_1x + a_2x^2 + a_3x^3$ forms a vector space denoted by P_3.
6. Consider the vector space P_3 of cubic polynomials $a_0 + a_1x + a_2x^2 + a_3x^3$ in Exercise 5. Give two examples of classes of cubic polynomials that belong to subspaces of P_3, and one example of a class of cubic polynomials that does not belong to a subspace of P_3, and explain why this is so.
7. Let V be the set of all real-valued continuous functions of a real variable x defined over the interval $a \leq x \leq b$, where addition and scaling are defined in the usual way. Show that V is a real vector space.
8. Let W be the set of all differentiable real-valued functions of a real variable x defined over the interval $a \leq x \leq b$, with addition and scaling defined in the usual way. Is W a subspace of the vector space P_3 in Exercise 5? Give reasons for your answer.
9. Using the ordinary definitions of addition and scaling, show the set of vectors V formed by all real and continuous functions of a real variable x such that their integral over the interval $a \leq x \leq b$ exists is a real vector space.
10. In Exercise 7 let the set of functions in V be replaced by the set of all real-valued discontinuous functions defined over the interval $a \leq x \leq b$. What

restriction, if any, must be imposed on the functions in V if their integrals over $a \leq x \leq b$ are to form a vector space ?

11. A function $f(x)$ defined over an interval $a \leq x \leq b$ is said to be *convex* over that interval if, for any two points P and Q on the graph of the function $y = f(x)$, all points on the chord between P and Q lie *above* the graph. Is the space V of all convex functions over the interval $a \leq x \leq b$ a vector space under the ordinary algebraic operations of addition and scaling? Give reasons for your answer.
12. Show, subject to the usual rules for multiplication and scaling, that the set of all functions of the form $f(x) = a + b\sin 2x + c\cos 2x$, with a, b and c arbitrary real numbers and $0 \leq x \leq \pi$, form a vector space V. Does the set W of all the derivatives $f'(x)$ of the functions $f(x)$ form a subspace of V ? Give reasons for your answer.
13. From amongst the set of vectors

$$\mathbf{v}_1 = [1, -2,\ 1,\ 3],\ \mathbf{v}_2 = [2,\ 1,\ 0, -1],\ \mathbf{v}_3 = [5,\ 0,\ 1,\ 1],$$
$$\mathbf{v}_4 = [1, -1,\ 1, -1],\ \mathbf{v}_5 = [1,\ 0,\ 2,\ 1],$$

find a set that forms a basis for R^4. Is your choice of four of the vectors in this set the only ones that form a basis? If not, find a different set from amongst the vectors $\mathbf{v}_1$ to $\mathbf{v}_5$ that will also serve as a basis for R^4.

14. In the standard ordered basis for R^4 a vector $\mathbf{v} = [3, 1, 2, 0]$. Find the form of $\mathbf{v}$ in terms of the basis vectors $\{\mathbf{v}_1, \mathbf{v}_2, \mathbf{v}_3, \mathbf{v}_4\}$, given that $\mathbf{v}_1 = [1, -1, 1, -1]$, $\mathbf{v}_2 = [1, 1, 0, 0]$, $\mathbf{v}_3 = [0, 1, 0, 1]$, $\mathbf{v}_4 = [1, -1, -1, 1]$.
15. In the standard ordered basis for R^5 a vector $\mathbf{v} = [1, 3, -2, 1, 2]$. Find the form of $\mathbf{v}$ in terms of the basis vectors $\{\mathbf{v}_1, \mathbf{v}_2, \mathbf{v}_3, \mathbf{v}_4, \mathbf{v}_5\}$, given that $\mathbf{v}_1 = [1, 0, 0, 0, 1]$, $\mathbf{v}_2 = [1, 1, 0, 0, 1]$, $\mathbf{v}_3 = [0, 1, 1, 0, 1]$, $\mathbf{v}_4 = [1, 0, 1, 0, 1]$, $\mathbf{v}_5 = [0, 1, 0, 1, 1]$.
16. Two other norms are often used when working with vectors in R_n, called the **1-norm** denoted by $\|.\|_1$ and the *infinity norm* denoted by $\|.\|_\infty$, where the dot is a placeholder for the vector quantity whose norm is required. These norms are defined for a vector $\mathbf{u} = [u_1, u_2, \ldots, u_n]$ by

$$\|\mathbf{u}\|_1 = |u_1| + |u_2| + \cdots + |u_n|$$

and

$$\|\mathbf{u}\|_\infty = \max\{u_1, u_2, \cdots, u_n\}.$$

Show these definitions satisfy N1 to N4 in Definition 7.1.4.

17. Use the axioms of Definition 7.1.3 to prove the following properties of an inner product:
(i) $\langle \mathbf{0}, \mathbf{v} \rangle = \langle \mathbf{v}, \mathbf{0} \rangle = 0$,
(ii) $\langle \mathbf{u}, \mathbf{v} + \mathbf{w} \rangle = \langle \mathbf{u}, \mathbf{v} \rangle + \langle \mathbf{u}, \mathbf{w} \rangle$,
(iii) $\langle \mathbf{u}, k\mathbf{v} \rangle = k\langle \mathbf{u}, \mathbf{v} \rangle$.

18. Let $\mathbf{u}$ and $\mathbf{v}$ be two arbitrary vectors in the vector space R^n. Give a mathematical justification of the fact that, in general, $\text{proj}_{\mathbf{v}}\mathbf{u} \neq \text{proj}_{\mathbf{u}}\mathbf{v}$, and illustrate this situation graphically when $\mathbf{u}$ and $\mathbf{v}$ are vectors in R^2. For what relationship between $\mathbf{u}$ and $\mathbf{v}$, if any, can the inequality sign between the two projections be replaced by an equality sign? Verify that $\text{proj}_{\mathbf{v}}\mathbf{u} \neq \text{proj}_{\mathbf{u}}\mathbf{v}$ when $\mathbf{u} = [1, 2, 3]$ and $\mathbf{v} = [1, 2, 1]$, and find the angle between these vectors.
19. The vectors $\mathbf{u}_1 = [1, -1,1]$, $\mathbf{u}_2 = [1, -1, -1]$ and $\mathbf{u}_3 = [1, 1, -1]$ are the vectors used in Example 7.1.1 arranged in a different order. Find an equivalent orthonormal set of vectors, and hence show these vectors are not those found in the example.
20. Check that the vectors $\mathbf{u}_1 = [1, -2, 1]$, $\mathbf{u}_2 = [1, 1, 1]$, $\mathbf{u}_3 = [-1, 0\ 1]$ are linearly independent, and find an equivalent orthonormal set of vectors.
21. Check that the vectors $\mathbf{u}_1 = [1, 1, 1, 1]$, $\mathbf{u}_2 = [1, 0, 1, 1]$, $\mathbf{u}_3 = [1, -1, -1, 0]$ and $\mathbf{u}_4 = [0, 1, 1, 0]$ are linearly independent and find an equivalent orthonormal set of vectors.
22. Let $\mathbf{u}_1$ and $\mathbf{u}_2$ be any two linearly independent three element vectors, and $\mathbf{u}_3$ be such that $\mathbf{u}_3 = \alpha\mathbf{u}_1 + \beta\mathbf{u}_2$, where α and β are arbitrary real numbers, not both of which are zero. Use the Gram–Schmidt orthogonalization process together with the definition of an orthogonal projection of a vector in the direction of another vector to show the process will generate two orthogonal vectors $\mathbf{v}_1$ and $\mathbf{v}_2$, and a third null vector $\mathbf{v}_3 = \mathbf{0}$. For the case $n = 3$ this justifies the result stated previously, that if only two of the three vectors $\mathbf{u}_1$, $\mathbf{u}_2$ and $\mathbf{u}_3$ are linearly independent, the Gram–Schmidt orthogonalization process will only generate two orthogonal vectors that span the same subspace as the one spanned by the vectors $\mathbf{u}_1$ and $\mathbf{u}_2$. The result extends immediately if, for $n > 3$, only $m < n$ of the vectors $\mathbf{u}_1, \mathbf{u}_2, \ldots, \mathbf{u}_n$ are linearly independent, because then only m linearly independent vectors will be generated.
23. Find the vector $\mathbf{q}$ with magnitude and direction equal to the vector projection of $\mathbf{p} = [2,\ 1,\ 4]$ onto the two-dimensional subspace W with basis $W = \{[1,-2,\ 1], [1, -1, 1]\}$, and hence find $\|\mathbf{q}\|$.
24. Find the vector $\mathbf{q}$ with magnitude and direction equal to the vector projection of $\mathbf{p} = [1, -1, -1,\ 2]$ onto the three-dimensional subspace W with basis $W = \{[1, 0, 1, -1], [0, 1, 0, 1], [1, 0, -1, 0]\}$, and hence find $\|\mathbf{q}\|$.
25. The position vector $\mathbf{p}$ on an ellipse in the (x_1, x_2)-plane, centred on the origin, with its axis of length a along the x_1-axis and its other axis of length b along the x_2-axis, has the parametric representation $\mathbf{p} = [a \cos t, b \sin t, 0]$, with $0 \leq t \leq 2\pi$. Find the parametric representation of the projection of this ellipse onto a plane Π that contains the x_1-axis and is rotated about it until it is inclined to the (x_1, x_2)-plane at an angle α, with $-\pi/2 \leq \alpha \leq \pi/2$. Name the shape of the projected curve.
26. The parametric form of the position vector on a space curve C in R^3 is $\mathbf{p} = [a \cos t, b \sin t, t^2]$, for $0 \leq t \leq 2\pi$. Find the parametric form of the equations describing the projection of C onto (a) the (x_1, x_2)-plane and (b) the (x_2, x_3)-plane.

27. Solve Example 7.7 using the geometrical unit vectors **i**, **j** and **k** together with geometrical reasoning. Compare the effort and geometrical insight that is required with the routine approach used in Example 7.7.
28. The differential equation $y'' + n^2y = 0$ has for its solutions the infinite set of functions $\sin x$, $\sin 2x$, $\sin 3x, \ldots, 1, \cos x, \cos 2x, \cos 3x, \ldots$, corresponding to different values of n. These form an orthogonal set of functions with respect to the weight function $\rho(x) \equiv 1$ over the interval $-\pi \le x \le \pi$, with the inner product $\langle \mathbf{u}_i, \mathbf{u}_j \rangle = \int_{-\pi}^{\pi} \mathbf{u}_i(x)\mathbf{u}_j(x)dx$, where vectors $\mathbf{u}_i(x)$ and $\mathbf{u}_j(x)$ are any two vectors (functions) belonging to the set. Prove the orthogonality of these functions with respect to the weight function $\rho(x) \equiv 1$ by showing that

$$\int_{-\pi}^{\pi} \sin mx \cos nx = 0, \text{ for all } m, n \quad \int_{-\pi}^{\pi} \sin mx \sin nx dx = \begin{cases} 0, & m \ne n \\ \pi, & m = n \end{cases}$$

and

$$\int_{-\pi}^{\pi} \cos mx \cos nx dx = \begin{cases} 0, & m \ne n \\ \pi, & m = n \ne 0 \\ 2\pi, & m = n = 0. \end{cases}$$

Find the norms $\|\sin nx\|$ and $\|\cos nx\|$, and hence an equivalent orthonormal set of functions. This system of orthogonal functions is used in the development of *Fourier series*.

29. The differential equation

$$(1 - x^2)y'' - xy' + n^2y = 0,$$

with $n = 0, 1, 2, \ldots$, is called the *Chebyshev equation of order n*. For each value of n the equation has a polynomial solution of degree n that is defined over the interval $-1 \le x \le 1$, and these form an orthogonal system with respect to the weight function $\rho(x) = 1/\sqrt{1 - x^2}$. Corresponding to $n = 0, 1, 2, 3$, the first four of these polynomial solutions, called *Chebyshev polynomials*, are

$$T_0(x) = 1, \quad T_1(x) = x, \quad T_2(x) = 2x^2 - 1, \text{ and } T_3(x) = 4x^3 - 3x.$$

Use the inner product $\langle \mathbf{u}_i, \mathbf{u}_j \rangle = \int_{-1}^{1} \mathbf{u}_i(x)\mathbf{u}_j(x)/\sqrt{1 - x^2}dx$ to prove the orthogonality of the polynomials $T_i(x)$ with respect to their weight function when they are considered as vectors $\mathbf{u}_0(x)$, $\mathbf{u}_1(x)$, $\mathbf{u}_2(x)$ and $\mathbf{u}_3(x)$.

30. The differential equation

$$(1 - x^2)y'' - 2xy' + n(n + 1)y = 0,$$

with $n = 0, 1, 2, \ldots$, is called the *Legendre equation of order **n***. For each value of n the equation has a polynomial solution of degree n defined over the interval $-1 \le x \le 1$, with respect to the weight function $\rho(x) \equiv 1$. As the equation is homogeneous, each of these solutions can be scaled arbitrarily. Corresponding to $n = 0, 1, 2, 3$, these polynomial solutions can be taken to be $u_0(x) = 1$, $u_1(x) = x$, $u_2(x) = 3x^2 - 1$, and $u_3(x) = 5x^3 - 3x$.

Prove that when these functions are considered as vectors $\mathbf{u}_0(x)$, $\mathbf{u}_1(x)$, $\mathbf{u}_2(x)$ and $\mathbf{u}_3(x)$ with inner product $\langle \mathbf{u}_i, \mathbf{u}_j \rangle = \int_{-1}^{1} \mathbf{u}_i(x)\mathbf{u}_j(x)dx$, the vectors are mutually orthogonal with respect to the weight function $\rho(x) \equiv 1$. These functions are scaled by a factor p_m to form the functions $P_m(x) = p_m u_m(x)$, called *Legendre polynomials of degree **m***, where the scale factors p_m are chosen such that

$$\|p_m \mathbf{u}_m\|^2 = \frac{2}{2m+1}, \quad m = 0,\ 1,\ 2, 3\ , \ \ldots \quad .$$

Find the form of the Legendre polynomials $P_0(x)$ to $P_3(x)$.

Chapter 8
Linear Transformations and the Geometry of the Plane

8.1 Rotation of Coordinate Axes

This chapter provides an introduction to the concept of a linear transformation, with initial motivation in this section provided by considering the rotation of orthogonal coordinate systems in two and three space dimensions. A more systematic study of linear transformations will be given in Section 8.3 though there, for simplicity, the discussion will be confined to linear transformations that are of importance when studying the geometry of the Euclidean plane. Even the simple transformations considered in this chapter are useful, because their geometrical interpretations find applications in topics as diverse as elasticity, crystallography and computer graphics.

A typical geometrical example of a linear transformation in the plane is the transformation encountered in connection with Fig. 3.1 of Chapter 3. There the effect on the coordinates of a point in the plane was considered when the (x, y)-axes were subjected to a counterclockwise rotation about the origin through an angle θ.

Such a rotation was shown to transform a general point P in the (x, y)-plane into a the corresponding point in the-(x', y') plane by means of the coordinate transformation

$$x' = x\cos\theta + y\sin\theta, \quad y' = -x\sin\theta + y\cos\theta. \tag{8.1}$$

In terms of matrices, this transformation becomes

$$\mathbf{x}' = \mathbf{A}\mathbf{x}, \tag{8.2}$$

where

$$\mathbf{A} = \begin{bmatrix} \cos\theta & \sin\theta \\ -\sin\theta & \cos\theta \end{bmatrix}, \quad \mathbf{x} = \begin{bmatrix} x \\ y \end{bmatrix}, \quad \mathbf{x}' = \begin{bmatrix} x' \\ y' \end{bmatrix}. \tag{8.3}$$

The coordinates in the rotated configuration are seen to be determined by the four elements in matrix $\mathbf{A}$, so it is appropriate to call $\mathbf{A}$ a *two-dimensional rotation*

A. Jeffrey, *Matrix Operations for Engineers and Scientists*,
DOI 10.1007/978-90-481-9274-8_8,

matrix. For later use, notice that $\sin\theta = \cos(\frac{1}{2}\pi - \theta)$ and $-\sin\theta = \cos(\frac{1}{2}\pi + \theta)$, so all of the coefficients of **A** can be interpreted as cosines of angles between the respective primed and unprimed axes, with the result that **A** can also be written

$$\mathbf{A} = \begin{bmatrix} \cos\theta & \cos(\frac{1}{2}\pi - \theta) \\ \cos(\frac{1}{2}\pi + \theta) & \cos\theta \end{bmatrix}. \tag{8.4}$$

In component form, the coordinate transformation $\mathbf{x}' = \mathbf{A}\mathbf{x}$ then becomes

$$x' = a_{11}x + a_{12}y, \quad y' = a_{21}x + a_{22}y, \tag{8.5}$$

where $a_{11} = \cos\theta$, $a_{12} = \cos(\frac{1}{2}\pi - \theta)$, $a_{21} = \cos(\frac{1}{2}\pi + \theta)$ and $a_{22} = \cos\theta$. The geometry of the situation in terms of the angles θ, $\frac{1}{2}\pi - \theta$ and $\frac{1}{2}\pi + \theta$ is shown in Fig. 8.1.

Notice that when $\theta = 0$, corresponding to there being no planar rotation of axes about the origin, matrix **A** reduces to the 2×2 identity matrix **I**. Consequently, the effect of this transformation is to leave the original configuration of axes unchanged. So when $\mathbf{A} = \mathbf{I}$, it is appropriate to call the transformation $\mathbf{x}' = \mathbf{I}\mathbf{x}$ an *identity transformation*. The effect of a clockwise rotation about the origin is obtained by reversing the sign of θ.

An examination of (8.4) shows that matrix **A** describes the nature of the transformation, while the equations in (8.5) show how the transformation relates the primed and unprimed coordinate systems.

In the context of the vector space R^2 where this transformation takes place, the point (x_p, y_p) can be considered to be the tip P of a space vector **r** in the (x, y)-plane with its base at the origin, so the result of the transformation is to keep the norm of **r** unchanged (see Section 3.1) while rotating the coordinate system through an angle θ. Geometrical reasoning shows that if the coordinate system (x, y) is subjected to the successive rotations θ_1 and θ_2, the result will be the same as a single rotation through the combined angle $\theta_1 + \theta_2$. The same geometrical

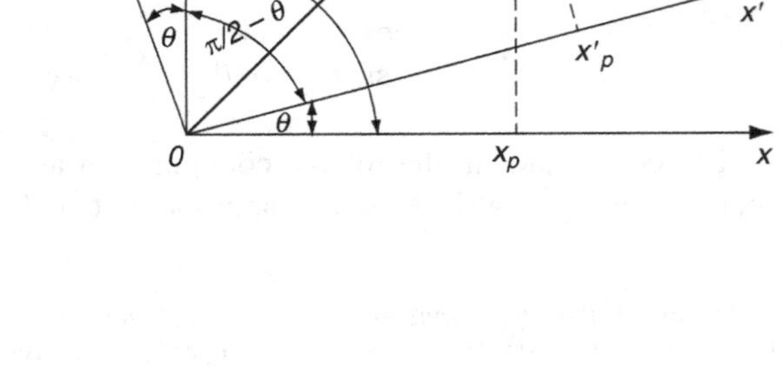

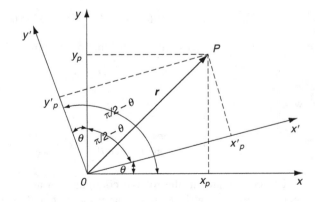

Fig. 8.1 The angles in a planar rotation of coordinates about the origin

reasoning asserts that the effect of a rotation through the angle $\lambda\theta$, with λ any number, will be the same as the effect of performing a rotation through the angle θ, and then scaling the angle of the resulting rotation by λ.

To make this argument more formal, let $\mathbf{T}(\theta)$ represent the effect on the coordinate system produced by its counterclockwise rotation about the origin through an angle θ. Now consider the two successive rotations about the origin in the plane described in matrix form by the transformations

$$\mathbf{x}' = \mathbf{A}\mathbf{x} \text{ and } \mathbf{X} = \mathbf{B}\mathbf{x}' \tag{8.6}$$

and when these rotations are performed in succession they are equivalent to

$$\mathbf{X} = \mathbf{B}\mathbf{A}\mathbf{x}, \tag{8.7}$$

where

$$\mathbf{A} = \begin{bmatrix} \cos\theta_1 & \sin\theta_1 \\ -\sin\theta_1 & \cos\theta_1 \end{bmatrix}, \quad \mathbf{B} = \begin{bmatrix} \cos\theta_2 & \sin\theta_2 \\ -\sin\theta_2 & \cos\theta_2 \end{bmatrix}, \quad \mathbf{x} = \begin{bmatrix} x \\ y \end{bmatrix}, \\ \mathbf{x}' = \begin{bmatrix} x' \\ y' \end{bmatrix}, \quad \mathbf{X} = \begin{bmatrix} X \\ Y \end{bmatrix}. \tag{8.8}$$

Evaluating the matrix product $\mathbf{BA}$, and using elementary trigonometric identities, it is easily shown that

$$\mathbf{BA} = \begin{bmatrix} \cos(\theta_1 + \theta_2) & \sin(\theta_1 + \theta_2) \\ -\sin(\theta_1 + \theta_2) & \cos(\theta_1 + \theta_2) \end{bmatrix}. \tag{8.9}$$

Recalling the interpretation of the rotation matrix in (8.4), and using the notation $\mathbf{T}(\theta)$ to indicate a counterclockwise rotation about the origin through an angle θ, this last result shows that the rotation operation $\mathbf{T}$ is *linear*, because it is equivalent to

$$\mathbf{T}(\theta_1 + \theta_2) = \mathbf{T}(\theta_1) + \mathbf{T}(\theta_2). \tag{8.10}$$

From the geometrical interpretation of $\mathbf{T}(\theta)$, it follows directly that $\mathbf{T}(\lambda\theta)$ represents a counterclockwise rotation of the coordinate system about the origin through an angle $\lambda\theta$, so in terms of this notation we have

$$\mathbf{T}(\lambda\theta) = \lambda\mathbf{T}(\theta), \tag{8.11}$$

which is another linear property of the rotation operation $\mathbf{T}$. The matrix product $\mathbf{BA}$ in (8.9) describes the nature of the successive transformations, while (8.7) describes the effect the transformation has on the respective primed and unprimed coordinate systems.

The two important properties just exhibited in (8.10) and (8.11) are not particular to this example, because it will be seen later that they are the two properties used to define a general linear transformation.

It will be useful to generalize the situation in Fig. 3.1 (equivalently Fig. 8.1) to a rotational transformation about the origin of an orthogonal system of axes in the three-dimensional Euclidean space R^3, corresponding to a transformation from R^3 to R^3. However before doing this, to permit generalization to the space R^n, we will switch to the more convenient notation used previously, where the coordinates x, y and z are replaced by x_1, x_2 and x_3.

A vector $\mathbf{r}$ is shown in Fig. 8.2 with its tip P at the point (r_1, r_2, r_3), relative to an orthogonal set of axes $O\{x_1, x_2, x_3\}$, with the unit basis vectors $\mathbf{e}_1$, $\mathbf{e}_2$ and $\mathbf{e}_3$ along the respective axes. The axes are then rotated about the origin to a new position, where they become the system $O\{x'_1, x'_2, x'_3\}$, with corresponding unit basis vectors $\mathbf{e}'_1$, $\mathbf{e}'_2$ and $\mathbf{e}'_3$ along the respective rotated coordinate axes. The position of the new coordinate system relative to the old one is determined by specifying the angles α_{11}, α_{22} and α_{33} between the corresponding primed and unprimed axes. These angles are shown in Fig. 8.2, together with some other angles that show how the

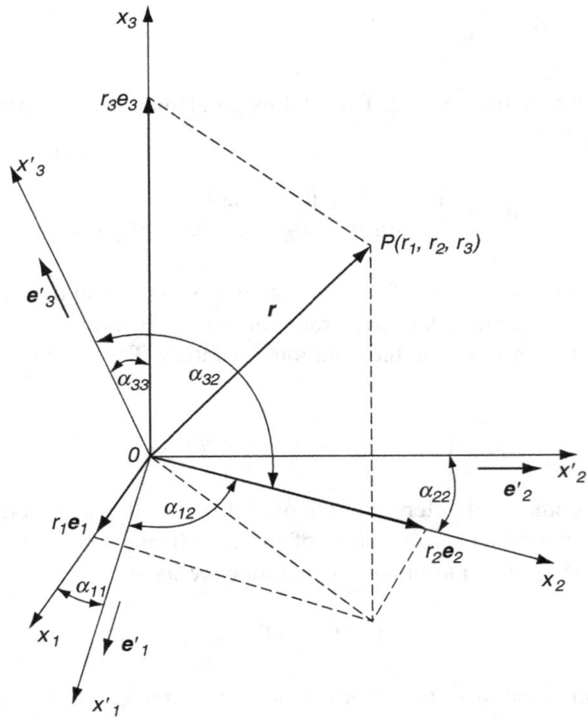

Fig. 8.2 The rotation of an orthogonal three-dimensional coordinate system $O\{x_1, x_2, x_3\}$ about the origin to form the system $O\{x'_1, x'_2, x'_3\}$

notation α_{mn} is used to signify the angle between the mth primed axis and the nth unprimed axis.

The general position vector $\mathbf{r}$ of a point in the first reference frame in terms of a matrix column vector becomes $\mathbf{r} = [r_1, r_2, r_3]^T$, or $\mathbf{r} = r_1\mathbf{e}_1 + r_2\mathbf{e}_2 + r_3\mathbf{e}_3$ in terms of the original geometrical unit vectors $\mathbf{e}_1$, $\mathbf{e}_2$ and $\mathbf{e}_3$. In the new reference frame it becomes $\mathbf{r}' = [r'_1, r'_2, r'_3]^T$, or $\mathbf{r}' = r_1\mathbf{e}'_1 + r_2\mathbf{e}'_2 + r_3\mathbf{e}'_3$ in terms of the unit vectors $\mathbf{e}'_1$, $\mathbf{e}'_2$ and $\mathbf{e}'_3$ in the new reference frame. Setting $a_{mn} = \cos\alpha_{mn}$, an examination of the Cartesian geometry involved shows, as in the two-dimensional case (8.5), that the transformation of coordinates in terms of cosines can again be described in terms of the angles α_{mn} through the coefficients a_{mn} as

$$\begin{aligned} x'_1 &= a_{11}x_1 + a_{12}x_2 + a_{13}x_3, \\ x'_2 &= a_{21}x_1 + a_{22}x_2 + a_{23}x_3, \\ x'_3 &= a_{31}x_1 + a_{32}x_2 + a_{33}x_3. \end{aligned} \tag{8.12}$$

In terms of matrices Eq. (8.12) become

$$\mathbf{x}' = \mathbf{A}\mathbf{x}, \tag{8.13}$$

with

$$\mathbf{A} = \begin{bmatrix} a_{11} & a_{12} & a_{13} \\ a_{21} & a_{22} & a_{23} \\ a_{31} & a_{32} & a_{33} \end{bmatrix}, \quad \mathbf{x} = \begin{bmatrix} x_1 \\ x_2 \\ x_3 \end{bmatrix}, \quad \mathbf{x}' = \begin{bmatrix} x'_1 \\ x'_2 \\ x'_3 \end{bmatrix}. \tag{8.14}$$

This shows that in the three-dimensional case the coordinates of $\mathbf{r}$ in the new configuration are determined by the nine coefficients (elements) a_{mn} in matrix $\mathbf{A}$, so $\mathbf{A}$ will be called a *three-dimensional rotation matrix*. It is instructive to discover the relationship between matrix $\mathbf{A}$ in (8.14), and matrix $\mathbf{A}$ in the two-dimensional case in (8.4). To simplify matters, let us consider a rotation that only takes place around the x_3-axis (that is around the old z-axis). Then, because each of the angles α_{13}, α_{23}, α_{31} and α_{32} between the primed and unprimed axes is equal to $\pi/2$, the terms a_{13}, a_{23}, a_{31} and a_{32} all vanish, while $a_{33} = 1$ because $\alpha_{33} = 0$. These results reduce the transformation matrix $\mathbf{A}$ in (8.14) to

$$\mathbf{A} = \begin{bmatrix} a_{11} & a_{12} & 0 \\ a_{21} & a_{22} & 0 \\ 0 & 0 & 1 \end{bmatrix}. \tag{8.15}$$

The presence of the element 1 in (8.15) means that $x'_3 = x_3$ (that is $z' = z$), so all points in any plane $x_3 =$ const. behave in exactly the same way as points in the (x_1, x_2)-plane when $x_3 = 0$, which is the (x, y)-plane in (8.4). So the three-dimensional case behaves like a *rigid body rotation* about its x_3-axis.

Fig. 8.3 The angles α_{mn} in terms of α_{11}

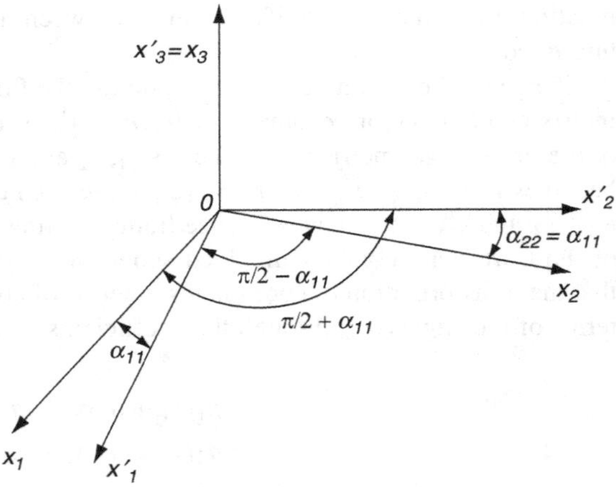

The meaning of matrix **A** in (8.15) can be clarified by considering Fig. 8.3, which shows the appropriate angles a_{mn} expressed in terms of the single rotation angle α_{11} about the x_3-axis.

An examination of Fig. 8.3 shows that $a_{11} = \cos\alpha_{11}$, $a_{12} = \cos(\frac{1}{2}\pi - \alpha_{11})$, $a_{21} = \cos(\frac{1}{2}\pi + \alpha_{11})$ and $a_{22} = \cos\alpha_{11}$, so the three-dimensional rotation matrix takes the form

$$\mathbf{A} = \begin{bmatrix} \cos\alpha_{11} & \cos(\frac{1}{2}\pi - \alpha_{11}) & 0 \\ \cos(\frac{1}{2}\pi + \alpha_{11}) & \cos\alpha_{11} & 0 \\ 0 & 0 & 1 \end{bmatrix}, \text{ when } \mathbf{x}' = \mathbf{A}\mathbf{x}. \tag{8.16}$$

Recalling that in (8.4) the angle $\theta = \alpha_{11}$, the rotation matrix **A** in (8.16) is seen to describe a rigid body rotation about the x_3-axis represented by the top 2×2 block of terms, together with an identity transformation with respect to the x_3-axis, represented by the single element 1.

By virtue of its construction, matrix **A** in (8.16), like matrix **A** in (8.4), is an *orthogonal matrix*, so $\mathbf{A}^{-1}$ always exists and is given by $\mathbf{A}^{-1} = \mathbf{A}^{\mathrm{T}}$. Consequently, $\mathbf{x} = \mathbf{A}^{-1}\mathbf{x}' = \mathbf{A}^{\mathrm{T}}\mathbf{x}'$ is the *inverse transformation* that always exists and reverses the effect of the rotation just described. Combining the original and inverse matrix transformations gives $\mathbf{x} = \mathbf{A}^{-1}\mathbf{A}\mathbf{x} = \mathbf{I}\mathbf{x} = \mathbf{x}$, as would be expected. So the effect produced by the rotation matrix **A** is reversed by an application of the inverse rotation matrix $\mathbf{A}^{-1}$, leading to a rotation matrix **I** that corresponds to the *identity transformation*, where the system is left unchanged after the successive transformations have been performed.

Similar reasoning shows that when the rotation represented by matrix **A** in (8.13) is about only the x_1-axis, or about the x_2-axis, the same linearity properties found in (8.10) and (8.11) again apply. For example, to show what happens when successive rotations occur about the x_3-axis, irrespective of the order in which they are performed, it will suffice for us to establish result (8.9) in terms of two rotations about the x_3-axis, one through an angle α_{11}, and another through an angle $\tilde{\alpha}_{11}$. The corresponding rotation matrices are

$$\mathbf{A} = \begin{bmatrix} \cos\alpha_{11} & \cos(\frac{1}{2}\pi - \alpha_{11}) & 0 \\ \cos(\frac{1}{2}\pi + \alpha_{11}) & \cos\alpha_{11} & 0 \\ 0 & 0 & 1 \end{bmatrix}$$

and

$$\mathbf{B} = \begin{bmatrix} \cos\tilde{\alpha}_{11} & \cos(\frac{1}{2}\pi - \tilde{\alpha}_{11}) & 0 \\ \cos(\frac{1}{2}\pi + \tilde{\alpha}_{11}) & \cos\tilde{\alpha}_{11} & 0 \\ 0 & 0 & 1 \end{bmatrix}.$$

Forming the matrix products **BA** and **AB**, and simplifying the result, gives

$$\mathbf{BA} = \mathbf{AB} = \begin{bmatrix} \cos(\alpha_{11} + \tilde{\alpha}_{11}) & \cos(\frac{1}{2}\pi - \alpha_{11} - \tilde{\alpha}_{11}) & 0 \\ \cos(\frac{1}{2}\pi + \alpha_{11} + \tilde{\alpha}_{11}) & \cos(\alpha_{11} + \tilde{\alpha}_{11}) & 0 \\ 0 & 0 & 1 \end{bmatrix},$$

confirming that property (8.10) holds with respect to these rotations about the x_3-axis independently of the order in which they occur. As would be expected, displacements along the x_3-axis are not affected by these rigid body rotations. Notice that property (8.11) is also true for the three-dimensional case for the same reason it is true for the two-dimensional case.

8.2 Linear Transformations

It is now necessary to give a formal definition of a linear transformation, though before doing so attention must be drawn to the fact that linear transformations do not necessarily have simple geometrical interpretations, because they are often between very general vector spaces.

Definition 8.1. *A Linear Transformation*

Let $\mathbf{x}_1$, $\mathbf{x}_2$ and $\mathbf{x}$ *be any vectors in a vector space X, and let* λ *be a scalar in the field of real numbers* $\mathcal{R}$. *Then a linear transformation* $\mathbf{T}$ *is a transformation between a vector space X and a vector space Y such that to each vector* $\mathbf{x}$ *in X there corresponds a unique vector* $\mathbf{y} = \mathbf{T}(\mathbf{x})$ *in Y and, in addition, the transformation* $\mathbf{T}$ *has the following fundamental properties*

$$\mathbf{T}(\mathbf{x}_1 + \mathbf{x}_2) = \mathbf{T}(\mathbf{x}_1) + \mathbf{T}(x_2) \ (\textit{linearity permits additivity})$$

and

$$\mathbf{T}(\lambda\mathbf{x}) = \lambda\mathbf{T}(\mathbf{x}). \ (\textit{linearity permits scaling}).$$

◇

Notice from this definition that a *general linear transformation*, denoted here by $\mathbf{T}$, is a transformation between two vector spaces X and Y that do not necessarily have the same dimension.

When examining the properties of coordinate rotations in Section 8.1, it was established that each possessed the additivity and scaling properties required of a linear transformation. Consequently, as both two and three-dimensional coordinate rotations satisfy the conditions of Definition 8.1, each is an example of a linear transformation.

Before considering an example of a general linear transformation with no particular geometrical significance, we will first make a direct application of Definition 8.1 to the projection operation defined in Chapter 7.

Example 8.1. Show that the projection operation defined in (7.35) of Chapter 7 is a linear transformation.

Solution. The projection operation involving the projection of a vector $\mathbf{p}$ in a vector space V onto a vector $\mathbf{u}$ in a subspace W of V was defined in (7.35) as

$$\mathrm{proj}_{V\to W(\mathbf{u})}(\mathbf{p})\frac{\mathbf{u}}{\|\mathbf{u}\|}=\frac{\langle\mathbf{p},\mathbf{u}\rangle}{\langle\mathbf{u},\mathbf{u}\rangle}\mathbf{u}. \tag{8.17}$$

Denoting the projection operation in (8.17) by $\mathbf{T(p)}$, and setting $p=\mathbf{p}_1+\mathbf{p}_2$, it follows directly from the properties of the inner product $\langle\mathbf{p},\mathbf{u}\rangle$ in $\mathbf{T(p)}$ that

$$\mathbf{T}(\mathbf{p}_1+\mathbf{p}_2)=\mathbf{T}(\mathbf{p}_1)+\mathbf{T}(\mathbf{p}_2), \tag{8.18}$$

while replacing $\mathbf{p}$ by $\lambda\mathbf{p}$, with λ a scalar, it also follows that

$$\mathbf{T}(\lambda\mathbf{p})=\lambda\mathbf{T}(\mathbf{p}). \tag{8.19}$$

Results (8.18) and (8.19) show the projection operation satisfies the two key properties of linearity required by Definition 8.1, so the projection operation is another example of a linear transformation.

The next Example of a linear transformation is of a more general nature, without the geometrical interpretation that was possible in the case of coordinate rotations and the projection operation.

◊

Example 8.2. The transformation $\mathbf{T}$ between a two-dimensional vector space X, containing vector $\mathbf{x}=[x_1,x_2]^{\mathrm{T}}$, and a three-dimensional vector space Y, containing the vector $\mathbf{y}$ corresponding to $\mathbf{x}$, is given by $\mathbf{y}=\mathbf{T}(\mathbf{x})=[x_1,\ 2x_1-x_2,\ x_1+2x_2]^{\mathrm{T}}$. Show how $\mathbf{T}$ transforms the vectors $\mathbf{x}_1=[1,-1]^{\mathrm{T}}$, $\mathbf{x}_2=[4,\ 3]^{\mathrm{T}}$, and $\mathbf{x}_3=\mathbf{0}=[0,\ 0]^{\mathrm{T}}$ in R^2 into vectors in R^3, and prove that $\mathbf{T}$ is a linear transformation.

Solution. First it is necessary to explain the notation that is used. It means that column vector $\mathbf{T}(\mathbf{x}) = [x_1,\ 2x_1 - x_2,\ x_1 + 2x_2]^{\mathrm{T}}$, with the three elements $x_1,\ 2x_1 - x_2$ and $x_1 + 2x_2$, is to be interpreted as a column vector in R^3 obtained by combining the two elements x_1 and x_2 of the vector $\mathbf{x} = [x_1, x_2]^{\mathrm{T}}$ in R^2. The first element of $\mathbf{T}(\mathbf{x}) = \mathbf{y}$ is the element x_1 of $\mathbf{x}$, the second element, namely $2x_1 - x_2$, is formed from the two elements x_1 and x_2 of $\mathbf{x}$, while the third element of, namely $x_1 + 2x_2$, is also formed from the two elements of $\mathbf{x}$.

We now show how $\mathbf{T}$ transforms the vectors $\mathbf{x}_1 = [1, -1]^{\mathrm{T}}$, $\boldsymbol{x}_2 = [4,\ 3]^{\mathrm{T}}$, and $\mathbf{x}_3 = \mathbf{0} = [0,\ 0]^{\mathrm{T}}$ in R^2 into vectors $\mathbf{T}(\mathbf{x}) = \mathbf{y}$ in R^3. The substitutions $x_1 = 1$ and $x_2 = -1$, show that $\mathbf{y}_1 = \mathbf{T}(\mathbf{x}_1) = [1,\ 3, -1]^{\mathrm{T}}$, while the substitutions $x_1 = 4$ and $x_2 = 3$ show that $\mathbf{y}_2 = \mathbf{T}(\boldsymbol{x}_2) = [4, 5, 10]^{\mathrm{T}}$. Similarly, the substitutions $x_1 = 0$ and $x_2 = 0$ show that $\mathbf{y}_3 = \mathbf{T}(\mathbf{x}_3) = \mathbf{T}(\mathbf{0}) = [0,\ 0,\ 0]^{\mathrm{T}}$, illustrating the fact that the zero vector $\mathbf{0}$ in R^2 is transformed into the zero vector $\mathbf{0}$ in R^3.

To prove that $\mathbf{T}$ is a linear transformation, we must show it possesses the two key properties of additivity and scaling in Definition 8.1. Let $\mathbf{x}_1 = [\xi_1,\ \xi_2]^{\mathrm{T}}$ and $\boldsymbol{x}_2 = [\eta_1,\ \eta_2]^{\mathrm{T}}$, then $\mathbf{x}_1 + \boldsymbol{x}_2 = [\xi_1 + \eta_1,\ \xi_2 + \eta_2]^{\mathrm{T}}$, so the first component of $\mathbf{x}_1 + \boldsymbol{x}_2$ is $\xi_1 + \eta_1$, while the second component is $\xi_2 + \eta_2$. Forming $\mathbf{T}(\mathbf{x}_1 + \boldsymbol{x}_2)$, separating out terms corresponding to $\mathbf{T}(\mathbf{x}_1)$ and $\mathbf{T}(\boldsymbol{x}_2)$, and using the property of vector addition, we find that

$$\begin{aligned}\mathbf{T}(\mathbf{x}_1 + \mathbf{x}_2) &= [\xi_1 + \eta_1, 2\xi_1 + 2\eta_1 - \xi_2 - \eta_2, \xi_1 + \eta_1 + 2\xi_2 + \eta_2]^{\mathrm{T}}\\ &\quad\times [\xi_1, 2\xi_1 - \xi_2, \xi_1 + 2\xi_2]^{\mathrm{T}} + [\eta_1, 2\eta_1 - \eta_2, \eta_1 + 2\eta_2]^{\mathrm{T}}\\ &= \mathbf{T}(\mathbf{x}_1) + \mathbf{T}(\mathbf{x}_2).\end{aligned}$$

Next, let $\mathbf{x}_1 = [\xi_1,\ \xi_2]^{\mathrm{T}}$ be an arbitrary vector, then $\lambda\mathbf{x}_1 = [\lambda\xi_1,\ \lambda\xi_1]^{\mathrm{T}}$. Forming $\mathbf{T}(\lambda\mathbf{x}_1)$, and using the scaling property of vectors, we find that $\mathbf{T}(\lambda\mathbf{x}_1) = \lambda\mathbf{T}(\mathbf{x}_1)$. Thus $\mathbf{T}$ possesses the properties of *additivity* and *scaling*, so $\mathbf{T}$ it is a linear transformation.

◊

A linear transformation is a special and very important example of a *mapping* between two vector spaces. Such mappings, or transformations, establish a procedure that assigns to every vector $\mathbf{x}$ in space X a vector $\mathbf{y}$ in space Y called the *image* of $\mathbf{x}$. To clarify the relationship between $\mathbf{x}$ and $\mathbf{y}$, when referring back to the space X from space Y, the vector $\mathbf{x}$ is called the *pre-image* of vector $\mathbf{y}$. The space X is called the *domain* of the mapping $\mathbf{T}$, and the space of vector images in Y corresponding to the mapping $\mathbf{y} = \mathbf{T}(\mathbf{x})$ of vectors $\mathbf{x}$ in X is called the *range* of the mapping $\mathbf{T}$, denoted by $\mathcal{R}(\mathbf{T})$. The range $\mathcal{R}(\mathbf{T})$ of $\mathbf{T}$ need not necessarily contain every vector in space Y, and this general situation is represented symbolically in Fig. 8.4a, where $\mathbf{T}$ maps space X onto only a part of space Y. However, when the mapping $\mathbf{T}$ is such that $\mathcal{R}(\mathbf{T})$ and Y coincide, $\mathbf{T}$ is said to map X *onto* Y and this situation is represented symbolically in Fig. 8.4b.

It is usual to adopt the standard notation when referring to transformations between vector spaces, and to do this the transformation of vector $\mathbf{x}$ in R^n to a

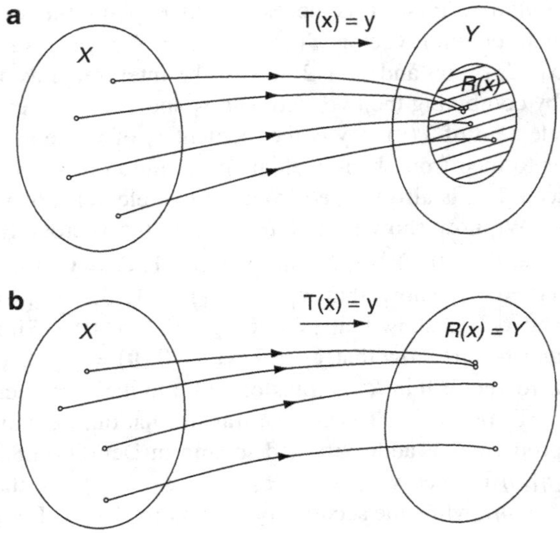

Fig. 8.4 (**a**) Mapping by **T** of vectors in space X onto the range $\mathcal{R}(\mathbf{T})$ that lies strictly within space Y. (**b**) Mapping by **T** of vectors in space X *onto* the range $\mathcal{R}(\mathbf{T})$, when $\mathcal{R}(\mathbf{T})$ and space Y are identical

vector $\mathbf{y}$ in R^m is denoted by writing $\mathbf{T}(\mathbf{x})$: $R^n \to R^m$. It is useful to display this transformation more clearly by writing it as

$$\mathbf{T}(\mathbf{x}) = \mathbf{T}\left(\underbrace{\begin{bmatrix} x_1 \\ x_2 \\ \vdots \\ x_n \end{bmatrix}}_{n \text{ elements}}\right) = \underbrace{\begin{bmatrix} y_1 \\ y_2 \\ \vdots \\ y_m \end{bmatrix}}_{m \text{ elements}}. \tag{8.20}$$

The effect of transformation **T** on the components of the vectors **x** and **y** depends on both **T** and the choice of the bases $\mathcal{X}$ and $\mathcal{Y}$; so it must be remembered that neither of these bases need necessarily be the standard ordered bases used above. Should it become necessary to emphasize this dependence on different bases this can be shown by writing $\mathbf{T}_{\mathcal{X}\mathcal{Y}}$. However, in the basic geometrical applications that are to follow, the transformations **T** will all be particularly simple, because they will be from R^2 to R^2, and the bases $\mathcal{X}$ and $\mathcal{Y}\mathcal{Y}$ used will each be the standard ordered basis for R^2. So in this case the bases $\mathcal{X} = \mathcal{Y} = \{\mathbf{e}_1, \mathbf{e}_2\}$ of the spaces X and Y coincide. Because of this simplification, the notation $\mathbf{T}_{\mathcal{X}\mathcal{X}}$ can be abbreviated to **T**.

When developing the linear transformations that describe coordinate rotations, matrices entered in a natural way. We now show that the use of matrices is not restricted to these examples, because a general linear transformations **T** from R^n to R^m can always be interpreted as the product of a suitable $m \times n$ matrix and an $n \times 1$ matrix column vector.

As usual, in what follows a vector $\mathbf{x}$ in R^n will be written as the n element matrix column vector $\mathbf{x} = [x_1, x_2, \ldots, x_n]^{\mathrm{T}}$, while a vector $\mathbf{y}$ in R^m will be written as the m element matrix column vector $\mathbf{y} = [y_1, y_2, \ldots, y_m]^{\mathrm{T}}$, where for simplicity in what follows the basis for R^n will be the usual standard ordered basis

$$\left.\mathbf{e}_1 = \begin{bmatrix} 1 \\ 0 \\ \vdots \\ 0 \end{bmatrix}, \mathbf{e}_2 = \begin{bmatrix} 0 \\ 1 \\ \vdots \\ 0 \end{bmatrix}, \ldots, \mathbf{e}_n = \begin{bmatrix} 0 \\ 0 \\ \vdots \\ 1 \end{bmatrix}\right\} \text{ all with } n \text{ elements,} \tag{8.21}$$

though this basis could be replaced by any equivalent basis. Similarly, for the basis $\mathcal{Y}$ of the space R^m to which $\mathbf{y}$ belongs, for simplicity we will again take the set of m element column matrices $\mathcal{Y} = \{\mathbf{e}'_1, \mathbf{e}'_2, \ldots, \mathbf{e}'_m\}$ that form the standard ordered basis for R^m, though this could also be replaced by any equivalent basis.

Any column vector $\mathbf{y}$ in R^m can be represented as a linear combination of the vectors

$$\left.\mathbf{e}'_1 = \begin{bmatrix} 1 \\ 0 \\ \vdots \\ 0 \end{bmatrix}, \mathbf{e}'_2 = \begin{bmatrix} 0 \\ 1 \\ \vdots \\ 0 \end{bmatrix}, \ldots, \mathbf{e}'_m = \begin{bmatrix} 0 \\ 0 \\ \vdots \\ 1 \end{bmatrix}\right\} \text{ all with } m \text{ elements,} \tag{8.22}$$

so the matrix column vector $\mathbf{y} = [y_1, y_2, \ldots, y_m]^{\mathrm{T}}$ can be written

$$\mathbf{y} = y_1\mathbf{e}'_1 + y_2\mathbf{e}'_2 + \cdots + y_m\mathbf{e}'_m. \tag{8.23}$$

An application of the linear transformation $\mathbf{T}$ to (8.21), will lead to the introduction of the transformed basis vectors $\mathbf{T}(\mathbf{e}_1)$, $\mathbf{T}(\mathbf{e}_2)$,..., $\mathbf{T}(\mathbf{e}_n)$, each with m elements. These transformed basis vectors in (8.21) will be written

$$\begin{gathered} \mathbf{T}(\mathbf{e}_1) = \begin{bmatrix} a_{11} \\ a_{21} \\ \vdots \\ a_{m1} \end{bmatrix}, \ \mathbf{T}(\mathbf{e}_2) = \begin{bmatrix} a_{12} \\ a_{22} \\ \vdots \\ a_{m2} \end{bmatrix}, \ \ldots, \\ \left.\mathbf{T}(\mathbf{e}_n) = \begin{bmatrix} a_{1n} \\ a_{2n} \\ \vdots \\ a_{mn} \end{bmatrix}\right\} \text{ all with } m \text{ elements.} \end{gathered} \tag{8.24}$$

For each i, the m components of the vector $\mathbf{T}(\mathbf{e}_i)$ are obtained by setting $\mathbf{x} = \mathbf{e}_i$ in $\mathbf{T}(\mathbf{x})$, where for the standard ordered basis this corresponds to setting $x_1 = 0$,

$x_2 = 0, \ldots, x_{i-1} = 0, x_i = 1, x_{i+1} = 0, \ldots, x_{n-1} = 0,\ x_n = 0$ in $\mathbf{T}(\mathbf{x})$. As a result, $\mathbf{y}$ can be written in terms of the $m \times n$ matrix

$$\mathbf{A} = [\mathbf{T}(\mathbf{e}_1),\ \mathbf{T}(\mathbf{e}_2),\ \ldots\ ,\ \mathbf{T}(\mathbf{e}_n)], \tag{8.25}$$

with its columns the m element column vectors $\mathbf{T}(\mathbf{e}_i)$. As a result, the m element column vector $\mathbf{y}$ is given by the matrix product

$$\mathbf{y} = \mathbf{A}\mathbf{x}. \tag{8.26}$$

Matrix $\mathbf{A}$ provides a unique representation of $\mathbf{T}(\mathbf{x})$, because the coefficients a_{ij} in $\mathbf{A}$ are uniquely determined once the basis for R^n has been chosen, and the nature of the linear transformation $\mathbf{T}$ has been specified. So we have succeeded in showing that a linear transformation from R^n to R^m can be represented as the product of an $m \times n$ matrix and an n element matrix column vector. The matrix $\mathbf{A}$, based on the use of the standard ordered basis in (8.21), is called the *standard matrix representation* of the linear transformation $\mathbf{T}$.

The advantage of matrix representations of linear transformations is that they allow linear transformations to be combined in a simple manner. For example, if $\mathbf{y} = \mathbf{A}\mathbf{x}$ and $\mathbf{x} = \mathbf{B}\mathbf{z}$ are general linear transformations, and not necessarily coordinate rotations, the linear transformation from $\mathbf{y}$ to $\mathbf{z}$ is given by $\mathbf{y} = \mathbf{A}\mathbf{B}\mathbf{z}$, without the necessity of first finding the transformation from $\mathbf{x}$ and $\mathbf{y}$, and then the effect of the transformation from $\mathbf{y}$ to $\mathbf{z}$, This process of successively combining transformations is called the *composition* of transformations.

Result (8.26) represents the fundamental way in which linear transformations between the spaces R^n and R^m can be represented, so this fundamental result will be stated as the following theorem.

Theorem 8.1 *Matrix representation of the transformation $T(x)$: $R^n \to R^m$*

All transformations $\mathbf{T}(\mathbf{x})$: $R^n \to R^m$, *where* $\mathbf{T}(\mathbf{x}) = \mathbf{y}$, *can be represented in the form of a matrix transformation* $\mathbf{T}(\mathbf{x}) = \mathbf{A}\mathbf{x} = \mathbf{y}$, *where* $\mathbf{A}$ *is an* $m \times n$ *matrix,* $\mathbf{x}$ *is an* $n \times 1$ *matrix column vector and* $\mathbf{y}$ *is an* $m \times 1$ *matrix column vector. If, instead of the standard ordered basis for* R^n, *an arbitrary ordered basis* $\mathcal{B}$ *is used, where the basis* $\mathcal{B} = \{\mathbf{b}_1,\ \mathbf{b}_2,\ \ldots, \mathbf{b}_n\}$, *then the* ith *column of* $\mathbf{A}$ *becomes* $\mathbf{T}(\mathbf{b}_i)$, *with* $i = 1, 2, \ldots, \mathrm{n}$.

◆

Of special interest are transformations from a vector space U onto a vector space V with the property that each vector $\mathbf{x}$ in U is mapped onto a unique vector $\mathbf{y}$ in V and, conversely, each vector $\mathbf{y}$ is the image of only one vector $\mathbf{x}$. These are called *one-one* transformations, or *mappings*, and this property is often shown by saying a transformation is **1:1** or *one-to-one*.

Definition 8.2. *One-One Transformations*

A linear transformation $\mathbf{T}(\mathbf{x})$: $U \to V$ *is said to be* one-one *if for any two vectors* $\mathbf{x}$ *and* $\mathbf{y}$ *in* U, $\mathbf{x} \neq \mathbf{y}$ *implies* $\mathbf{T}(\mathbf{x}) \neq \mathbf{T}(\mathbf{y})$ *or, equivalently, the linear transformation is* one-one *if* $\mathbf{T}(\mathbf{x}) = \mathbf{T}(\mathbf{y})$ *implies that* $\mathbf{x} = \mathbf{y}$.

□

In what follows, interest will be confined to the case when the vector spaces U and V in Definition 8.2 are both the space R^n, so the transformation is of the form $\mathbf{T}(\mathbf{x})$: $R^n \to R^n$. In this case the transformation converts a vector $\mathbf{x}$ in the space R^n into a vector $\mathbf{y}$ also in R^n. Theorem 8.1 allows such a transformation to be represented by the matrix equation $\mathbf{Ax} = \mathbf{y}$, where $\mathbf{A}$ is an $n \times n$ matrix, with $\mathbf{x}$ and $\mathbf{y}$ $n \times 1$ column vectors. A unique vector $\mathbf{x}$ in R^n will determine a unique vector $\mathbf{y}$ that is also in R^n, but for such a transformation to be one-one it is necessary that the pre-image of any vector $\mathbf{y}$ must determine a unique vector $\mathbf{x}$. Matrix $\mathbf{A}$ will determine a unique vector $\mathbf{y}$ through the matrix equation $\mathbf{Ax} = \mathbf{y}$, and provided $\mathbf{A}$ is nonsingular the transformation will have a unique inverse $\mathbf{A}^{-1}$, in which case vector $\mathbf{y}$ will have a unique pre-image $\mathbf{x}$ given $\mathbf{x} = \mathbf{A}^{-1}\mathbf{y}$. So, in this case, the condition for the transformation to be one-one is simply that the $n \times n$ matrix $\mathbf{A}$ is nonsingular.

Two examples of this type that have already been encountered arose when two and three-dimensional rotation matrices were introduced. Each rotation matrix was an orthogonal matrix, but when $\mathbf{A}$ is orthogonal $\mathbf{A}^{\mathrm{T}} = \mathbf{A}^{-1}$, so that both of these transformations are one-one. In summary, if $\mathbf{Ax} = \mathbf{y}$, the *inverse transformation* is given by $\mathbf{x} = \mathbf{A}^{-1}\mathbf{y}$, provided $\det\mathbf{A} \neq 0$.

When a transformation is between spaces of different dimensions it is harder to decide if the transformation is one-one. However, when the spaces are each of low dimension, a direct approach is all that is necessary to establish the explicit relationships between the vector $\mathbf{x}$, its image $\mathbf{y}$ and, when it exists, the unique pre-image of $\mathbf{y}$.

Example 8.3. Determine the nature of the transformation $\mathbf{Ax} = \mathbf{y}$ from R^3 to R^2, where

$$\mathbf{A} = \begin{bmatrix} 2 & -1 & 2 \\ 1 & -2 & -3 \end{bmatrix},$$

when $\mathbf{x} = [x_1, x_2, x_3]^{\mathrm{T}}$ and $\mathbf{y} = [y_1, y_2]^{\mathrm{T}}$.

Solution. As $\mathbf{A}$ is a 2×3 matrix it has no inverse, so the transformation cannot be one-one. To examine the situation more closely, notice that any vector $\boldsymbol{x}$ in R^3 will always determine a vector $\mathbf{y}$ in R^2, but to determine when an arbitrary vector $\mathbf{y} = [y_1, y_2]^{\mathrm{T}}$ in R^2 has a pre-image in R^3, and if that pre-image is unique, requires a careful examination of the transformation, which when expanded becomes

$$2x_1 - x_2 + 2x_3 = y_1, \quad \text{and} \quad x_1 - x_2 - 3x_3 = y_2.$$

Only two equations for the three variables x_1 to x_3 are involved, so solving for x_1 and x_2 in terms of x_3, y_1 and y_2 as parameters gives

$$x_1 = -5x_3 + y_1 - y_2, \quad x_2 = -8x_3 + y_1 - 2y_2.$$

This shows that for any fixed vector $\mathbf{y}$, the components x_1 and x_2 of vector $\mathbf{x}$ also depend on x_3, which is arbitrary. So, as vector $\mathbf{x}$ is not determined uniquely in terms of the vector $\mathbf{y}$, the transformation is *not* one-one.

◊

Example 8.4. A transformation from R^3 to R^4 is represented by the matrix equation $\mathbf{Ax} = \mathbf{y}$, where

$$\mathbf{A} = \begin{bmatrix} 1 & 2 & 1 \\ 1 & -1 & 2 \\ 1 & 1 & -3 \\ 2 & 0 & -1 \end{bmatrix}, \text{ where } \mathbf{x} = \begin{bmatrix} x_1 \\ x_2 \\ x_3 \end{bmatrix} \text{ and } \mathbf{y} = \begin{bmatrix} y_1 \\ y_2 \\ y_3 \\ y_4 \end{bmatrix}.$$

Show that not every vector $\mathbf{y}$ in R^4 is the image of a vector $\mathbf{x}$ in R^3, and determine the relationship between the components of $\mathbf{y}$ in order that it is the image of a vector $\mathbf{x}$ in R^3.

Solution. An arbitrary vector $\mathbf{x} = [x_1, x_2, x_3]^{\mathrm{T}}$ in R^3 will always give rise to a vector $\mathbf{y} = [y_1, y_2, y_3, y_4]^{\mathrm{T}}$ in R^4, but the system $\mathbf{Ax} = \mathbf{y}$ represents four equations in the three unknowns x_1, x_2 and x_3, and so is overdetermined. Thus, in general, an arbitrary vector $\mathbf{y}$ in R^4 will *not* be the image of a vector $\mathbf{x}$ in R^3.

It will be recalled that when examining system of equations of the form $\mathbf{Ax} = \mathbf{b}$ in Chapter 4, it was found that for a solution to exist it is necessary that $\text{rank}(\mathbf{A}) = \text{rank}(\mathbf{A}|\mathbf{b})$, where $\mathbf{A}|\mathbf{b}$ is the augmented matrix. Thus for a solution $\mathbf{y}$ to exist for the given $\mathbf{A}$ it is necessary that $\text{rank}(\mathbf{A}|\mathbf{y}) = \text{rank}(\mathbf{A})$. A simple calculation shows the reduced echelon form $\mathbf{A_E}$ of $\mathbf{A}$ is

$$\mathbf{A_E} = \begin{bmatrix} 1 & 0 & 0 \\ 0 & 1 & 0 \\ 0 & 0 & 1 \\ 0 & 0 & 0 \end{bmatrix},$$

showing that $\text{rank}(\mathbf{A}) = 3$. Consequently the necessary condition for $\mathbf{y}$ to be the image of a vector $\mathbf{x}$ in R^3 is that $\text{rank}(\mathbf{A}|\mathbf{y}) = 3$. To proceed further we write out in full the matrix equation $\mathbf{Ax} = \mathbf{y}$,

$$\begin{aligned} x_1 + 2x_2 + x_3 &= y_1, \\ x_1 - x_2 + 2x_3 &= y_2, \\ x_1 + x_2 - 3x_3 &= y_3, \\ 2x_1 - x_3 &= y_4. \end{aligned}$$

The reduced echelon form of $\mathbf{A}$ shows the fourth row of $\mathbf{A}$ is linearly dependent on the first three rows, and it also tells us how to construct vectors $\mathbf{y}$ in order that each such vector is the image of a vector $\mathbf{x}$. It can be seen from the structure of $\mathbf{A_E}$

that while the first three components y_1, y_2 and y_3 can be assigned arbitrarily, the fourth component y_4 must be compatible with fourth equation in the system of equations $\mathbf{Ax} = \mathbf{y}$, so it is necessary that $y_4 = 2x_1 - x_3$.
Solving the first three equations for x_1 to x_3 gives

$$x_1 = (y_1 + 7y_2 + 5y_3)/13, \quad x_2 = (5y_1 - 4y_2 - y_3)/13,$$
$$x_3 = (2y_1 + y_2 - 3y_3)/13,$$

from which it follows that for y_4 to be compatible we must set $y_4 = y_2 + y_3$ corresponding to $y_4 = (7y_1 - 3y_2 - 4y_3)/13$.

So the way to construct a vector $\mathbf{y}$ that is the image of a vector $\mathbf{x}$ is seen to require two steps. The first involves assigning y_1 to y_3 arbitrarily, and then using these values to solve the first three equations for x_1 to x_3. The second step, having found x_1 to x_3, is to use the fourth equation to find y_4 from the result $y_4 = y_2 + y_3$.

◇

Example 8.5. Let $\mathbf{T}(\mathbf{x})$: $R^2 \to R^4$ be a linear transformation, with $\mathbf{T}(\mathbf{x})$ defined as $\mathbf{T}(x_1, x_2) = [x_1 + x_2, x_1 - x_2, 3x_1 + x_2, x_2]$. Find the matrix representation of $\mathbf{T}$ using the standard ordered basis $\chi = \{[0, 1]^{\mathrm{T}}, [1, 0]^{\mathrm{T}}\}$ for R^2, and hence find how the vector $\mathbf{x} = [2, 3]^{\mathrm{T}}$ is transformed.

Solution. Setting $x_1 = 1, x_2 = 0$, we find that $\mathbf{T}(1, 0) = [1, 1, 3, 0]^{\mathrm{T}}$, while setting $x_1 = 0$, $x_2 = 1$, we find that $\mathbf{T}(0, 1) = [1, -1, 1, 1]^{\mathrm{T}}$, so in terms of the chosen basis vectors the matrix representation of $\mathbf{T}$ determined by Theorem 8.1 becomes

$$\mathbf{A} = \begin{bmatrix} 1 & 1 \\ 1 & -1 \\ 3 & 1 \\ 0 & 1 \end{bmatrix}.$$

Substituting $\mathbf{x} = [2, 3]^{\mathrm{T}}$ in $\mathbf{y} = \mathbf{T}(\mathbf{x}) = \mathbf{Ax}^{\mathrm{T}}$, shows that $\mathbf{y} = \mathbf{Ax}^{\mathrm{T}} = [5, -1, 9, 3]^{\mathrm{T}}$.

◇

Example 8.6. Using the standard ordered basis, find the standard matrix representation for the linear transformation $\mathbf{T}(\mathbf{x})$: $R^3 \to R^4$, given that

$$\mathbf{T}(x_1, x_2, x_3) = [x_1 + 2x_3, x_1 - x_2, x_2 + x_3, x_3].$$

Solution. Proceeding as in Example 8.5, and using the standard basis for R^3, we have

$$\mathbf{A} = \begin{bmatrix} 1 & 0 & 2 \\ 1 & -1 & 0 \\ 0 & 1 & 1 \\ 0 & 0 & 1 \end{bmatrix}.$$

This is easily checked, because $\mathbf{Ax} = \begin{bmatrix} 1 & 0 & 2 \\ 1 & -1 & 0 \\ 0 & 1 & 1 \\ 0 & 0 & 1 \end{bmatrix} \begin{bmatrix} x_1 \\ x_2 \\ x_3 \end{bmatrix} = \begin{bmatrix} x_1 + 2x_3 \\ x_1 - x_2 \\ x_2 + x_3 \\ x_3 \end{bmatrix}$ $= [x_1 + 2x_3, x_1 - x_2, x_2 + x_3, x_3]^{\mathrm{T}} = \mathbf{T}(\mathbf{x})$. ◇

The matrix representation of a linear transformation makes the following basic properties of a linear transformations $\mathbf{T}(\mathbf{x})$: $R^n \to R^m$ almost self-evident:

$$\text{(i)}\,\mathbf{T}(\mathbf{0}) = \mathbf{0}, \tag{8.27}$$

$$\text{(ii)}\,\mathbf{T}(-\mathbf{x}) = -\mathbf{T}(\mathbf{x}) \text{ for every vector } \mathbf{x} \text{ in } R^m, \tag{8.28}$$

$$\text{(iii)}\,\mathbf{T}(\mathbf{x} - \mathbf{y}) = \mathbf{T}(\mathbf{x}) - \mathbf{T}(\mathbf{y}) \text{ for all vectors } \mathbf{x} \text{ and } \mathbf{y} \text{ in } R^m. \tag{8.29}$$

It is left as an exercise to show these properties can be established directly from the definition of a linear transformation, without appealing to the equivalent matrix representation of the transformation.

Example 8.7. Let $\mathbf{T}$ be a linear transformation from R^3 to R^3, with $\mathbf{e}_1$ to $\mathbf{e}_3$ the standard ordered basis vectors for R^3. Find the matrix representation for this transformation if

$$\mathbf{T}(2\mathbf{e}_1 + \mathbf{e}_2) = \begin{bmatrix} 2 \\ 3 \\ 0 \end{bmatrix}, \quad \mathbf{T}(\mathbf{e}_1 - \mathbf{e}_2) = \begin{bmatrix} 1 \\ 0 \\ 3 \end{bmatrix}, \quad \mathbf{T}(\mathbf{e}_2 + \mathbf{e}_3) = \begin{bmatrix} 1 \\ 1 \\ 2 \end{bmatrix}.$$

Solution. As the standard matrix $\mathbf{A} = [\mathbf{T}(\mathbf{e}_1), \mathbf{T}(\mathbf{e}_2), \mathbf{T}(\mathbf{e}_3)]$, to reconstruct it from the given information it is necessary to find $\mathbf{T}(\mathbf{e}_1)$, $\mathbf{T}(\mathbf{e}_2)$ and $\mathbf{T}(\mathbf{e}_3)$. The linearity of the transformation means that $\mathbf{T}(2\mathbf{e}_1 + \mathbf{e}_2) = 2\mathbf{T}(\mathbf{e}_1) + \mathbf{T}(\mathbf{e}_2)$, $\mathbf{T}(\mathbf{e}_1 - \mathbf{e}_2) = \mathbf{T}(\mathbf{e}_1) - \mathbf{T}(\mathbf{e}_2)$ and $\mathbf{T}(\mathbf{e}_2 + \mathbf{e}_3) = \mathbf{T}(\mathbf{e}_2) + \mathbf{T}(\mathbf{e}_3)$. Adding the first two of these rewritten equations gives

$$3\mathbf{T}(\mathbf{e}_1) = \begin{bmatrix} 2 \\ 3 \\ 0 \end{bmatrix} + \begin{bmatrix} 1 \\ 0 \\ 3 \end{bmatrix} = \begin{bmatrix} 3 \\ 3 \\ 3 \end{bmatrix}, \quad \text{so } \mathbf{T}(\mathbf{e}_1) = \begin{bmatrix} 1 \\ 1 \\ 1 \end{bmatrix}.$$

Using this result in the second equation $\mathbf{T}(\mathbf{e}_1) - \mathbf{T}(\mathbf{e}_2) = \mathbf{T}(\mathbf{e}_1) - \mathbf{T}(\mathbf{e}_2)$ gives

$$\mathbf{T}(\mathbf{e}_2) = \begin{bmatrix} 1 \\ 1 \\ 1 \end{bmatrix} - \begin{bmatrix} 1 \\ 0 \\ 3 \end{bmatrix} = \begin{bmatrix} 0 \\ 1 \\ -2 \end{bmatrix},$$

while from the third equation $\mathbf{T}(\mathbf{e}_2 + \mathbf{e}_3) = \mathbf{T}(\mathbf{e}_2) + \mathbf{T}(\mathbf{e}_3)$. we have

$$\mathbf{T}(\mathbf{e}_3) = \begin{bmatrix} 1 \\ 1 \\ 2 \end{bmatrix} - \begin{bmatrix} 0 \\ 1 \\ -2 \end{bmatrix} = \begin{bmatrix} 1 \\ 0 \\ 4 \end{bmatrix}.$$

Thus the matrix representation of $\mathbf{T}$ is

$$\mathbf{A} = \begin{bmatrix} 1 & 0 & 1 \\ 1 & 1 & 0 \\ 1 & -2 & 4 \end{bmatrix}.$$

◇

It should be mentioned that any linear transformation $\mathbf{T}(\mathbf{x}): U \to V$ with the property that $\mathbf{T}(\mathbf{x}) = \mathbf{0}$ for *every* vector $\mathbf{x}$ in U is called the *zero transformation.*

8.3 The Null Space of a Linear Transformation and Its Range

We now make a brief mention of some definitions and their consequences that are of importance to all linear transformations, and in particular when the general structure of linear transformations is studied. However, in this introductory account it would be inappropriate to discuss some of the ways in which these results are used, because this information belongs more properly to a more advanced account of linear algebra.

Definition 8.3. *The Null Space of a Linear Transformation and Its Range*

Let $\mathbf{T}(\mathbf{x}): U \to V$ *be a linear transformation of the space* U *in* R^n *containing the* $n \times 1$ *vector* $\mathbf{x}$ *and the space* V *in* R^m*, where the transformation* $\mathbf{T}$ *is determined by the multiplication of* $\mathbf{x}$ *by an* $m \times n$ *matrix* $\mathbf{A}$*. Then the set of vectors* $\mathbf{x}$ *in the vector space* U *that are mapped by* $\mathbf{A}$ *into the zero vector* $\mathbf{0}$ *in* V *is called the* nullspace *of* $\mathbf{A}$*, or the* kernel *of* $\mathbf{A}$*, and denoted either by* $\mathcal{N}(\mathbf{A})$ *or by* ker$(\mathbf{A})$*. The dimension* $\nu_{\mathbf{A}}$ *of the nullspace is called the* nullity *of* $\mathbf{A}$*. The* range *of the linear transformation* $\mathbf{T}$ *determined by* $\mathbf{A}$*, denoted by* $\mathcal{R}(\mathbf{A})$*, is the set of all vectors in* V *that correspond to the image of at least one vector in* U*, so that* $\mathcal{R}(\mathbf{A})$ = *the column space of* $\mathbf{A}$.

□

Two properties of a linear transformation $\mathbf{T}$ determined by $\mathbf{A}$ that follow almost immediately from Definition 8.3 are:

$$\text{(a) The nullspace } \mathcal{N}(\mathbf{A}) \text{ of } \mathbf{A} \text{ is a subspace of } U. \tag{8.30}$$

$$\text{(b) The range } \mathcal{R}(\mathbf{A}) \text{ of } \mathbf{A} \text{ is a subspace of } V. \tag{8.31}$$

That $\mathcal{N}(\mathbf{A})$ belongs to U is obvious, but to show it is a subspace of U it is necessary to show $\mathcal{N}(\mathbf{A})$ satisfies the conditions for a linear transformation given in Definition 8.1. This follows because the linearity of the transformation means that for any two vectors $\mathbf{x}_1$ and $\boldsymbol{x}_2$ in U, $\mathbf{A}(\mathbf{x}_1 + \boldsymbol{x}_2) = \mathbf{A}\mathbf{x}_1 + \mathbf{A}\mathbf{x}_2$, but if $\mathbf{x}_1$ and $\boldsymbol{x}_2$ belong to $\mathcal{N}(\mathbf{A})$ we have $\mathbf{A}\mathbf{x}_1 = \mathbf{A}\mathbf{x}_2 = \mathbf{0}$, so $\mathbf{A}(\mathbf{x}_1 + \boldsymbol{x}_2) = \mathbf{0}$. Similarly, if k is an arbitrary real number, $\mathbf{A}(k\mathbf{x}_1) = k\mathbf{A}\mathbf{x}_1$, but as $\mathbf{x}_1$ belongs to $\mathcal{N}(\mathbf{A})$ we have $\mathbf{A}\mathbf{x}_1 = \mathbf{0}$, so $\mathbf{A}(k\mathbf{x}_1) = \mathbf{0}$, and result (a) is established. The proof of result (b) is left as an exercise.

When $\mathbf{T}(\mathbf{x})$: $R^n \rightarrow R^m$ is represented by a matrix $\mathbf{A}$, the range $\mathcal{R}(\mathbf{A})$ must coincide with the column space of $\mathbf{A}$, but the row and column ranks of $\mathbf{A}$ are the same, with each denoted by rank($\mathbf{A}$), so we arrive at the result

$$\operatorname{rank}(\mathbf{T}) = \operatorname{rank}(\mathbf{A}). \tag{8.32}$$

As $\mathcal{N}(\mathbf{T}) =$ the dimension of the null space of$\mathbf{A}$, which is equal to $\mathcal{N}(\mathbf{A})$, it also follows that

$$\mathcal{N}(\mathbf{T}) = \mathcal{N}(\mathbf{A}). \tag{8.33}$$

We are now in a position to establish an important connection between the rank of a transformation represented by a matrix $\mathbf{A}$, and its nullity $\nu_{\mathbf{A}}$. Let $\mathbf{A}$ be an $m \times n$ matrix. Then if rank($\mathbf{A}$) $= r$, precisely r rows of $\mathbf{A}$ are linearly independent. Consequently, the remaining rows of $\mathbf{A}$ that are solutions of $\mathbf{A}\mathbf{x} = \mathbf{0}$ must belong to the nullspace $\mathcal{N}(\mathbf{A})$ of $\mathbf{A}$ with dimension equal to the nullity $\nu_{\mathbf{A}}$. However, the sum $\operatorname{rank}(\mathbf{A}) + \nu_{\mathbf{A}}$ must equal the number of vectors in the basis for the space, which in turn must equal the dimension of the space R^n, so we have proved the following important relationship.

Theorem 8.2 *The Rank Nullity Theorem*

Let $\mathbf{A}$ *be an arbitrary real* $m \times n$ *matrix, then*

$$\operatorname{rank}(\mathbf{A}) + \nu_{\mathbf{A}} = n.$$

◆

Example 8.8. A linear transformation $\mathbf{A}(\mathbf{x})$: $R^4 \rightarrow R^3$ of a vector $\mathbf{x}$ in R^4 to a vector $\mathbf{y}$ in R^3 is described by the matrix equation $\mathbf{A}\mathbf{x} = \mathbf{y}$, where

$$\mathbf{A} = \begin{bmatrix} -1 & 2 & -1 & -2 \\ 0 & 2 & 2 & 1 \\ 2 & 1 & 0 & 3 \end{bmatrix}, \quad \mathbf{x} = \begin{bmatrix} x_1 \\ x_2 \\ x_3 \\ x_4 \end{bmatrix} \quad \text{and } \mathbf{y} = \begin{bmatrix} y_1 \\ y_2 \\ y_3 \end{bmatrix}.$$

(i) Find the nullspace $\mathcal{N}(\mathbf{A})$ of $\mathbf{A}$, and hence the nullity $\nu_{\mathbf{A}}$ of $\mathbf{A}$.
(ii) Verify Theorem 8.2.
(iii) Given that $\mathbf{x} = [a, b, c, d]^{\mathrm{T}}$, for what values of a, b, c and d, if any, will the vectors $\mathbf{y}_1 = [1, 2, 3]^{\mathrm{T}}$, $\mathbf{y}_2 = [0, 2, 4]^{\mathrm{T}}$ and $\mathbf{y}_3 = [-1, 2, -5]^{\mathrm{T}}$ belong to $\mathcal{R}(\mathbf{A})$.

Solution. (i) The four component column vectors $\mathbf{x}$ will be in $\mathcal{N}(\mathbf{A})$ if $\mathbf{A}\mathbf{x} = \mathbf{0}$, so setting $\mathbf{x} = \begin{bmatrix} a \\ b \\ c \\ d \end{bmatrix}$ we must solve the matrix equation $\begin{bmatrix} -1 & 2 & -1 & -2 \\ 0 & 2 & 2 & 1 \\ 2 & 1 & 0 & 3 \end{bmatrix} \begin{bmatrix} a \\ b \\ c \\ d \end{bmatrix} = \begin{bmatrix} 0 \\ 0 \\ 0 \end{bmatrix}$ for a, b, c and d. This reduces to solving the algebraic system of equations

$$
\begin{aligned}
&-1 + 2b - c - 2d = 0, \\
&2b + 2c + d = 0, \\
&2a + b + 3d = 0.
\end{aligned}
$$

The solution of these homogeneous equations in terms of c as an arbitrary parameter is $a = 3c$, $b = 0$, $c = c$ (arbitrary) and $d = -2c$, so the vector in $\mathcal{N}(\mathbf{A})$ must be of the form $\mathbf{x} = [3c \quad 0 \quad c \quad -2c]^{\mathrm{T}}$. The vector $\mathbf{x}$ is the solution of a homogeneous set of equations and so can be scaled arbitrarily, so setting $c = 1$ we find the only vector in the nullspace $\mathcal{N}(\mathbf{A})$ is$\mathbf{x} = [3 \quad 0 \quad 1 \quad -2]^{\mathrm{T}}$. Consequently the dimension of the nullspace $\mathcal{N}(\mathbf{A})$ is $\nu_{\mathbf{A}} = 1$.

(ii) We have $n = 4$, and a check shows that rank($\mathbf{A}$) $= 3$, so as $\nu_{\mathbf{A}} = 1$ the result of Theorem 8.2 is confirmed, because $\mathrm{rank}(\mathbf{A}) + \nu_{\mathbf{A}} = 3 + 1 = 4 = n$.

(iii) The image $\mathbf{y} = [p, q, r]^{\mathrm{T}}$ in R^3 of an arbitrary column vector $\mathbf{x} = [a, b, c, d]^{\mathrm{T}}$ in R^4 is determined by the matrix equation $\mathbf{A}\mathbf{x} = \mathbf{y}$, so that

$$
\begin{aligned}
&-1 + 2b - c - 2d = p, \\
&2b + 2c + d = q, \\
&2a + b + 3d = r.
\end{aligned}
$$

As there are three equations connecting four unknowns, this system may always be solved for any p, q and r, in terms of c as an arbitrary parameter. So $\mathbf{y}_1$, $\mathbf{y}_2$ and $\mathbf{y}_3$ must all belong to $\mathcal{R}(\mathbf{A})$. It follows directly that all vectors $\mathbf{y}$ belong to $\mathcal{R}(\mathbf{A})$. ◇

8.4 Linear Transformations and the Geometry of the Plane

In this final section, linear transformations described by matrices will be used to establish some simple results concerning the geometry of the plane. The linear transformation $\mathbf{T}(\mathbf{x})\colon X \to Y$ from R^2 to R^2 will be in the form $\mathbf{y} = \mathbf{A}\mathbf{x}$, where $\mathbf{A}$ is the matrix $\mathbf{A} = \begin{bmatrix} a_{11} & a_{12} \\ a_{21} & a_{22} \end{bmatrix}$ with real elements, vector $\mathbf{x} = [x_1, x_2]^{\mathrm{T}}$ belongs to the vector space X, and vector $\mathbf{y} = [y_1, y_2]^{\mathrm{T}}$ belongs to the vector space Y. Specifically, the transformation will map vectors $\mathbf{x} = [\tilde{x}_1, \tilde{x}_2]^T$ in space X, that is points in the (x_1, x_2)-plane, into vectors $\mathbf{y} = [\tilde{y}_1, \tilde{y}_2]^T$ in space Y containing points in the (y_1, y_2)-plane. The elements of $\mathbf{x}$ will represent the Cartesian coordinates (x_1, x_2) of a point in the space X, while the elements of $\mathbf{y}$ will represent the Cartesian coordinates (y_1, y_2) of a corresponding image point in space Y. When displayed explicitly, the linear transformation becomes

$$\begin{bmatrix} y_1 \\ y_2 \end{bmatrix} = \begin{bmatrix} a_{11} & a_{12} \\ a_{21} & a_{22} \end{bmatrix} \begin{bmatrix} x_1 \\ x_2 \end{bmatrix}, \tag{8.34}$$

or in scalar form

$$\begin{aligned} y_1 &= a_{11}x_1 + a_{12}x_2, \\ y_2 &= a_{21}x_1 + a_{22}x_2. \end{aligned} \tag{8.35}$$

Theorem 8.3 *Geometrical Properties of Transformation (8.34)*

1. *Provided* $\det \mathbf{A} \neq 0$, *the origin in the* (x_1, x_2)*-plane (the space X) is mapped into the origin in the* (y_1, y_2)*-plane (the space Y).*
2. *When* $\det \mathbf{A} \neq 0$, *the transformation maps arbitrary straight lines in the* (x_1, x_2)*-plane into straight lines in the* (y_1, y_2)*-plane.*
3. *When* $\det \mathbf{A} \neq 0$, *the transformation maps parallel straight lines in the* (x_1, x_2)*-plane into parallel straight lines in the* (y_1, y_2)*-plane.*
4. *When* $\det \mathbf{A} \neq 0$, *straight lines in the* (x_1, x_2)*-plane that pass through the origin are mapped into straight lines in the* (y_1, y_2)*-plane that also pass through the origin.*

Proof.

1. The proof is trivial, because $\mathbf{0} = [0, 0]^T$, so $\mathbf{A}[0, 0]^T = [0, 0]^T$, showing the origin in X space is mapped to the origin in Y space.
2. Let the straight line pass through the arbitrary point (α, β) in the (x_1, x_2)-plane, then α and β must be such that $\beta = m\alpha + c$. From (8.35), the image of this point in the (y_1, y_2)-plane, say the point (α', β'), is determined by the equations

$$\begin{aligned} \alpha' &= (a_{11} + a_{12}m)\alpha + a_{12}c, \\ \beta' &= (a_{21} + a_{22}m)\alpha + a_{22}c. \end{aligned}$$

Eliminating α between these two equations gives

$$(a_{11} + ma_{12})\beta' = (a_{21} + ma_{22})\alpha' - (a_{11}a_{22} - a_{12}a_{21})c.$$

However $a_{11}a_{22} - a_{12}a_{21} = \det \mathbf{A} \neq 0$ (by hypothesis), so this becomes

$$(a_{11} + ma_{12})\beta' = (a_{21} + ma_{22})\alpha' - (\det \mathbf{A})c.$$

The point (α, β) in the (x_1, x_2)-plane was arbitrary, with (α', β') its image in the (y_1, y_2)-plane. So replacing α' by y_1 and β' by y_2, the image of the line $x_2 = mx_1 + c$ is seen to be described by the equation

$$(a_{11} + ma_{12})y_2 = (a_{21} + ma_{22})y_1 - (\det \mathbf{A})c.$$

This is the equation of a straight line in the (y_1, y_2)-plane, so we have shown that if $\det \mathbf{A} \neq 0$, any straight line in the (x_1, x_2)-plane that is not parallel to the x_2-axis maps to a straight line in the (y_1, y_2)-plane. This conclusion forms the main part of result 2. There remains the question of how a straight line parallel to the x_2-axis is mapped by (8.35).

Setting $x_1 = K$ in Eq. (8.35), where K is arbitrary, and eliminating x_2, shows that $a_{12}y_2 = a_{22}y_1 - (\det \mathbf{A})K$, which is again the equation of a straight line. So we have proved all of the assertions in 2; namely that if $\det \mathbf{A} \neq 0$, then arbitrary straight lines in the (x_1, x_2)-plane, including those parallel to the x_2-axis, map to straight lines in the (y_1, y_2)-plane.

1. The proof of result 3 follows directly from the result

$$(a_{11} + ma_{12})\beta' = (a_{21} + ma_{22})\alpha - (\det \mathbf{A})c.$$

in 1, because parallel straight lines in the (x_1, x_2)-plane all have the same slope m, so they will map to straight lines in the (y_1, y_2)-plane corresponding to different values of c, and hence the image lines must also be parallel.

2. Result 4 follows from the fact that straight lines in the (x_1, x_2)-plane will only pass through the origin if $c = 0$, where from the result in 1 we see their images must also pass through the origin.

◆

Apart from requiring $\det \mathbf{A} \neq 0$ for the general properties 1 through 4 in Theorem 8.3 to be true, the value of $\det \mathbf{A}$ in transformation (8.34) has two other important consequences that are described in the following theorem.

In preparation for the theorem it is necessary to introduce the term *orientation* in relation to the way a point moves around the boundary of an area in the plane. By convention, when a point moves around the boundary of an area in the *counterclockwise* sense the *orientation* of the trajectory described by the point is considered to be *positive*, whereas if it moves around the boundary in the *clockwise* sense, the orientation of the trajectory is considered to be *negative*. Thus movement around the boundary of the unit square in Fig. 8.5a is *positively* oriented if it follows the path $OABC$, while it is *negatively* oriented if it follows

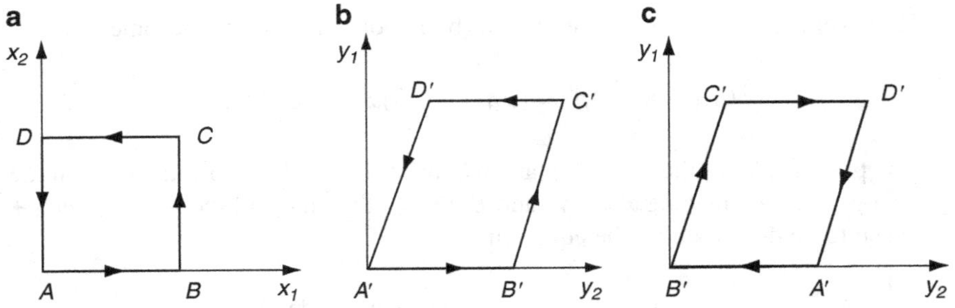

Fig. 8.5 (**a**) A *positive orientation OABC* of the unit square in the (x_1, x_2)-plane. (**b**) A *positively oriented* image of the unit square in the (y_1, y_2)-plane. (**c**) A *negatively oriented* image of the unit square in the (y_1, y_2)-plane

the path *OCBA*. Figure 8.5b shows a *positively oriented* image of the unit square in Fig. 8.5a under the transformation $\mathbf{y} = \mathbf{A}\mathbf{x}$, while Fig. 8.5c shows an image of the unit square that is *negatively oriented*. By Theorem 8.3(3) both of the images must be a parallelogram.

It will be seen from Theorem 8.4 that the positive orientation in Fig. 8.5b follows when $\det \mathbf{A} > 0$ in the transformation $\mathbf{A}\mathbf{x} = \mathbf{y}$, while the negative orientation in Fig. 8.5c follows when $\det \mathbf{A} < 0$.

Theorem 8.4 *Area Magnification and the Orientation of an Image*

1. *Let a triangle or parallelogram in the (x_1, x_2)-plane have area S, and let* $\det \mathbf{A} \neq 0$. *Then if* $\mathbf{A}\mathbf{x} = \mathbf{y}$, *the area S of the image of the corresponding triangle or parallelogram in the (y_1, y_2)-plane is* $S = |\det \mathbf{A}|s$. *So the absolute value of* $\det \mathbf{A}$ *is the* area magnification factor *when the shape is transformed from the (x_1, x_2)-plane to the (y_1, y_2)-plane.*
2. *Let the boundaries of a triangle or parallelogram in the (x_1, x_2)-plane be positively oriented. Then if* $\det \mathbf{A} > 0$ *the boundary of the corresponding image triangle or parallelogram in the (y_1, y_2)-plane will be positively oriented, while if* $\det \mathbf{A} < 0$ *the boundary of the corresponding image triangle or parallelogram will be negatively oriented.*

Proof.

1. To determine the magnification factor, transformation (8.34) will be applied to the unit square *OABC* with unit area shown in Fig. 8.5a. The square is seen to be located in the first quadrant of the (x_1, x_2)-plane with a corner located at the origin and its sides parallel to the axes. Properties 1 and 2 show the transformation will change the square into a parallelogram $O'A'B'C'$ with area *S*, as in Fig. 8.5b, with its corner O' located at the origin of the (y_1, y_2)-plane, because O' is the image of *O*. From (8.35), the respective coordinates of $O'A'B'C'$ are $(0, 0)$, (a_{11}, a_{21}), $(a_{11}+a_{12}, a_{21}+a_{22})$ and (a_{12}, a_{22}). The linearity of the transformation ensures that the magnification factor will remain the same between any triangle

or parallelogram in the (x_1, x_2)-plane and its image in the (y_1, y_2)-plane, while the square of the area S is given by

$$S^2 = (O'C')^2(O'A')^2 \sin^2\theta,$$

but $\sin^2\theta = 1 - \cos^2\theta$, and so

$$S^2 = (O'C')^2(O'A')^2(1 - \cos^2\theta). \tag{8.36}$$

However,

$$\cos^2\theta = \frac{\langle \mathbf{u}, \mathbf{v}\rangle^2}{\|\mathbf{u}\|^2\|\mathbf{v}\|^2}, \tag{8.37}$$

so from (8.36) and (8.37) we find that

$$S^2 = \|\mathbf{u}\|^2\|\mathbf{v}\|^2 \sin^2\theta = \|\mathbf{u}\|^2\|\mathbf{v}\|^2(1 - \cos^2\theta) = \|\mathbf{u}\|^2\|\mathbf{v}\|^2 - \langle \mathbf{u}, \mathbf{v}\rangle^2.$$

As $\mathbf{u} = [a_{12}, a_{22}]$ and $\mathbf{v} = [a_{11}, a_{21}]$, substituting for these vectors in the above expression for S^2, and simplifying, gives

$$S^2 = (a_{11}a_{22} - a_{12}a_{21}) = (\det \mathbf{A})^2, \tag{8.38}$$

showing that area $S = |\det \mathbf{A}|$.

Thus matrix $\mathbf{A}$ has been shown to determine the scaling when triangles and parallelograms are mapped from the (x_1, x_2)-plane to the (y_1, y_2)-plane, with $|\det \mathbf{A}|$ as the magnification factor, so result 1 of the theorem has been established.

To prove result 2, associate the positive sign of det $\mathbf{A}$ with a positive orientation around the parallelogram image of the unit square. Now suppose the rows of $\mathbf{A}$ are interchanged. Then an examination of Fig. 8.5b shows this corresponds to an interchange of the points A' and C', which has the effect of reversing the orientation around the parallelogram.. However, if the rows of $\mathbf{A}$ are interchanged the sign of det $\mathbf{A}$ is reversed, and as a point moves with a positive orientation around a boundary of the square in the (x_1, x_2)-plane, so the image point will move with a negative orientation around its image in the (y_1, y_2)-plane. Consequently the orientation of the parallelogram that forms the image of the unit square is positive when det $\mathbf{A} > 0$, and negative when det $\mathbf{A} < 0$, and result 2 of the theorem has been established, so the proof is complete.

◆

Having established some important general properties of linear transformations, this section will close with a discussion of some specific transformations that form a basis for further study of linear algebra applied to the geometry of the plane.

8.4.1 Rotation About the Origin

The linear transformation $\mathbf{y} = \mathbf{A}\mathbf{x}$, with

$$\mathbf{A} = \begin{bmatrix} \cos\theta & \sin\theta \\ -\sin\theta & \cos\theta \end{bmatrix}, \quad \mathbf{x} = \begin{bmatrix} x_1 \\ x_2 \end{bmatrix}, \quad \mathbf{y} = \begin{bmatrix} y_1 \\ y_2 \end{bmatrix}, \tag{8.39}$$

describes a counterclockwise rotation about the origin of a vector $\mathbf{x}$ through an angle θ to form vector $\mathbf{y}$ in the same plane, where $\mathbf{A}$ is the *rotation matrix*.

This transformation has already been considered in Section 8.1, and the reasoning that gave rise to it can be used to determine some related transformations, as we now show.

8.4.2 A Reflection in a Line L Through the Origin

A point P is said to be reflected in a line L to form an image P' if the line PP' is perpendicular to L, with P and P' equidistant from L. Consider Fig. 8.6 in which the line L about which reflection is to take place is inclined to the x_1-axis at an angle α. Point A located at the tip of the unit vector along the x_1-axis has the coordinates $(1, 0)$ while point B located at the tip of the unit vector along the x_2-axis has the coordinates $(0, 1)$. Point A' is the reflected image of point A, and it is seen to have the coordinates $(\cos 2\alpha, \sin 2\alpha)$, and as the angle BOA is equal to $\frac{1}{2}\pi - 2\alpha$, the coordinates of A' are $\left(\cos(\frac{1}{2}\pi - 2\alpha), -\sin(\frac{1}{2}\pi - 2\alpha)\right) = (\sin 2\alpha, -\cos 2\alpha)$. The *reflection matrix* describing reflection in line L is

$$\mathbf{A} = \begin{bmatrix} \cos 2\alpha & \sin 2\alpha \\ \sin 2\alpha & -\cos 2\alpha \end{bmatrix}. \tag{8.40}$$

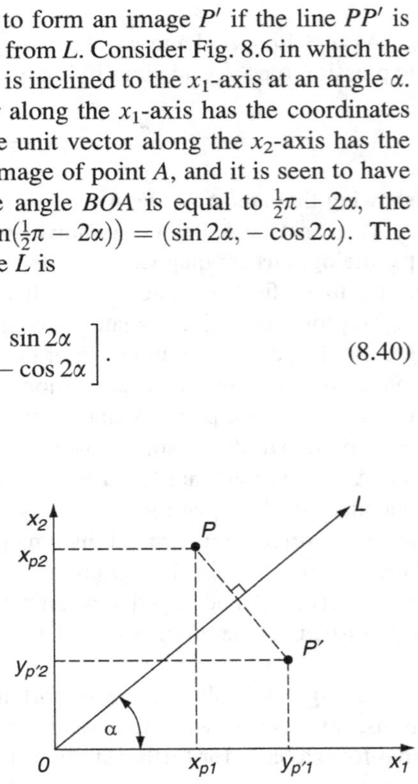

Fig. 8.6 Reflection of a point P in a line L through the origin

Thus the transformation from vector $\mathbf{x} = [x_{P1}, x_{P2}]^{\mathrm{T}}$ that defines a point P, to vector $\mathbf{y} = [y_{P'1}, y_{P'2}]$ which defines its reflection in line L, becomes

$$\begin{bmatrix} y_{P''1} \\ y_{P'2} \end{bmatrix} = \begin{bmatrix} \cos 2\alpha & \sin 2\alpha \\ \sin 2\alpha & -\cos 2\alpha \end{bmatrix} \begin{bmatrix} x_{P1} \\ x_{P2} \end{bmatrix}. \tag{8.41}$$

This result has been obtained from Fig. 8.6 assuming α to be an acute angle, though the result remains true when α is obtuse. The justification for this assertion is left as an exercise.

8.4.3 The Orthogonal Projection of a Point P Onto a Line L Through the Origin

This situation is illustrated in Fig. 8.7, where the coordinates of P in the (x_1, x_2)-plane are (x_{P1}, x_{P2}). Point P' with coordinates (y_{P1}, y_{P2}) is the orthogonal projection of P onto the line L, and the line L is inclined to the x_1-axis at an angle α.

The unit vector along the x_1-axis has its tip at point A with coordinates (1, 0), so if point B is the orthogonal projection of the tip of this unit vector onto L, the length OB is cos α, so its component along the x_1-axis is $OB \cos\alpha = \cos^2\alpha$, while its component along the x_2-axis is $OB \sin\alpha = \cos\alpha \sin\alpha$. So after scaling these components by x_1 and x_2, respectively, the horizontal coordinate $y_{P'1}$ of P' is seen to be $y_{P'1} = x_{P1}\cos^2\alpha + x_{P2}\cos\alpha\sin\alpha$. After projecting D onto L, where the unit vector along the x_2-axis with its tip at D has coordinates (0, 1), similar reasoning shows that the vector normal to L has an x_1-component equal to cos α sin α, and an x_2-component equal to $\sin^2\alpha$. So after scaling these components by x_{P1} and x_{P2}, the vertical y_2 coordinate of P' is found to be $y_{P'2} = x_{P1}\cos\alpha\sin\alpha + x_{P2}\sin^2\alpha$ Thus the transformation matrix describing this procedure is seen to be

$$\mathbf{A} = \begin{bmatrix} \cos^2\alpha & \cos\alpha\sin\alpha \\ \cos\alpha\sin\alpha & \sin^2\alpha \end{bmatrix}, \tag{8.42}$$

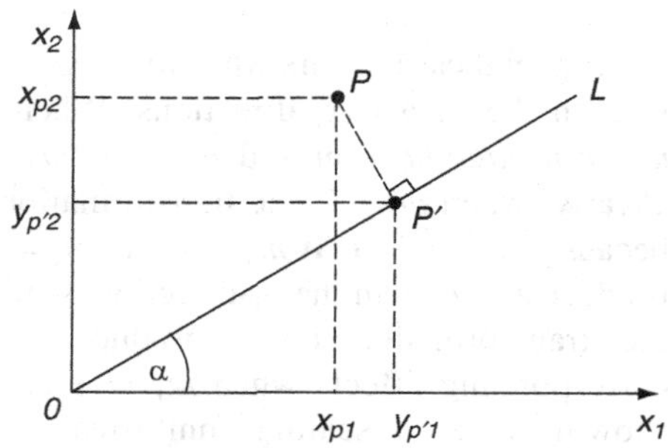

Fig. 8.7 The orthogonal projection of point P onto a line L through the origin

with the coordinates $(y_{P'1}, y_{P'2})$ of point P', the perpendicular projection of P onto the line L inclined at an angle α to the x_1-axis, given by the *projection matrix*

$$\begin{bmatrix} y_{P'1} \\ y_{P'2} \end{bmatrix} = \begin{bmatrix} \cos^2\alpha & \cos\alpha\sin\alpha \\ \cos\alpha\sin\alpha & \sin^2\alpha \end{bmatrix} \begin{bmatrix} x_{P1} \\ x_{P2} \end{bmatrix}. \tag{8.43}$$

As a check, setting $\alpha = 0$ causes line L, and so also point P', to lie on the x_1-axis with $y_{P'1} = x_{P1}$ and $y_{P'2} = 0$, and this is indeed the case, because (8.32) reduces to

$$\begin{bmatrix} y_{P'1} \\ y_{P'2} \end{bmatrix} = \begin{bmatrix} 1 & 0 \\ 0 & 0 \end{bmatrix} \begin{bmatrix} x_{P1} \\ x_{P2} \end{bmatrix},$$

confirming that $y_{P'1} = x_{P1}$ and $y_{P'2} = 0$.

8.4.4 Scaling in the $\mathbf{x}_1$ and $\mathbf{x}_2$ Directions

The rule for matrix multiplication shows that scaling in the x_1-direction with scale factor m_1, and scaling in the x_2-direction with scale factor m_2, is represented by the *scaling matrix*

$$\mathbf{A} = \begin{bmatrix} m_1 & 0 \\ 0 & m_2 \end{bmatrix}. \tag{8.44}$$

In this case $\mathbf{y} = \mathbf{Ax}$ takes the form

$$\begin{bmatrix} y_1 \\ y_2 \end{bmatrix} = \begin{bmatrix} m_1 & 0 \\ 0 & m_2 \end{bmatrix} \begin{bmatrix} x_1 \\ x_2 \end{bmatrix}, \tag{8.45}$$

and after expansion this becomes

$$y_1 = m_1 x_1 \text{ and } y_2 = m_2 x_2. \tag{8.46}$$

A special case occurs when $m_1 = m_2 = m$, because then the scaling is the same in both the x_1 and x_2-directions. When $m > 1$ this situation corresponds to a *uniform magnification*, and when $0 < m < 1$ it corresponds to a *uniform shrinkage*. Clearly, when $m = 1$, the transformation reduces to the identity transformation because then $\mathbf{A} = \mathbf{I}$. If $m_1 > 1$ and $m_2 = 1$ the effect of the transformation is to produce a *stretch* in the x_1-direction, while if $0 < m_1 < 1$ and $m_2 = 1$ the effect of the transformation is to produce a *shrinkage* in the x_1-direction, with corresponding effects when $m_1 = 1$ and $m_2 \neq 1$. Some typical examples that show the effect of scaling a unit square are given in Figs. 8.8a–c, where Fig. 8.8a shows the unit square before scaling.

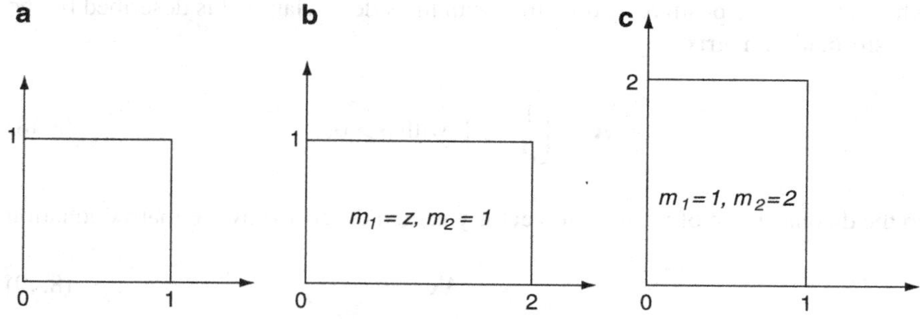

Fig. 8.8 Some typical examples of the scaling transformation

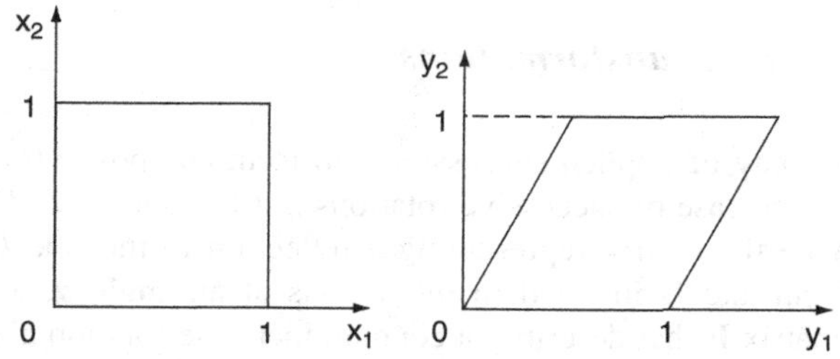

Fig. 8.9 The effect on a unit square of a shear parallel to the x_1-axis

8.4.5 A Shear

The geometrical effect of a *shear* is easily understood if it is applied to a unit square, the base of which is fixed, while the shear takes place parallel to the base line. To be more specific, let the unit square have one corner located at the origin and, before the shear is applied, let two of its sides coincide with the x_1 and x_2-axes. Then the geometrical effect of a shear applied in the x_1-direction is to cause each line $x_2 = k$ that pass through the square to be displaced to the right by an amount proportional to k. This is illustrated in Fig. 8.9, where the undistorted unit square is shown in the diagram at the left, while the effect of the shear is shown in the diagram at the right. Because the amount of shear is proportional to the perpendicular distance of the shear line from the x_1-axis, and so is it is proportional to k, the effect of the shear is to transform the unit square into a parallelogram.

It is easy to see that shear in the positive x_1-direction with the side $x_1 = 0$ clamped is described by the *shear matrix*

$$\mathbf{A} = \begin{bmatrix} 1 & \alpha \\ 0 & 1 \end{bmatrix}, \text{ with } \alpha > 0, \tag{8.47}$$

while shear in the positive x_2-direction with the side x_2 clamped is described by the transformation matrix

$$\mathbf{A} = \begin{bmatrix} 1 & 0 \\ \beta & 1 \end{bmatrix}, \text{ with } \beta > 0, \tag{8.48}$$

so the displacement of vector $\mathbf{x}$ to vector $\mathbf{y}$ is then described by the matrix equation

$$\mathbf{y} = \mathbf{A}\mathbf{x}. \tag{8.49}$$

Reversing the signs of α and β simply reverses the directions of the shear.

8.4.6 *Composite Transformations*

Transformations may be applied successively to form composite transformations, as was shown in the case of successive rotations in (8.8) and (8.9). For a different example, let $\mathbf{A}$ be the matrix representing a reflection in the line L_1 that passes through the origin and is inclined to the x_1-axis at an angle α, and let this be followed by a matrix $\mathbf{B}$ that describes a counterclockwise rotation about the origin through an angle β. This is described by the matrix product

$$\mathbf{A} = \begin{bmatrix} \cos 2\alpha & \sin 2\alpha \\ \sin 2\alpha & -\cos 2\alpha \end{bmatrix} \text{ and } \mathbf{B} = \begin{bmatrix} \cos\theta & -\sin\theta \\ \sin\theta & \cos\theta \end{bmatrix}. \tag{8.50}$$

Then a point represented by a vector $\mathbf{x}$ is mapped by $\mathbf{A}$ into a point $\mathbf{y}$, where $\mathbf{y} = \mathbf{A}\mathbf{x}$. The point represented by vector $\mathbf{y}$ is then mapped by $\mathbf{B}$ into a point $\mathbf{z}$, where $\mathbf{z} = \mathbf{B}\mathbf{y}$. Combining these results to form a composite transformation then gives $\mathbf{z} = \mathbf{B}\mathbf{A}\mathbf{x}$. Notice that $\mathbf{A}$ precedes $\mathbf{B}$, because it is the first operator to act on $\mathbf{x}$, after which $\mathbf{B}$ acts on the vector $\mathbf{A}\mathbf{x}$. Thus the effect of this composite transformation on a vector $\mathbf{x}$ to produce a vector $\mathbf{y}$ is given by

$$\begin{bmatrix} y_1 \\ y_2 \end{bmatrix} = \begin{bmatrix} \cos\theta & -\sin\theta \\ \sin\theta & \cos\theta \end{bmatrix} \begin{bmatrix} \cos 2\alpha & \sin 2\alpha \\ \sin 2\alpha & -\cos 2\alpha \end{bmatrix} \begin{bmatrix} x_1 \\ x_2 \end{bmatrix}. \tag{8.51}$$

As would be expected, the geometrical effect of applying these transformations in the reverse order will be different, and this is reflected by the fact that in general matrix products are not commutative.

To illustrate this last remark consider the case when the reflection is about the line L inclined to the x_1-axis at an angle $\pi/8$, after which a counterclockwise rotation about the origin is made through an angle $\pi/4$, so $\alpha = \pi/8$ and $\beta = \pi/4$. The transformation matrices in (8.51) become

$$\mathbf{BA} = \begin{bmatrix} 1/\sqrt{2} & -1/\sqrt{2} \\ 1/\sqrt{2} & 1/\sqrt{2} \end{bmatrix} \begin{bmatrix} 0 & 1 \\ 1 & 0 \end{bmatrix} = \begin{bmatrix} -1/\sqrt{2} & 1/\sqrt{2} \\ 1/\sqrt{2} & 1/\sqrt{2} \end{bmatrix},$$

whereas, if the transformations are performed in the reverse order, the product **AB** of the transformation matrices becomes

$$\mathbf{AB} = \begin{bmatrix} 0 & 1 \\ 1 & 0 \end{bmatrix} \begin{bmatrix} 1/\sqrt{2} & -1/\sqrt{2} \\ 1/\sqrt{2} & 1/\sqrt{2} \end{bmatrix} = \begin{bmatrix} 1/\sqrt{2} & 1/\sqrt{2} \\ 1/\sqrt{2} & -1/\sqrt{2} \end{bmatrix}.$$

8.4.7 The Transformation of Curves

The transformation $\mathbf{y} = \mathbf{Ax}$ will be one-one when $\det \mathbf{A} \neq 0$, because then $\mathbf{A}^{-1}$ exists and $\mathbf{x} = \mathbf{A}^{-1}\mathbf{y}$. So in this case, if a curve C in the (x_1, x_2)-plane is continuous, it will be mapped by this linear transformation onto a unique continuous image curve C' in the (y_1, y_2)-plane. Conversely, a continuous curve C' in the (y_1, y_2)-plane will have as its pre-image a unique continuous curve C in the (x_1, x_2)-plane. In general, the effect of such a transformation will be to distort the image curve C' relative to curve C, and a typical result of a mapping $\mathbf{y} = \mathbf{Ax}$, with

$$\mathbf{A} = \begin{bmatrix} 1 & 2 \\ 3 & -1 \end{bmatrix}$$

is shown in Fig. 8.10, where the unit circle centered on the origin in Fig. 8.10a is seen to be mapped into the ellipse in Fig. 8.10b.

Exercises

1. Is matrix $\mathbf{A} = \begin{bmatrix} \cos\theta & \sin\theta \\ \sin\theta & -\cos\theta \end{bmatrix}$ orthogonal? Give a reason for your answer.
2. Using the matrices **A** and **B** in (8.8), verify the form of the matrix product **BA** in (8.9).
3. Identify the nature of the two-dimensional rotations produced by the matrices:

$$\text{(a)} \begin{bmatrix} -\frac{1}{2} & \frac{\sqrt{3}}{2} \\ -\frac{\sqrt{3}}{2} & -\frac{1}{2} \end{bmatrix} \quad \text{(b)} \begin{bmatrix} -\frac{1}{\sqrt{2}} & \frac{1}{\sqrt{2}} \\ -\frac{1}{\sqrt{2}} & -\frac{1}{\sqrt{2}} \end{bmatrix} \quad \text{(c)} \begin{bmatrix} -\frac{1}{2} & -\frac{\sqrt{3}}{2} \\ \frac{\sqrt{3}}{2} & -\frac{1}{2} \end{bmatrix} \quad \text{(d)} \begin{bmatrix} \frac{1}{\sqrt{2}} & -\frac{1}{\sqrt{2}} \\ \frac{1}{\sqrt{2}} & \frac{1}{\sqrt{2}} \end{bmatrix}.$$

4. Given that $\mathbf{x} = [x_1, x_2, x_3]$, show from first principles that the transformation $\mathbf{T}(\mathbf{x}) = [2x_1, x_1 + 3x_3, 2x_1 - x_3]$ is linear.

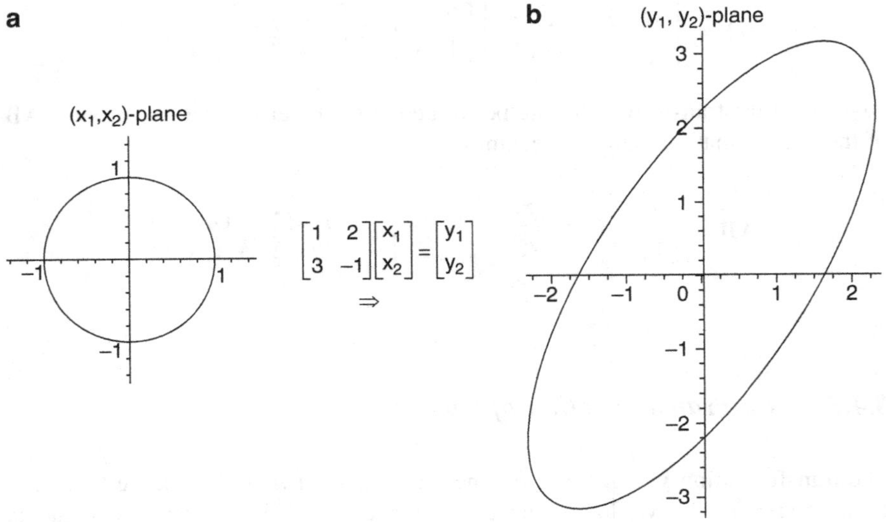

Fig. 8.10 (**a**) A unit circle. (**b**) The elliptical image of the unit circle

5. Given that $\mathbf{x} = [x_1, x_2, x_3, x_4]$, show from first principles that the transformation $\mathbf{T}(\mathbf{x}) = [3x_1, x_2 + x_3, x_2 - 2x_3, x_1 + x_4]$ is linear.
6. Write down the matrix representations of the linear transformations

$$\text{(a) } \mathbf{T}\begin{bmatrix} x_1 \\ x_2 \\ x_3 \end{bmatrix} = \begin{bmatrix} x_1 - x_2 + 3x_3 \\ 2x_1 + x_2 - x_3 \\ -x_1 - x_2 + 2x_3 \end{bmatrix} \quad \text{(b) } \mathbf{T}\begin{bmatrix} x_1 \\ x_2 \\ x_3 \\ x_4 \end{bmatrix} = \begin{bmatrix} x_2 + 3x_3 - x_4 \\ x_1 - x_2 + x_3 - x_4 \\ x_3 + x_4 \\ x_1 - 2x_2 + 3x_3 + 2x_4 \end{bmatrix}.$$

An *integral transformation*, also called an *integral transform*, used in the solution of ordinary and partial differential equations, is a correspondence between two functions $f(t)$ and $F(s)$ determined by an integral of the form

$$F(s) = \int_{-\infty}^{\infty} K(s,t)f(t)dt.$$

The function $F(s)$ is called the *transform* of $f(t)$, the variable s is called the transform variable, and the function $K(s, t)$ is called the *kernel* of the transformation. The interval $-\infty \leq t \leq \infty$ in which the function $f(t)$ is defined is called the *domain* of the space to which $f(t)$ belongs, and the transform $F(s)$ is then said to be defined in a domain called the *image space*, which is usually in the complex plane. In general, the variable t is real, but the transform variable is complex with $s = \sigma + i\omega$. The Laplace transform is a particular case of an integral transform in which the kernel

$$K(s,t) = \begin{cases} 0, & t<0, \\ e^{-st}, & t>0, \end{cases}$$

where the integral is taken over the interval $0 \leq t < \infty$ to which $f(t)$ belongs. As in Chapter 6, the Laplace transform of $F(s)$ is indicated by writing $F(s) = \mathcal{L}\{f(t)\}$.

7. When the improper integral defining an integral transformation given above is defined, show that the integral transform is a linear transformation.
8. Three integral transforms that are used when solving ordinary and partial differential equations are:
   (a) The **Mellin transform** $\mathcal{M}\{f(t)\} = \int_0^\infty t^{p-1} f(t)dt$,
   (b) The **Fourier transform** $\mathcal{F}\{f(t)\} = \int_{-\infty}^\infty e^{-i\omega t} f(t)dt$,
   (c) The **Hankel transform of order** v is $\mathcal{H}_v\{f(t)\} = \int_0^\infty tJ_v(\sigma t) f(t)dt$, where $J_v(\sigma t)$ is the Bessel function of the first kind of order v.

   Why are these linear transformations, what is the space to which $f(t)$ belongs, and, what are the kernels of the transforms?

9. Given that $\mathbf{T}$ is a linear transformation from $R^3 \to R^3$, and $\mathbf{e}_1$ to $\mathbf{e}_3$ are the normal ordered basis vectors for R^3, find the matrix representation $\mathbf{A}$ of $\mathbf{T}$ if

$$\mathbf{T}(2\mathbf{e}_1 + \mathbf{e}_3) = \begin{bmatrix} 1 \\ -1 \\ 0 \end{bmatrix}, \quad \mathbf{T}(-\mathbf{e}_2 + \mathbf{e}_3) = \begin{bmatrix} -2 \\ 1 \\ 1 \end{bmatrix}, \quad \mathbf{T}(\mathbf{e}_2 + \mathbf{e}_3) = \begin{bmatrix} 0 \\ 1 \\ 1 \end{bmatrix}.$$

10. Given that $\mathbf{T}$ is a linear transformation from $R^4 \to R^3$, and $\mathbf{e}_1$ to $\mathbf{e}_4$ are the normal ordered basis vectors for R^4, find the matrix representation $\mathbf{A}$ of $\mathbf{T}$ if

$$\mathbf{T}(\mathbf{e}_1 - \mathbf{e}_3) = \begin{bmatrix} 1 \\ 0 \\ 0 \end{bmatrix}, \ \mathbf{T}(\mathbf{e}_2 + \mathbf{e}_3) = \begin{bmatrix} -2 \\ 1 \\ 1 \end{bmatrix}, \ \mathbf{T}(\mathbf{e}_2 - \mathbf{e}_4) = \begin{bmatrix} 2 \\ -2 \\ 0 \end{bmatrix},$$

$$\mathbf{T}(\mathbf{e}_2 + \mathbf{e}_4) = \begin{bmatrix} 2 \\ -4 \\ 2 \end{bmatrix}.$$

11. Which, if any, of the following matrices $\mathbf{A}$ makes the matrix representation of a linear transformation $\mathbf{Ax} = \mathbf{y}$ one-to-one? If this is not the case, what must be

the relationship between the components of the matrix vector **y** in order that **y** is a unique image of a vector **x**?

(a) $\mathbf{A} = \begin{bmatrix} 2 & 1 & 3 \\ -1 & 1 & 2 \\ -6 & -6 & -16 \end{bmatrix}$ (b) $\mathbf{A} = \begin{bmatrix} 1 & 4 & 3 \\ 2 & 1 & 1 \\ 1 & -1 & 2 \end{bmatrix}$ $\mathbf{A} = \begin{bmatrix} 1 & -1 & 1 \\ 1 & 1 & 1 \\ 1 & 0 & -1 \\ 1 & -2 & 1 \end{bmatrix}$.

In Exercises 12 through 15, given the matrix **A**, find rank(**A**), the nullspace $\mathcal{N}(\mathbf{A})$ of **A**, and the nullity $\nu_{\mathbf{A}}$ of **A**. Verify the result of Theorem 8.2.

12.
$$\mathbf{A} = \begin{bmatrix} 2 & -1 & 2 \\ 1 & 1 & -1 \\ 6 & 0 & 2 \end{bmatrix}$$

13.
$$\mathbf{A} = \begin{bmatrix} 1 & 1 & 3 \\ 0 & -1 & 2 \\ 2 & -1 & 4 \end{bmatrix}$$

14.
$$\mathbf{A} = \begin{bmatrix} 1 & 2 & 0 & 1 \\ 1 & 1 & 1 & 3 \\ 4 & 6 & 2 & 8 \\ 0 & -1 & 1 & 2 \end{bmatrix}$$

15.
$$\mathbf{A} = \begin{bmatrix} 1 & 1 & 1 & 0 \\ 1 & -1 & 0 & 1 \\ 2 & -1 & 1 & 1 \\ 4 & -1 & 2 & 2 \end{bmatrix}$$

In Exercises 16 through 19 identify the geometrical effect of the transformation involved.

16. (a) $\mathbf{A} = \begin{bmatrix} \frac{1}{2} & \frac{\sqrt{3}}{2} \\ \frac{\sqrt{3}}{2} & -\frac{1}{2} \end{bmatrix}$ (b) $\mathbf{A} = \begin{bmatrix} \frac{1}{2} & -\frac{\sqrt{3}}{2} \\ \frac{\sqrt{3}}{2} & \frac{1}{2} \end{bmatrix}$.

17. (a) $\mathbf{A} = \begin{bmatrix} 0 & 1 \\ 1 & 0 \end{bmatrix}$ (b) $\mathbf{A} = \begin{bmatrix} \frac{1}{2} & \frac{1}{2} \\ \frac{1}{2} & \frac{1}{2} \end{bmatrix}$.

18. (a) $\mathbf{A} = \begin{bmatrix} \frac{1}{\sqrt{2}} & \frac{1}{\sqrt{2}} \\ -\frac{1}{\sqrt{2}} & \frac{1}{\sqrt{2}} \end{bmatrix}$ (b) $\mathbf{A} = \begin{bmatrix} -1 & 0 \\ 0 & 1 \end{bmatrix}$.

19. (a) $\mathbf{A} = \begin{bmatrix} \frac{1}{2} & -\frac{1}{2} \\ -\frac{1}{2} & \frac{1}{2} \end{bmatrix}$ (b) $\mathbf{A} = \begin{bmatrix} -\frac{1}{\sqrt{2}} & -\frac{1}{\sqrt{2}} \\ \frac{1}{\sqrt{2}} & -\frac{1}{\sqrt{2}} \end{bmatrix}$.

20. Write down the matrix **A** that describes the reflection in a line L through the origin inclined at an angle $\theta = -\pi/3$. to the $x_1 =$ axis. The corners of a

rectangle in the (x_1, x_2)-plane lie at the points A (1, 1), B (1, 3), C (0, 3) and D (0,1). Sketch to scale this rectangle and its reflection in line L, where the images of the corners A, B, C, D occur at the points A', B', C', D'. Use matrix $\mathbf{A}$ to find the coordinates of the corners of the image rectangle, and check that these agree with the location of the corners of the sketch of the image.

21. Find the coordinates of the orthogonal projection of the point (3, 5) in the (x_1, x_2)-plane onto the line L through the origin inclined to the x_1-axis at an angle $\theta = \pi/4$. Check the result with a sketch drawn to scale.
22. Find the coordinates of the orthogonal projection of the point (2, 6) in the (x_1, x_2)-plane onto the line L through the origin inclined to the x_1-axis at an angle $\theta = \pi/6$. Check the result with a sketch drawn to scale.
23. Prove that when a two-dimensional geometrical shape is reflected in a straight line L through the origin, and the image is then again reflected through the same line L, the original shape is reproduced.
24. Let a two-dimensional geometrical shape be reflected in the x_2-axis, and then let the resulting image be reflected in the x_1-axis. Prove that this final image can also be obtained by first reflecting the shape in a line inclined to the x_1-axis at an angle $\alpha = \pi/4$, and then reflecting this image in a line inclined at an angle $\alpha = -\pi/4$ to the x_1-axis.
25. Prove that a reflection in a line L_1 through the origin inclined to the x_1-axis at an angle α, followed by a reflection in a line L_2 through the origin inclined at an angle β to the x_1-axis, is equivalent to a rotation about the origin through an angle $2(\beta - \alpha)$. Taking $\alpha = \pi/12$ and $\beta = \pi/6$, find how the point $\mathbf{x} = [1,\ 1]^{\mathrm{T}}$ is mapped by this composite transformation, and sketch the result to scale. Does the result satisfy the rotation condition through an angle $2(\beta - \alpha)$, and if this appears not to be the case what is wrong with your interpretation of the rotation condition?
26. Write down the composite transformation matrix that describes first a uniform magnification by a factor 2, followed by a counterclockwise rotation about the origin through an angle $\theta = \pi/6$ measured from the x_1-axis, and finally a shear in the positive x_1-direction with parameter $\alpha = 1$. Using geometrical arguments, sketch the effect of this transformation on the rectangle A, B, C, D with its corners at the respective points (1.5, 0), (1.5, 1), (0, 1) and (0, 0), where these map to the corresponding image points A', B', C' and D'. Apply the composite transformation to find the coordinates of the respective image points A', B', C' and D' and check them against the sketch drawn to scale.
27. Construct a rectangle R in the (x_1, x_2)-plane of your own choice by specifying the coordinates of its corners. Use the transformation $\mathbf{Ax} = \mathbf{y}$, with $\mathbf{A} = \begin{bmatrix} 1 & 2 \\ 3 & 2 \end{bmatrix}$, to find the images in the (y_1, y_2)-plane of the corners of rectangle R. Verify that R maps to a rectangle $\widehat{R}$ in the (y_1, y_2)-plane, and confirm that the area of $\tilde{R}$ is $|\det \mathbf{A}|$ times the area of R.
28. Construct a rectangle R of your choice in the (x_1, x_2)-plane by specifying the coordinates of its corners, and a matrix $\mathbf{A}$ that will produce a counterclockwise rotation about the origin through an angle $\frac{1}{6}\pi$. By applying the transformation

$\mathbf{Ax} = \mathbf{y}$ to R, find how the corners of rectangle R map to the corners of the image $\tilde{R}$ of R in the (y_1, y_2)-plane. Hence confirm that the transformation has had the desired effect, and that the areas of R and $\tilde{R}$ are identical. How should matrix $\mathbf{A}$ be modified if the area of $\tilde{R}$ is to be three times that of R ?

29. Construct a triangle R of your choice in the (x_1, x_2)-plane by specifying the coordinates of its vertices. Construct a matrix $\mathbf{A}$ that will reflect the triangle R about a line L through the origin into an image $\tilde{R}$, where line L is inclined to the x_1-axis at an angle of your choice. Verify that the transformation has the desired effect, and give a mathematical reason why the areas R and $\tilde{R}$ are identical.
30. Write down (a) a matrix $\mathbf{A}$ that will stretch a rectangle or triangle of your choice in the (x_1, x_2)-plane in the x_1- and x_2-directions by the respective amounts a and b. (b) Write down a matrix $\mathbf{B}$ that will produce a shear in the x_1-direction by an amount k. Describe the geometrical effect on a rectangle or triangle if it is mapped (i) by the transformation $\mathbf{ABx} = \mathbf{y}$, and (ii) by the transformation $\mathbf{BAx} = \mathbf{y}$.
31. Matrix $\mathbf{A}$ is said to be a *singular transformation* if $\det \mathbf{A} = 0$. Describe the effect of a singular transformation $\mathbf{Ax} = \mathbf{y}$ when it maps points in the (x_1, x_2)-plane onto the (y_1, y_2)-plane.
32. Explain why changing the sign in an element of a matrix $\mathbf{A}$ that produces a stretch of a rectangle or triangle in the (x_1, x_2)-plane, causes a reflection of a stretched image rectangle or triangle in the direction of one of the axes.
33. Explain the geometrical effect on a transformation $\mathbf{Ax} = \mathbf{y}$ that maps a rectangle or triangle in the (x_1, x_2)-plane onto a corresponding rectangle or triangle in the (y_1, y_2)-plane if $\mathbf{A}$ is replaced by $\mathbf{PA}$ where $\mathbf{P} = \begin{bmatrix} 0 & 1 \\ 1 & 0 \end{bmatrix}$.

Solutions for All Exercises

Solutions 1

1. It is necessary to arrange entries in each equation in the *same* order before writing down the coefficient matrix **A**. Taking the order to be x_1, x_2, x_3 and x_4 gives

$$\mathbf{A} = \begin{bmatrix} 3 & 2 & -4 & 5 \\ 3 & 2 & 4 & -1 \\ -2 & 4 & 1 & 5 \\ 3 & 2 & 6 & 0 \end{bmatrix} \text{ and } \mathbf{B} = \begin{bmatrix} 4 \\ 3 \\ 2 \\ 1 \end{bmatrix}.$$

2.
$$\mathbf{A} + 2\mathbf{B} = \begin{bmatrix} 0 & 4 & 11 \\ -3 & 11 & 13 \end{bmatrix}, \ 3\mathbf{A} - 4\mathbf{B} = \begin{bmatrix} 10 & -8 & 3 \\ 11 & -7 & -21 \end{bmatrix}.$$

3. $a = 0$, $b = 6$, $c = 4$.
4.
$$3\mathbf{A} - \mathbf{B}^{\mathrm{T}} = \begin{bmatrix} 2 & 10 \\ 17 & 6 \\ 3 & 8 \end{bmatrix}, \quad 2\mathbf{A}^{\mathrm{T}} + 4\mathbf{B} = \begin{bmatrix} 20 & 16 & -12 \\ 16 & -10 & 10 \end{bmatrix}.$$

5.
$$\mathbf{A}^{\mathrm{T}} + \mathbf{B} = \begin{bmatrix} 3 & 5 & 6 \\ 2 & 9 & 2 \\ 4 & 1 & 4 \end{bmatrix}, \quad 2\mathbf{A} + 3(\mathbf{B}^{\mathrm{T}})^{\mathrm{T}} = 2\mathbf{A} + 3\mathbf{B} = \begin{bmatrix} 6 & 12 & 5 \\ 8 & 23 & 9 \\ 19 & -4 & 10 \end{bmatrix}.$$

6. The result follows directly from the definitions of matrix addition and transposition. $\mathbf{A} + \mathbf{B} = [a_{ij}] + [b_{ij}] = [a_{ij} + b_{ij}]$ and so $(\mathbf{A} + \mathbf{B})^{\mathrm{T}} = [a_{ij} + b_{ij}]^{\mathrm{T}} = [a_{ji} + b_{ji}] = \mathbf{A}^{\mathrm{T}} + \mathbf{B}^{\mathrm{T}}$.
7. (a) **A** is symmetric if the a_{ii} for $i = 1, 2, 3, 4$ are arbitrary, but $a_{14} = 1$, $a_{21} = 4$, $a_{31} = -3$, $a_{23} = 6$, $a_{24} = a_{42} = \alpha$ (arbitrary), $a_{43} = 7$.
   (b) **A** is skew symmetric if $a_{ii} = 0$ for $i = 1, 2, 3, 4$, $a_{14} = -1$, $a_{21} = -4$, $a_{31} = 3$, $a_{23} = -6$, $a_{24} = -\,a_{42}$ with $a_{24} = \alpha$ (arbitrary), $a_{43} = -\,7$.

8. Denote the symmetric matrix by $\mathbf{M} = [m_{ij}]$ and the skew symmetric matrix by $\mathbf{S} = [s_{ij}]$. Then if $\mathbf{A} = [a_{ij}]$ is to be decomposed into the sum $\mathbf{A} = \mathbf{M} + \mathbf{S}$, we must have $a_{ij} = m_{ij} + s_{ij}$ and $a_{ji} = m_{ij} - s_{ij}$. Solving these equations gives m_{ij} and s_{ij}, and hence $\mathbf{M}$ and $\mathbf{S}$. A more sophisticated approach uses the following argument: $\mathbf{A} = \frac{1}{2}(\mathbf{A} + \mathbf{A}^{\mathrm{T}}) + (\mathbf{A} - \mathbf{A}^{\mathrm{T}})$, but $\frac{1}{2}(\mathbf{A} + \mathbf{A}^{\mathrm{T}})^{\mathrm{T}} = \frac{1}{2}(\mathbf{A}^{\mathrm{T}} + \mathbf{A}) = \frac{1}{2}(\mathbf{A} + \mathbf{A}^{\mathrm{T}})$, so $\mathbf{M} = \frac{1}{2}(\mathbf{A} + \mathbf{A}^{\mathrm{T}})$. A similar argument shows $\frac{1}{2}(\mathbf{A} - \mathbf{A}^{\mathrm{T}})^{\mathrm{T}} = -\frac{1}{2}(\mathbf{A} - \mathbf{A}^{\mathrm{T}})$, and so $\mathbf{S} = \frac{1}{2}(\mathbf{A} - \mathbf{A}^{\mathrm{T}})$. For example, if

$$\mathbf{A} = \begin{bmatrix} 8 & 1 & 3 \\ 2 & 2 & 1 \\ 0 & 1 & 2 \end{bmatrix}, \quad \text{then} \quad \mathbf{M} = \begin{bmatrix} 8 & 3/2 & 3/2 \\ 3/2 & 2 & 1 \\ 3/2 & 1 & 2 \end{bmatrix},$$

$$\mathbf{S} = \begin{bmatrix} 0 & -1/2 & 3/2 \\ 1/2 & 0 & 0 \\ -3/2 & 0 & 0 \end{bmatrix}.$$

9. The solution set for system (a) with $x_2 = p$ arbitrary is $\{\frac{1}{2}(1-p), p, \frac{1}{2}(1-p)\}$, while the solution set for system (b) with $x_1 = q$ arbitrary is $\{q, 1-2q, q\}$. The solution set in the text with $x_3 = k$ arbitrary was $\{k, 1-2k, k\}$. Replacing q by k shows that the solution set for (b) is the same as the solution set found in the text. Setting $k = \frac{1}{2}(1-p)$ in the solution set for (a) it becomes $\{k, 1-2k, k\}$, which is again the solution set found in the text. Thus all three solution sets are equivalent. This demonstrates, as would be expected, that it is immaterial which variable is chosen as the arbitrary parameter.
10. System (a) has no solution. This can be shown in more than one way. The most elementary way being to solve the first three equations for x_1, x_2 and x_3, and then to substitute these values into the last equation to show that they do *not* satisfy it. Thus the last equation contradicts the other three, so there can be no solution set.
In system (b) the third equation is the sum of the first equation and twice the second equation, while the fourth equation is the difference between the first and second equations. Thus the last two equations are redundant. Setting $x_3 = k$ in the first two equations with k arbitrary parameter, and solving for x_1 and x_2 by elimination, gives $x_1 = 1 - k$ and $x_3 = k$, so a nonunique solution set exists given by $\{1 - k, k, k\}$.

Solutions 2

1. (a) $\det \mathbf{A} = 7$, (b) $\det \mathbf{A} = 0$, (c) $\det \mathbf{A} = -1$.
2. (a) $\det \mathbf{A} = -e^t \sin t$, (b) $\det \mathbf{A} = e^{-2t} + \cos t$.
3. The proof for an nth-order determinant is by induction. First, direct expansion shows that for a second-order determinant $\det \mathbf{A} = \det \mathbf{A}^{\mathrm{T}}$, showing that this result is true for $n = 2$. Now suppose the result is true for $n = k + 1$, then as the cofactor of a_{1j} in $\mathbf{A}$ is simply the cofactor of a_{j1} in $\mathbf{A}^{\mathrm{T}}$, expanding $\mathbf{A}$ in terms of elements of its first row is the same as expanding $\mathbf{A}^{\mathrm{T}}$ in terms of elements of its

first column, so det $\mathbf{A}$ = det $\mathbf{A}^{\mathrm{T}}$ when $n = k+1$. However the result is true for $k = 2$, so it is also true for $n = 2, 3, \ldots$, and the result is proved. In fact the result is true for $n = 1, 2, \ldots$, because the result is trivial when $n = 1$ since then det $\mathbf{A}$ is simply the single element a_{11}.

4. Subtract row 1 from row 2 and remove a factor $(\cos x - \sin x)$. Add column 2 to column 3, and then subtract column 3 from column 1. Finally, subtract $(1 - a)$ times column 3 from column 2 to obtain

$$\det\mathbf{A} = (\cos x - \sin x)\begin{vmatrix} e^x - \cos x - \sin x & \cos x - a(\cos x + \sin x) & \cos x + \sin x \\ 0 & 1 & 0 \\ 0 & 0 & 1 \end{vmatrix}$$
$$= (\cos x - \sin x)(e^x - \cos x - \sin x)\,.$$

5. Add rows 2 and 3 to row 1 and then subtract column 1 from columns 2 and 3 to obtain

$$\begin{vmatrix} 1+a & a & a \\ b & 1+b & b \\ b & b & 1+b \end{vmatrix} = \begin{vmatrix} 1+a+2b & 0 & 0 \\ b & 1 & 0 \\ b & 0 & 1 \end{vmatrix} = (1+a+2b)\,.$$

6. Subtract row 3 from row 1 and row 3 from row 2. Remove factors x^3 from rows 1 and 2. Subtract row 1 from row 3 and then row 2 from the new row 3 to obtain

$$x^6\begin{vmatrix} 1 & 0 & -1 \\ 0 & 1 & -1 \\ 0 & 0 & (x^3+3) \end{vmatrix} = x^6(x^3+3)\,.$$

7. Subtract 3/2 times row 1 from row 2 and ½ times row 1 from row 3 to obtain $\Delta = \begin{vmatrix} 2 & 1 & 0 & 1 \\ 3 & 2 & 4 & 2 \\ 1 & 2 & 1 & 3 \\ 0 & 3 & 1 & 1 \end{vmatrix} = \begin{vmatrix} 2 & 1 & 0 & 1 \\ 0 & \frac{1}{2} & 4 & \frac{1}{2} \\ 0 & \frac{3}{2} & 1 & \frac{5}{2} \\ 0 & 3 & 1 & 1 \end{vmatrix}$. Subtract 3 times row 2 from row 3 and 6 times row 2 from row 4 to obtain $\Delta = \begin{vmatrix} 2 & 1 & 0 & 1 \\ 0 & \frac{1}{2} & 4 & \frac{1}{2} \\ 0 & \frac{3}{2} & 1 & \frac{5}{2} \\ 0 & 3 & 1 & 1 \end{vmatrix} = \begin{vmatrix} 2 & 1 & 0 & 1 \\ 0 & \frac{1}{2} & 4 & \frac{1}{2} \\ 0 & 0 & -11 & 1 \\ 0 & 0 & -23 & -2 \end{vmatrix}$. Finally subtract 23/11 times row 3 from row 4 to obtain

$$\Delta = \begin{vmatrix} 2 & 1 & 0 & 1 \\ 0 & \frac{1}{2} & 4 & \frac{1}{2} \\ 0 & 0 & -11 & 1 \\ 0 & 0 & -23 & -2 \end{vmatrix} = \begin{vmatrix} 2 & 1 & 0 & 1 \\ 0 & \frac{1}{2} & 4 & \frac{1}{2} \\ 0 & 0 & -11 & 1 \\ 0 & 0 & 0 & -\frac{45}{11} \end{vmatrix} = 2 \times (\tfrac{1}{2}) \times (-11) \times (-\tfrac{45}{11})$$
$$= 45\,.$$

8. $\Delta = 63,\ \Delta_1 = 204,\ \Delta_2 = 183,\ \Delta_3 = 3$. Thus $x_1 = 68/21, x_2 = 61/21, x_3 = 1/21$.
9. (a) det $\mathbf{A} = 0$, so the equations are linearly dependent
   (b) det $\mathbf{A} = 26$, so the equations are linearly independent.
10. No, because det $\mathbf{A} = 0$, so the equations are linearly dependent.
11. $\det[\mathbf{A} - \lambda\mathbf{I}] = 0$ becomes $(\lambda - 1)(\lambda - 2)(\lambda - 3) = 0$, so the eigenvalues of $\mathbf{A}$ are $\lambda = 1, 2, 3$. The new matrix is $\mathbf{B} = \mathbf{A} + k\mathbf{I}$, so det $\mathbf{B} = \det[\mathbf{A} - (\lambda - k)\mathbf{I}] = 0$ and when expanded this becomes $(\lambda - 1 + k)(\lambda - 2 + k)(\lambda - 3 + k) = 0$, so the eigenvalues of $\mathbf{B}$ are $\lambda = 1 - k$, $\lambda = 2 - k$ and $\lambda = 3 - k$, confirming the statement in the exercise. The result *could* have been deduced directly from the form of the modified matrix $\mathbf{B}$, because the eigenvalues of $\mathbf{B}$ are solutions of $\det|\mathbf{A} - (\lambda - k)\mathbf{I}| = 0$ showing the eigenvalues of $\mathbf{B}$ are simply the eigenvalues of matrix $\mathbf{A}$ from each of which has been subtracted the constant k. The result is true for *all* square matrices, because in each case the modified matrix has the same property as the 3×3 matrix in the exercise.
12. Self checking
13. $J = r$. The Jacobian vanishes if $r = 0$, and the transformation fails because then the angle θ has no meaning.
14. $J = r^2 \sin\theta$. The Jacobian fails if $r = 0$ or $\theta = 0$. When $r = 0$ the angles $= 0$ θ and $\boldsymbol{\phi}$ have no meaning, and when $\theta = 0$ the angle ϕ has no meaning.

Solutions 3

1.
$$\mathbf{xy} = -15, \mathbf{yx} = \begin{bmatrix} 2 & -4 & 8 & 6 \\ 4 & -8 & 16 & 12 \\ -3 & 6 & -12 & -9 \\ 1 & -2 & 4 & 3 \end{bmatrix}.$$

2. $\mathbf{xA} = [15, -4, 23, 16]$.
3.
$$\mathbf{AB} = \begin{bmatrix} 2 & -6 & 7 \\ 20 & 2 & 11 \\ 7 & -4 & 8 \\ 17 & -13 & 16 \end{bmatrix}.$$

4.
$$(\mathbf{AB})^{\mathrm{T}} = \mathbf{B}^{\mathrm{T}}\mathbf{A}^{\mathrm{T}} = \begin{bmatrix} 56 & 54 \\ -18 & -20 \end{bmatrix}.$$

5.
$$Q(\mathbf{x}) = 2x_1^2 - 2x_1x_2 + x_1x_4 + x_2^2 + 5x_2x_3 + 9x_2x_4 + 3x_1x_3 + 4x_3^2 + 2x_3x_4 - x_4^2.$$

$$\mathbf{A} = \begin{bmatrix} 2 & -1 & \frac{3}{2} & \frac{1}{2} \\ -1 & 1 & \frac{5}{2} & \frac{9}{2} \\ \frac{3}{2} & \frac{5}{2} & 4 & 1 \\ \frac{1}{2} & \frac{9}{2} & 1 & -1 \end{bmatrix}.$$

6.

$$\mathbf{A}^{-1} = \frac{1}{ad-bc}\begin{bmatrix} d & -b \\ -c & a \end{bmatrix} = \begin{bmatrix} d/(ad-bc) & -b/(ad-bc) \\ -c/(ad-bc) & a/(ad-bc) \end{bmatrix},$$

$ad - bc \neq 0$.

7.

$$\mathbf{A}^{-1} = \begin{bmatrix} 1 & -1 & 3 \\ 2 & -3 & 8 \\ 5 & -7 & 18 \end{bmatrix}.$$

8.

$$\mathbf{A}^{-1} = \begin{bmatrix} -\frac{3}{7} & \frac{4}{7} & \frac{2}{7} \\ \frac{4}{7} & -\frac{3}{7} & \frac{2}{7} \\ -\frac{3}{7} & \frac{4}{7} & -\frac{5}{7} \end{bmatrix}.$$

9.

$$\mathbf{A}^{-1} = \begin{bmatrix} \frac{13}{121} & \frac{2}{11} & -\frac{10}{11} \\ \frac{2}{11} & 0 & \frac{1}{11} \\ -\frac{10}{11} & \frac{1}{11} & \frac{17}{121} \end{bmatrix}.$$

10.

$$\operatorname{adj}(\mathbf{A}) = \begin{bmatrix} 5 & 9 & 12 \\ 2 & 19 & -26 \\ -11 & 11 & -11 \end{bmatrix}, \quad \operatorname{adj}(\mathbf{A}^{-1}) = \begin{bmatrix} \frac{1}{77} & \frac{3}{77} & -\frac{6}{77} \\ \frac{4}{77} & \frac{1}{77} & \frac{2}{77} \\ \frac{3}{77} & -\frac{2}{77} & \frac{1}{77} \end{bmatrix},$$

$$\operatorname{adj}(\mathbf{A})\operatorname{adj}(\mathbf{A}^{-1}) = \mathbf{I}$$

11. Let $\mathbf{A}_1, \mathbf{A}_2, \ldots, \mathbf{A}_n$ be n nonsingular $m \times m$ matrices, the repeated application of the result $(\mathbf{AB})^{-1} = \mathbf{B}^{-1}\mathbf{A}^{-1}$ gives $(\mathbf{A}_1\mathbf{A}_2 \ldots \mathbf{A}_n)^{-1} = \mathbf{A}_n{}^{-1}\mathbf{A}_{n-1}{}^{-1} \ldots \mathbf{A}_1{}^{-1}$. The result then follows by setting $\mathbf{A}_1 = \mathbf{A}_2 = \cdots = \mathbf{A}_n$.
12. Setting $\mathbf{B} = \mathbf{A}^{-1}$ we have $\det \mathbf{I} = 1 = \det \mathbf{A} \det \mathbf{A}^{-1}$, from which the required result follows immediately.
13.

$$x_1 = \frac{12}{7}, \quad x_2 = \frac{20}{7}, \quad x_3 = \frac{27}{7}, \quad x_4 = \frac{9}{7}.$$

14.

$$x_1 = \frac{7}{11}, \quad x_2 = 1, \quad x_3 = -\frac{28}{11}, \quad x_4 = -\frac{23}{11}.$$

15. **PA** interchanges rows two and three, while **AP** interchanges columns two and three. If **PA** is to interchange the first and last rows of **A**, then **P** is obtained from **I** by interchanging its first and last rows. If, however, **AP** is to interchange the second and fourth columns of **A**, then **P** is obtained from **I** by interchanging its second and fourth columns.
16. If **P** is any permutation matrix, it interchanges rows in a certain order. The effect of $\mathbf{P}^{\mathrm{T}}$, where rows are transposed, is to reverse the effect of **P** by changing back the altered rows to their original order, so $\mathbf{P}\mathbf{P}^{\mathrm{T}} = \mathbf{P}^{\mathrm{T}}\mathbf{P} = \mathbf{I}$.
17. Equations one and two must be interchanged, and equations three and four must be interchanged, so

$$\mathbf{P} = \begin{bmatrix} 0 & 1 & 0 & 0 \\ 1 & 0 & 0 & 0 \\ 0 & 0 & 0 & 1 \\ 0 & 0 & 1 & 0 \end{bmatrix}.$$

18. Interchanging rows of **I** to form **P** will cause det **P** to equal -1 if an *odd* number of row interchanges have been made, and to equal 1 if an *even* number of row interchanges have been made. So from the result det (**AB**) = det **A** det **B** in Exercise 12 it follows that det **AP** = det **A** × (sign of det **P**). Pre-multiplication of a conformable matrix **A** by **P** to form **PA** interchanges its rows in a particular way, so $\mathbf{P}^2\mathbf{A}$ returns them to their original positions, consequently $\mathbf{P}^2\mathbf{A} = \mathbf{A}$, showing that $\mathbf{P}^2 = \mathbf{I}$. Thus $\mathbf{P}^2 = \mathbf{P}^{-1}\mathbf{P} = \mathbf{P}\mathbf{P}^{-1}$, confirming that $\mathbf{P}^{-1} = \mathbf{P}$.
19. (a) Orthogonal (b) not orthogonal (c) orthogonal.
20. **Q** is orthogonal because $\mathbf{Q}\mathbf{Q}^{\mathrm{T}} = \mathbf{I}$. To prove the last property notice first that if $\mathbf{q}_i$ and $\mathbf{q}_j$ are columns of **Q**, then $\mathbf{q}_i^{\mathrm{T}}$ and $\mathbf{q}_j$ are orthogonal. Next, permuting the columns of **Q** to produce $\mathbf{Q}_1$ results in a corresponding permutation of the rows of $\mathbf{Q}_1^{\mathrm{T}}$, so it remains true that $\mathbf{Q}_1^{\mathrm{T}}\mathbf{Q}_1 = \mathbf{I}$.
21. $$\mathbf{A} = \begin{bmatrix} 1 & 3 & -1 \\ 2 & -1 & 1 \\ -1 & 1 & 2 \end{bmatrix}, \quad \mathbf{b} = \begin{bmatrix} -5 \\ 9 \\ 5 \end{bmatrix}, \quad \mathbf{x} = \begin{bmatrix} x_1 \\ x_2 \\ x_3 \end{bmatrix} \quad \text{and}$$

$$\mathbf{A}^{-1} = \begin{bmatrix} \frac{3}{19} & \frac{7}{19} & -\frac{2}{19} \\ \frac{5}{19} & -\frac{1}{19} & \frac{3}{19} \\ -\frac{1}{19} & \frac{4}{19} & \frac{7}{19} \end{bmatrix}$$

so as $\mathbf{x} = \mathbf{A}^{-1}\mathbf{b}$, $x_1 = 2$, $x_2 = -1$ and $x_3 = 4$.

22. $\det[\mathbf{A} - \lambda\mathbf{I}] = \begin{vmatrix} 1-\lambda & 2 \\ 2 & 1-\lambda \end{vmatrix} = \lambda^2 - 2\lambda - 3$. So the eigenvalues of **A** are the roots of $\lambda^2 - 2\lambda - 3 = 0$; namely $\lambda_1 = -1$ and $\lambda_2 = 3$.

To find the eigenvector $\mathbf{x}^{(1)} = \left[x_1^{(1)}, x_2^{(1)}\right]^{\mathrm{T}}$ we must set $\lambda = \lambda_1$ in $[\mathbf{A} - \lambda_1 \mathbf{I}]\mathbf{x}^{(1)} = 0$.

The matrix equation becomes $\begin{bmatrix} 2 & 2 \\ 2 & 2 \end{bmatrix}\begin{bmatrix} x_1^{(1)} \\ x_2^{(1)} \end{bmatrix} = \mathbf{0}$, so this reduces to solving the single equation $x_1^{(1)} + x_1^{(2)} = 0$. Setting $x_1{}^{(2)} = k_1$ (arbitrary) we find that $x_1{}^{(1)} = -k_1$. Thus the eigenvector $\mathbf{x}^{(1)}$ corresponding to $\lambda = \lambda_1 = -1$ is seen to be $\mathbf{x}^{(1)} = k_1[-1, 1]^{\mathrm{T}}$, where k_1 is arbitrary. A similar argument with $\lambda = \lambda_2$ shows that the eigenvector $\mathbf{x}^{(2)}$ corresponding to $\lambda = \lambda_2 = 3$ is $\mathbf{x}^{(2)} = k_2[1,\ 1]^{\mathrm{T}}$, where k_2 is arbitrary.

23. $\det[\mathbf{A} - \lambda\mathbf{I}] = 0$ becomes $\lambda^3 + \lambda^2 - 10\lambda + 8 = 0$, so its roots (the eigenvalues of **A**) are $\lambda_1 = 1$, $\lambda_2 = 2$ and $\lambda_3 = -4$. The eigenvector $\mathbf{x}^{(1)} = [x_1{}^{(1)}, x_2{}^{(1)}, x_3{}^{(1)}]^{\mathrm{T}}$ is the solution of $[\mathbf{A} - \lambda_1\mathbf{I}]\mathbf{x}^{(1)} = \mathbf{0}$. Writing out this system in full, as $\det[\mathbf{A} - \lambda\mathbf{I}] = 0$, the system must be linearly dependent. Solving by elimination shows that the third equation is redundant, and setting $x_3{}^{(1)} = k_3$ (arbitrary) it turns out that the eigenvector $\mathbf{x}^{(1)}$ corresponding to $\lambda = \lambda_1 = 1$ is $\mathbf{x}^{(1)} = k_1[0,\ 0,\ 1]^{\mathrm{T}}$. Similar reasoning shows that the eigenvector $\mathbf{x}^{(2)}$ corresponding to $\lambda = \lambda_2 = 2$ has the form $x^{(2)} = k_2[1,\ 1,\ -4]^{\mathrm{T}}$, where k_2 is arbitrary, while the eigenvector $\mathbf{x}^{(3)}$ corresponding to $\lambda = \lambda_3 = -4$ has the form $\mathbf{x}^{(3)} = k_3[-1,\ 1,\ 0]^{\mathrm{T}}$, where k_3 is arbitrary.

24. Self-checking.

25.

$$\mathbf{AB} = \left[\mathbf{A}_1\mathbf{B}_1 + \mathbf{A}_2\mathbf{B}_2 + \cdots + \mathbf{A}_n\mathbf{B}_n\right],\ \mathbf{BA} = \left[\begin{array}{c|c|c|c} \mathbf{B}_1\mathbf{A}_1 & \mathbf{B}_1\mathbf{A}_2 & \cdots & \mathbf{B}_1\mathbf{A}_n \\ \hline \mathbf{B}_2\mathbf{A}_1 & \mathbf{B}_2\mathbf{A}_2 & \cdots & \mathbf{B}_2\mathbf{A}_n \\ \hline \vdots & \vdots & \vdots & \vdots \\ \hline \mathbf{B}_n\mathbf{A}_1 & \mathbf{B}_n\mathbf{A}_2 & \cdots & \mathbf{B}_n\mathbf{A}_n \end{array}\right].$$

26. $$\mathbf{AB} = \left[\begin{array}{c|c} \sum_{i=1}^{2}\mathbf{A}_{1i}\mathbf{B}_{i1} & \sum_{i=1}^{2}\mathbf{A}_{1i}\mathbf{B}_{i2} \\ \hline \sum_{i=1}^{2}\mathbf{A}_{2i}\mathbf{B}_{i1} & \sum_{i=1}^{2}\mathbf{A}_{i2}\mathbf{B}_{i2} \end{array}\right] \quad \text{so} \quad (\mathbf{AB})^{\mathrm{T}} = \left[\begin{array}{c|c} \sum_{i=1}^{2}\mathbf{A}_{1i}\mathbf{B}_{i1} & \sum_{i=1}^{2}\mathbf{A}_{2i}\mathbf{B}_{i1} \\ \hline \sum_{i=1}^{2}\mathbf{A}_{1i}\mathbf{B}_{i2} & \sum_{i=1}^{2}\mathbf{A}_{i2}\mathbf{B}_{i2} \end{array}\right] = \mathbf{B}^{\mathrm{T}}\mathbf{A}^{\mathrm{T}}.$$

27. The argument proceeds in the same way as in the text, but partitioning **A** and $\mathbf{B} = \mathbf{A}^{-1}$ into 2×2 block matrices. Solving the four matrix equations for the $\mathbf{B}_{ij}$ then gives

$$\mathbf{A}^{-1} = \left[\begin{array}{c|c} \mathbf{A}_{11}^{-1} & -\mathbf{A}_{11}^{-1}\mathbf{A}_{12}\mathbf{A}_{22}^{-1} \\ \hline \mathbf{0} & \mathbf{A}_{22}^{-1} \end{array}\right].$$

This result follows from the resulting the text by taking for $\mathbf{A}^{-1}$ the results in the first 2×2 block matrix, because the entries outside this block do not influence the block.

28. $$\mathbf{AB} = \left[\begin{array}{c|c} \mathbf{A}_{11}\mathbf{B}_{11} & \mathbf{A}_{11}\mathbf{B}_{12} + \mathbf{A}_{12}\mathbf{B}_{22} \\ \hline \mathbf{0} & \mathbf{A}_{22}\mathbf{B}_{22} \end{array}\right],$$ so from the result of Exercise 27,

$$(\mathbf{AB})^{-1} = \left[\begin{array}{c|c} (\mathbf{A}_{11}\mathbf{B}_{11})^{-1} & -(\mathbf{A}_{11}\mathbf{B}_{11})^{-1}[\mathbf{A}_{11}\mathbf{B}_{12} + \mathbf{A}_{12}\mathbf{B}_{22}](\mathbf{A}_{22}\mathbf{B}_{22})^{-1} \\ \hline \mathbf{0} & (\mathbf{A}_{22}\mathbf{B}_{22})^{-1} \end{array}\right].$$

The products $\mathbf{A}_{11}\mathbf{B}_{11}$ and $\mathbf{A}_{22}\mathbf{B}_{22}$ are ordinary matrix products, so $(\mathbf{A}_{11}\mathbf{B}_{11})^{\mathrm{T}} = \mathbf{B}_{11}^{T}\mathbf{A}_{11}^{\mathrm{T}}$ and $(\mathbf{A}_{22}\mathbf{B}_{22})^{\mathrm{T}} = \mathbf{B}_{22}^{\mathrm{T}}\mathbf{A}_{22}^{\mathrm{T}}$. Inserting these results into the expression for $(\mathbf{AB})^{\mathrm{T}}$and examining the product $\mathbf{B}^{\mathrm{T}}\mathbf{A}^{\mathrm{T}}$ shows that $(\mathbf{AB})^{\mathrm{T}} = \mathbf{B}^{\mathrm{T}}\mathbf{A}^{\mathrm{T}}$.

29. The result in Exercise 4 shows $\mathbf{A}^{-1} = \left[\begin{array}{c|c} \mathbf{B}^{-1} & -\mathbf{B}^{-1} \\ \hline \mathbf{0} & \mathbf{I} \end{array}\right]$. For the given matrix

$$\mathbf{A}^{-1} = \begin{bmatrix} 1 & 0 & -1 & -2 \\ 0 & 1 & 3 & -1 \\ 0 & 0 & 1 & 0 \\ 0 & 0 & 0 & 1 \end{bmatrix}.$$

30. (a) $\mathbf{A}$ is idempotent, so $\mathbf{A}^2 = \mathbf{A}$. The proof is by induction. Assuming $\mathbf{A}^k = \mathbf{A}$, multiplying by $\mathbf{A}$ gives $\mathbf{A}^{k+1} = \mathbf{A}^2 = \mathbf{A}$, so the result for $k + 1$ follows from the result for k. However, the result is true for $k = 2$, so it is true for $k \geq 2$.
(b) $\mathbf{A}^2 = \mathbf{A}$, so $\det \mathbf{A} \det \mathbf{A} = \det \mathbf{A}$, so either $\det \mathbf{A} = 0$ or $\det \mathbf{A} = 1$.
(c) If $\mathbf{D}$ is idempotent, then $\mathbf{D}^2 = \mathbf{D}$, but $\mathbf{D}^2 = \{\lambda_{12}^2, \lambda_2^2, \ldots, \lambda_n^2\}$. So $\mathbf{D}^2 = \mathbf{D}$ if and only if $\lambda_i = \lambda_i^2$ for $i = 1, 2, \ldots, n$ which is only possible if the λ_i are 0 or 1.
(d) $(\mathbf{A} + \mathbf{B})^2 = \mathbf{A}^2 + \mathbf{AB} + \mathbf{BA} + \mathbf{B}^2$. If $\mathbf{AB} = \mathbf{BA} = \mathbf{0}$, then $(\mathbf{A} + \mathbf{B})^2 = \mathbf{A}^2 + \mathbf{B}^2 = \mathbf{A} + \mathbf{B}$ establishing that $\mathbf{A} + \mathbf{B}$ is idempotent.
(e) $(\mathbf{I} - \mathbf{A})^2 = \mathbf{I} - 2\mathbf{A} + \mathbf{A}^2$, but $\mathbf{A}^2 = \mathbf{A}$, so $(\mathbf{I} - \mathbf{A})^2 = \mathbf{I} - \mathbf{A}$.
(f) $\mathbf{A}^2 = \mathbf{A}$, so $\mathbf{A}(\mathbf{I} - \mathbf{A}) = \mathbf{0}$. Hence, $\det \mathbf{A} \det(\mathbf{I} - \mathbf{A}) = 0$. As both factors cannot vanish, either $\det \mathbf{A} = 0$ or $\det(\mathbf{A} - \mathbf{I}) = 0$.

31. $a = d, b = d(d - 1)/c,\ c = c,\ d = d$, with c and d arbitrary. Four other cases are possible that cannot be deduced from these results. The first two are the trivial results $\mathbf{A} = \mathbf{I}$ and $\mathbf{A} = \mathbf{0}$, while the others are $a = 1, b = b, c = 0$ and $d = 0$, and $a = 0, b = b, c = 0$ and $d = 1$.

32. $\det \mathbf{A} = 106$ and the Hadamard overestimate of $|\det \mathbf{A}|$ is 174.24.

33.

$$\mathbf{X} = \begin{bmatrix} 1 & 0 \\ 1 & 1 \\ 1 & 2 \\ 1 & 3 \\ 1 & 4 \end{bmatrix}, \quad \mathbf{X}^T = \begin{bmatrix} 1 & 1 & 1 & 1 & 1 \\ 0 & 1 & 2 & 3 & 4 \end{bmatrix}, \quad \mathbf{X}^{\mathrm{T}}\mathbf{X} = \begin{bmatrix} 5 & 10 \\ 10 & 30 \end{bmatrix},$$

$$(\mathbf{X}^{\mathrm{T}}\mathbf{X})^{-1} = \begin{bmatrix} \frac{3}{5} & -\frac{1}{5} \\ -\frac{1}{5} & \frac{1}{10} \end{bmatrix}$$

$$\mathbf{y} = [-0.8,\ 0.3,\ 0.3,\ 1.3,\ 1.7]^{\mathrm{T}}, \quad \mathbf{a} = (\mathbf{X}^{\mathrm{T}}\mathbf{X})^{-1}\mathbf{X}^{\mathrm{T}}\mathbf{y} = \begin{bmatrix} -0.64 \\ 0.6 \end{bmatrix}.$$

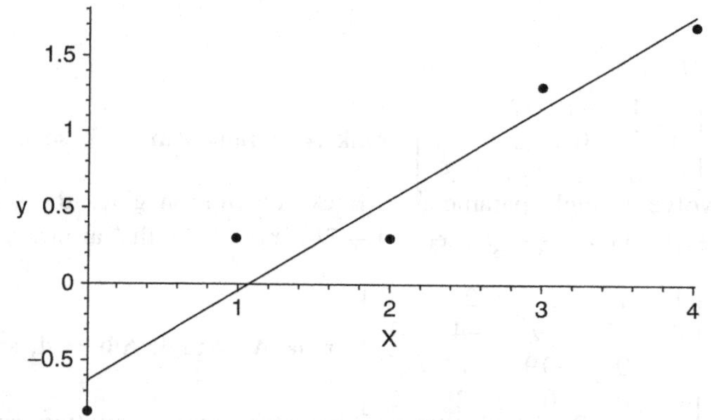

Fig. S3.33

The least-squares straight-line approximation is $\eta = -0.64 + 0.6x, \quad 0 \le x \le 4$.

34.

$$\mathbf{X} = \begin{bmatrix} 1 & -2 \\ 1 & -1 \\ 1 & 0 \\ 1 & 1 \\ 1 & 2 \\ 1 & 3 \end{bmatrix}, \quad \mathbf{X}^{\mathrm{T}} = \begin{bmatrix} 1 & 1 & 1 & 1 & 1 & 1 \\ -2 & -1 & 0 & 1 & 2 & 3 \end{bmatrix}, \quad \mathbf{X}^{\mathrm{T}}\mathbf{X} = \begin{bmatrix} 6 & 3 \\ 3 & 19 \end{bmatrix},$$

$$\left(\mathbf{X}^{\mathrm{T}}\mathbf{X}\right)^{-1} = \begin{bmatrix} \frac{19}{105} & -\frac{1}{35} \\ -\frac{1}{35} & \frac{2}{35} \end{bmatrix}.$$

$$\mathbf{y} = [1.93,\ 1.63,\ 0.75,\ 0.71,\ 0.47, -0.27]^{\mathrm{T}}, \quad \mathbf{a} = \left(\mathbf{X}^{\mathrm{T}}\mathbf{X}\right)^{-1}\mathbf{X}^{\mathrm{T}}\mathbf{y} = \begin{bmatrix} 1.077 \\ -0.415 \end{bmatrix}.$$

The least-squares straight-line approximation is $\eta = 1.077 - 0.415x, -2 \le x \le 3$.

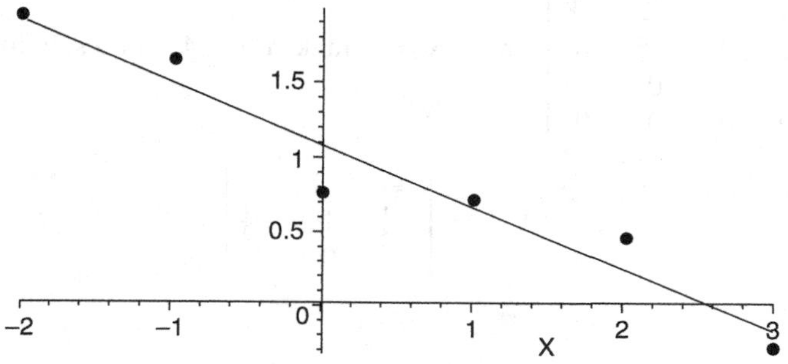

Fig. S3.34

Solutions 4

1. $\mathbf{A}|\mathbf{b} \sim \begin{bmatrix} 1 & 1 & -1 & 2 & 1 \\ 0 & 3 & 0 & 1 & -1 \\ 0 & 0 & 3 & -6 & 3 \end{bmatrix}$. rank $\mathbf{A}$ = rank $\mathbf{A|b}$ = 3, so the solutions will involve a single parameter k. Back substitution gives the solution set: $x_1 = \frac{7}{3} + \frac{1}{3}k$, $x_2 = -\frac{1}{3} - \frac{1}{3}k$, $x_3 = 1 + 2k$, $x_4 = k$, with k arbitrary.

2. $\mathbf{A}|\mathbf{b} \sim \begin{bmatrix} 1 & 1 & 3 & 2 & -2 \\ 0 & -2 & -9 & -4 & 9 \\ 0 & 0 & -19 & -12 & 17 \\ 0 & 0 & 0 & 12 & 2 \end{bmatrix}$. rank $\mathbf{A}$ = rank $\mathbf{A|b}$ = 4, so a unique solution exists. Back substitution gives the unique solution set $x_1 = 1$, $x_2 = -\frac{1}{3}$, $x_3 = -1$, $x_4 = \frac{1}{6}$.

3. $\mathbf{A}|\mathbf{b} \sim \begin{bmatrix} 1 & 4 & 2 & 3 \\ 0 & -5 & -3 & -5 \\ 0 & 0 & -3 & -10 \\ 0 & 0 & 0 & -16 \end{bmatrix}$. rank $\mathbf{A} = 3$, rank $\mathbf{A|b} = 4$, so no solution exists.

4. $\mathbf{A}|\mathbf{b} \sim \begin{bmatrix} 1 & -1 & 0 & 2 \\ 0 & 4 & -1 & -2 \\ 0 & 0 & 3 & -2 \\ 0 & 0 & 0 & 0 \\ 0 & 0 & 0 & 0 \end{bmatrix}$. rank $\mathbf{A}$ = rank $\mathbf{A|b}$ = 3, and there are only 3 linearly independent solutions, so back substitution gives the unique solution set $x_1 = \frac{4}{3}$, $x_2 = -\frac{2}{3}$, $x_3 = -\frac{2}{3}$.

5. $\mathbf{A}|\mathbf{b} \sim \begin{bmatrix} 2 & 0 & 1 & 4 \\ 0 & 8 & 4 & -4 \\ 0 & 0 & 20 & -8 \\ 0 & 0 & 0 & 0 \end{bmatrix}$. rank $\mathbf{A}$ = rank $\mathbf{A|b}$ = 3, and there are only 3 linearly independent solutions, so back substitution gives the unique solution set $x_1 = \frac{11}{5}$, $x_2 = -\frac{3}{10}$, $x_3 = -\frac{2}{5}$.

6. $\mathbf{A}|\mathbf{b} \begin{bmatrix} 1 & 2 & 1 & 5 \\ 0 & -3 & -2 & -8 \\ 0 & 0 & -6 & 6 \\ 0 & 0 & 0 & -2 \\ 0 & 0 & 0 & 0 \end{bmatrix}$. rank $\mathbf{A}$ = 3, rank $\mathbf{A|b}$ = 4, so no solution exists.

7.

$$\mathbf{A}^{-1} = \begin{bmatrix} \frac{2}{3} & -1 & \frac{2}{3} \\ -\frac{1}{3} & 1 & -\frac{1}{3} \\ -\frac{2}{3} & 0 & \frac{1}{3} \end{bmatrix}.$$

8.

$$\mathbf{A}^{-1} = \begin{bmatrix} -\frac{1}{4} & \frac{3}{4} & 1 \\ \frac{3}{4} & -\frac{5}{4} & -1 \\ \frac{1}{2} & -\frac{1}{2} & -1 \end{bmatrix}.$$

9.

$$\det(\mathbf{A} - \lambda\mathbf{I}) = -20 + 3\lambda^2 - \lambda^3 .$$

10.

$$\det(\mathbf{A} - \lambda\mathbf{I}) = 6 + 29\lambda - 11\lambda^2 - 3\lambda^3 + \lambda^4 .$$

11. (a) $\alpha > 0$ (b) $\alpha = -1/2$ (c) $\alpha < -1/2$
12. Self-checking
13.

$$\mathbf{L} = \begin{bmatrix} 1 & 0 & 0 \\ 2 & 1 & 0 \\ 1 & 0 & 1 \end{bmatrix}, \quad \mathbf{U} = \begin{bmatrix} 1 & 2 & 3 \\ 0 & -5 & -5 \\ 0 & 0 & -4 \end{bmatrix},$$

(a) $x_1 = \frac{1}{4}$, $x_2 = -\frac{3}{4}$, $x_3 = -\frac{1}{4}$, (b) $x_1 = \frac{7}{20}$, $x_2 = \frac{19}{20}$, $x_3 = \frac{1}{4}$.

14.

$$\mathbf{L} = \begin{bmatrix} 1 & 0 & 0 \\ 1 & 1 & 0 \\ -2 & -\frac{5}{2} & 1 \end{bmatrix}, \quad \mathbf{U} = \begin{bmatrix} 1 & -1 & 2 \\ 0 & 3 & -3 \\ 0 & 0 & -2 \end{bmatrix}, \quad x_1 = \tfrac{19}{6},\ x_2 = -\tfrac{3}{2},\ x_3.$$

15.

$$\mathbf{L} = \begin{bmatrix} 1 & 0 & 0 & 0 \\ \frac{1}{2} & 1 & 0 & 0 \\ \frac{1}{2} & -1 & 1 & 0 \\ -\frac{1}{2} & 3 & 0 & 1 \end{bmatrix}, \quad \mathbf{U} = \begin{bmatrix} 2 & 1 & 1 & 2 \\ 0 & \frac{1}{2} & \frac{1}{2} & -1 \\ 0 & 0 & -1 & -1 \\ 0 & 0 & 0 & 5 \end{bmatrix},$$

(a) $x_1 = \frac{3}{5}$, $x_2 = -\frac{2}{5}$, $x_3 = \frac{9}{5}$, $x_4 = -\frac{4}{5}$, (b) $x_1 = \frac{4}{5}$, $x_2 = -\frac{1}{5}$, $x_3 = -\frac{8}{5}$, $x_4 = \frac{3}{5}$.

16. This exercise requires a row interchange between rows 1 and 2, so **P**, **L** and **U** are

$$\mathbf{P} = \begin{bmatrix} 0 & 1 & 0 & 0 \\ 1 & 0 & 0 & 0 \\ 0 & 0 & 1 & 0 \\ 0 & 0 & 0 & 1, \end{bmatrix}, \quad \mathbf{L} = \begin{bmatrix} 1 & 0 & 0 & 0 \\ 0 & 1 & 0 & 0 \\ \frac{1}{2} & \frac{5}{2} & 1 & 0 \\ \frac{1}{2} & \frac{3}{2} & \frac{1}{3} & 1 \end{bmatrix}, \quad \mathbf{U} = \begin{bmatrix} 2 & -1 & -1 & -1 \\ 0 & 1 & -1 & 1 \\ 0 & 0 & 3 & -1 \\ 0 & 0 & 0 & -\frac{5}{3} \end{bmatrix},$$

$$x_1 = -1,\ x_2 = \tfrac{1}{2},\ x_3 = -\tfrac{5}{2},\ x_4 = -1.$$

17. Self checking.

Solutions 5

Remember that eigenvectors are indeterminate up to an arbitrary scale factor, and they can be arranged in any order when constructing a diagonalizing matrix or an orthogonal diagonalizing matrix.

1.
$$\lambda^3 - 3\lambda^2 + 2\lambda = 0; \quad \lambda_1 = 2, \quad \mathbf{x}_1 = \begin{bmatrix} 1 \\ -1 \\ -1 \end{bmatrix}, \quad \lambda_2 = 0, \quad \mathbf{x}_2 = \begin{bmatrix} 1 \\ 1 \\ 1 \end{bmatrix},$$
$$\lambda_3 = 1, \quad \mathbf{x}_3 = \begin{bmatrix} 0 \\ 1 \\ 0 \end{bmatrix}.$$

2.
$$\lambda^3 - 5\lambda^2 + 7\lambda - 3 = 0; \ \lambda_1 = \lambda_2 = 1, \ \mathbf{x}_1 = \mathbf{x}_2 = \begin{bmatrix} -1 \\ 0 \\ 1 \end{bmatrix}, \ \lambda_3 = 3, \ \mathbf{x}_3 = \begin{bmatrix} 1 \\ 2 \\ 1 \end{bmatrix}.$$

There are three eigenvalues, but only two linearly independent eigenvectors. The algebraic multiplicity of the eigenvalue $\lambda = 1$ is 2, but its geometric multiplicity is 1.

3.
$$\lambda^3 - \lambda^2 - \lambda + 1 = 0; \quad \lambda_1 = -1,$$
$$\mathbf{x}_1 = \begin{bmatrix} -1 \\ -2 \\ 1 \end{bmatrix}, \lambda_2 = \lambda_3 = 1, \ \mathbf{x}_2 = \mathbf{x}_3 = \begin{bmatrix} -1 \\ 0 \\ 1 \end{bmatrix}.$$

There are three eigenvalues, but only two linearly independent eigenvectors. The algebraic multiplicity of the eigenvalue $\lambda = 1$ is 2, but its geometric multiplicity is 1.

4.
$$\lambda^3 - 3\lambda = 0; \quad \lambda_1 = \sqrt{3}, \quad \mathbf{x}_1 = \begin{bmatrix} 1 \\ 1 + \sqrt{3} \\ 2 + \sqrt{3} \end{bmatrix}, \quad \lambda_2 = -\sqrt{3},$$
$$\mathbf{x}_2 = \begin{bmatrix} 1 \\ 1 - \sqrt{3} \\ 2 - \sqrt{3} \end{bmatrix}, \quad \lambda_3 = 0, \quad \mathbf{x}_3 = \begin{bmatrix} -1 \\ -1 \\ 1 \end{bmatrix}.$$

5.
$$\lambda^3 + 2\lambda^2 - 7\lambda - 2 = 0; \quad \lambda_1 = 2, \quad \mathbf{x}_1 = \begin{bmatrix} 1 \\ -4 \\ -9 \end{bmatrix}, \quad \lambda_2 = -2 + \sqrt{3},$$
$$\mathbf{x}_2 = \begin{bmatrix} \sqrt{3} - 1 \\ 1 \\ -1 \end{bmatrix}, \quad \lambda_3 = -2 - \sqrt{3}, \quad \mathbf{x}_3 = \begin{bmatrix} -1 - \sqrt{3} \\ 1 \\ -1 \end{bmatrix}.$$

6.

$$\lambda^3 - 4\lambda^2 + 5\lambda - 2 = 0; \quad \lambda_1 = 2, \quad \mathbf{x}_1 = \begin{bmatrix} 1 \\ 0 \\ 1 \end{bmatrix}, \quad \lambda_2 = 1, \quad \mathbf{x}_2 = \begin{bmatrix} -1 \\ 1 \\ 0 \end{bmatrix},$$

$$\lambda_3 = 1, \quad \mathbf{x}_3 = \begin{bmatrix} 0 \\ 0 \\ 1 \end{bmatrix}.$$

Three linearly independent eigenvectors. The repeated eigenvalue $\lambda = 1$ has the algebraic multiplicity = geometric multiplicity = 2.

7.

$$\lambda^3 - 2\lambda^2 - \lambda + 2 = 0; \quad \lambda_1 = -1, \quad \mathbf{x}_1 = \begin{bmatrix} 1 \\ 1 \\ 1 \end{bmatrix}, \quad \lambda_2 = 2, \quad \mathbf{x}_2 = \begin{bmatrix} 0 \\ 1 \\ 1 \end{bmatrix},$$

$$\lambda_3 = 1, \quad \mathbf{x}_3 = \begin{bmatrix} 1 \\ 0 \\ 1 \end{bmatrix}.$$

8.

$$\lambda^3 - 2\lambda^2 + \lambda = 0; \quad \lambda_1 = 0, \quad \mathbf{x}_1 = \begin{bmatrix} 0 \\ 1 \\ 2 \end{bmatrix}, \quad \lambda_2 = 1, \quad \mathbf{x}_2 = \begin{bmatrix} 1 \\ 0 \\ 1 \end{bmatrix}, \quad \lambda_3 = 1,$$

$$\mathbf{x}_3 = \begin{bmatrix} 1 \\ 1 \\ 0 \end{bmatrix}.$$

Three linearly independent eigenvectors. The repeated eigenvalue $\lambda = 1$ has the algebraic multiplicity = geometric multiplicity = 2.

9.

$$\lambda^3 + 2\lambda^2 - \lambda - 2 = 0; \quad \mathbf{P} = \begin{bmatrix} 0 & 1 & 2 \\ -1 & 0 & 1 \\ 1 & 1 & 0 \end{bmatrix}.$$

10. $\lambda^3 + 2\lambda^2 - \lambda - 2 = 0; \quad \mathbf{P} = \begin{bmatrix} -1 & 1 & 0 \\ 1 & 0 & 1 \\ 1 & 1 & 1 \end{bmatrix}$. Notice that Exercise 9 has the same characteristic equation as here, but the matrices **A** and **P** are different.

11.

$$\lambda^3 + 2\lambda^2 - \lambda - 2 = 0; \quad \mathbf{P} = \begin{bmatrix} 2 & -2 & 1 \\ -1 & 1 & 0 \\ 1 & 0 & -1 \end{bmatrix}.$$

12.

$$\lambda^3 - 2\lambda^2 - 5\lambda + 6 = 0; \quad \mathbf{P} = \begin{bmatrix} 1 & -1 & 2 \\ 0 & 1 & -1 \\ -1 & -1 & 1 \end{bmatrix}.$$

13.

$$\lambda^3+\lambda^2+\lambda=0;\quad \lambda_1=0,\quad \mathbf{x}_1=\begin{bmatrix}2\\1\\-2\end{bmatrix},\quad \lambda_2=-\tfrac{1}{2}(1+i\sqrt{3}),$$

$$\mathbf{x}_2=\begin{bmatrix}1\\1\\-\tfrac{1}{2}(1-i\sqrt{3})\end{bmatrix},$$

$$\lambda_3=-\tfrac{1}{2}(1-i\sqrt{3}),\quad \mathbf{x}_3=\begin{bmatrix}1\\1\\-\tfrac{1}{2}(1+i\sqrt{3})\end{bmatrix}.\quad \text{Diagonalizable.}$$

14.

$$\lambda_1=1,\ \mathbf{x}_1=\begin{bmatrix}1\\5\\4\end{bmatrix},\ \lambda_2=\lambda_3=-2,\ \mathbf{x}_2=\mathbf{x}_3=\begin{bmatrix}1\\-1\\1\end{bmatrix}.\text{Non}-\text{diagonalizable.}$$

There are three eigenvalues but only two linearly independent eigenvectors. The algebraic multiplicity of the eigenvalue $\lambda=-2$ is 2, but its geometric multiplicity is 1.

15.

$$\lambda^3-\lambda^2-2\lambda=0,\quad \mathbf{Q}=\begin{bmatrix}0&\frac{1}{\sqrt{2}}&-\frac{1}{\sqrt{2}}\\1&0&0\\0&\frac{1}{\sqrt{2}}&\frac{1}{\sqrt{2}}\end{bmatrix}.$$

16.

$$\lambda^3+\lambda^2-17\lambda+15=0,\quad \mathbf{Q}=\begin{bmatrix}-\frac{1}{\sqrt{3}}&0&\frac{2}{\sqrt{6}}\\\frac{1}{\sqrt{3}}&-\frac{1}{\sqrt{2}}&\frac{1}{\sqrt{6}}\\\frac{1}{\sqrt{3}}&\frac{1}{\sqrt{2}}&\frac{1}{\sqrt{6}}\end{bmatrix}.$$

In Exercises 17 through 20 remember that matrix $\mathbf{Q}$ is not unique, because the order in which the eigenvectors are used to form its columns is arbitrary. However, the final reduction will always be the same whatever the ordering of the columns of $\mathbf{Q}$.

17.

$$\mathbf{Q}=\begin{bmatrix}-\frac{1}{\sqrt{2}}&\frac{1}{\sqrt{3}}&\frac{1}{\sqrt{6}}\\\frac{1}{\sqrt{2}}&\frac{1}{\sqrt{3}}&\frac{1}{\sqrt{6}}\\0&-\frac{1}{\sqrt{3}}&\frac{2}{\sqrt{6}}\end{bmatrix},\quad x_1=-\tfrac{1}{\sqrt{2}}y_1+\tfrac{1}{\sqrt{3}}y_2+\tfrac{1}{\sqrt{6}}y_3,$$

$$x_2=\tfrac{1}{\sqrt{2}}y_1+\tfrac{1}{\sqrt{3}}y_2+\tfrac{1}{\sqrt{6}}y_3,\quad x_3=-\tfrac{1}{\sqrt{3}}y_2+\tfrac{2}{\sqrt{6}}y_2.$$

Eigenvalues are $-3, 2, -1$. $\mathbf{Q}(\mathbf{y}) = -3y_1^2 + 2y_2^2 - y_3^2$, so the quadratic form is indefinite. The classification is obvious from the values of the eigenvalues without further calculation.

18.

$$\mathbf{Q} = \begin{bmatrix} \frac{1}{\sqrt{3}} & \frac{1}{\sqrt{6}} & -\frac{1}{\sqrt{2}} \\ \frac{1}{\sqrt{3}} & -\frac{2}{\sqrt{6}} & 0 \\ \frac{1}{\sqrt{3}} & \frac{1}{\sqrt{6}} & \frac{1}{\sqrt{2}} \end{bmatrix}, \quad x_1 = \frac{1}{\sqrt{3}}y_1 + \frac{1}{\sqrt{6}}y_2 - \frac{1}{\sqrt{2}}y_3, \quad x_2 = \frac{1}{\sqrt{3}}y_1 - \frac{2}{\sqrt{6}}y_2,$$

$$x_3 = \frac{1}{\sqrt{3}}y_1 + \frac{1}{\sqrt{6}}y_2 + \frac{1}{\sqrt{2}}y_3.$$

Eigenvalues are $2, -1, 1$. $\mathbf{Q}(\mathbf{y}) = 2y_1^2 - y_2^2 + y_3^2$, so the quadratic form is indefinite. The classification is obvious from the values of the eigenvalues without further calculation.

19.

$$\mathbf{Q} = \begin{bmatrix} -\frac{2}{\sqrt{5}} & \frac{1}{\sqrt{5}} & 0 \\ 0 & 0 & 1 \\ \frac{1}{\sqrt{5}} & \frac{2}{\sqrt{5}} & 0 \end{bmatrix}, \quad x_1 = -\frac{2}{\sqrt{5}}y_1 + \frac{1}{\sqrt{5}}y_2, \quad x_2 = y_3, \quad x_3 = \frac{1}{\sqrt{5}}y_1 + \frac{2}{\sqrt{5}}y_2.$$

Eigenvalues are $-2, 3, -1$. $\mathbf{Q}(\mathbf{y}) = -y_1^2 - 2y_2^2 + 3y_3^2$, so the quadratic form is indefinite. The classification is obvious from the values of the eigenvalues without further calculation.

20.

$$\mathbf{Q} = \begin{bmatrix} \frac{2}{\sqrt{6}} & -\frac{1}{\sqrt{3}} & 0 \\ -\frac{1}{\sqrt{6}} & -\frac{1}{\sqrt{3}} & \frac{1}{\sqrt{2}} \\ \frac{1}{\sqrt{6}} & \frac{1}{\sqrt{3}} & \frac{1}{\sqrt{2}} \end{bmatrix}, \quad x_1 = \frac{2}{\sqrt{6}}y_1 - \frac{1}{\sqrt{3}}y_2, \quad x_2 = -\frac{1}{\sqrt{6}}y_1 - \frac{1}{\sqrt{3}}y_2 + \frac{1}{\sqrt{2}}y_3,$$

$x_3 = \frac{1}{\sqrt{6}}y_1 + \frac{1}{\sqrt{3}}y_2 + \frac{1}{\sqrt{2}}y_3$. Eigenvalues are $3, 0, 1$. $\mathbf{Q}(\mathbf{y}) = 3y_1^2 + y_3^2$, so the quadratic form is positive definite. The classification is obvious from the values of the eigenvalues without further calculation.

21. If the characteristic equation is $\lambda^n + c_1\lambda^{n-1} + \cdots + c_{n-1}\lambda + c_n = 0$, the Cayley–Hamilton equation becomes $\mathbf{A}^n + c_1\mathbf{A}^{n-1} + c_2\mathbf{A}^{n-2} + \cdots + c_{n-1}\mathbf{A} + c_n\mathbf{I} = \mathbf{0}$. When $\det \mathbf{A} \neq 0$ matrix $\mathbf{A}^{-1}$ exists, and pre-multiplication of the Cayley–Hamilton equation by $\mathbf{A}^{-1}$ followed by a rearrangement of terms gives $\mathbf{A}^{-1} = (1/c_n)\left(\mathbf{A}^{n-1} + c_1\mathbf{A}^{n-2} + \cdots + c_{n-1}\mathbf{A}\right)$, so $\mathbf{A}^{-1}$ is expressed in terms of powers of $\mathbf{A}$. If $\mathbf{A}$ is singular, $\mathbf{A}^{-1}$ does not exist, and the matrix form of the characteristic equation will not contain a multiple of the unit matrix $\mathbf{I}$. So the *formal* result of the pre-multiplication such a matrix characteristic equation by $\mathbf{A}^{-1}$ will not yield an expression for $\mathbf{A}^{-1}$, so the inverse matrix cannot be found. In addition, the characteristic equation will have a root $\lambda = 0$, and when it inserted in $|\mathbf{A} - \lambda\mathbf{I}| = 0$, it will give $|\mathbf{A}| = 0$, showing that $\mathbf{A}$ is singular. The purpose of the Cayley–Hamillton equation in this exercise is to provide a simple application of this general result, and also to help develop experience combining matrices. When this method is

used with the given matrix **A** to find $\mathbf{A}^{-1}$, to ensure the result is correct it is necessary check that $\mathbf{AA}^{-1} = \mathbf{I}$. The matrix **A** in this exercise is not singular, because $\det\mathbf{A} = -2$, so $\mathbf{A}^{-1}$ exists. The characteristic equation is $\lambda^3 - 3\lambda^2 + 4\lambda + 4 = 0$, so using the above method, but omitting the details of the calculations, gives

$$\mathbf{A}^{-1} = \begin{bmatrix} \frac{1}{2} & -\frac{1}{2} & \frac{1}{2} \\ -\frac{1}{2} & \frac{1}{2} & \frac{1}{2} \\ \frac{1}{2} & \frac{1}{2} & -\frac{1}{2} \end{bmatrix}.$$

22. Matrix **A** has the repeated eigenvalue $\lambda = 2$ and the single eigenvector $[0, 1]^{\mathrm{T}}$. Set $\mathbf{C} = \begin{bmatrix} 0 & 1 \\ 1 & 1 \end{bmatrix}$ when $\mathbf{C}^{-1} = \begin{bmatrix} -1 & 1 \\ 1 & 0 \end{bmatrix}$, so $\mathbf{C}^{-1}\mathbf{AC} = \begin{bmatrix} 2 & 2 \\ 0 & 2 \end{bmatrix}$.

$\mathbf{M} = \begin{bmatrix} 1 & 0 \\ 0 & \frac{1}{2} \end{bmatrix}$, so $\mathbf{Q} = \mathbf{CM} = \begin{bmatrix} 0 & \frac{1}{2} \\ 1 & \frac{1}{2} \end{bmatrix}$ and $\mathbf{Q}^{-1} = \begin{bmatrix} -1 & 1 \\ 2 & 0 \end{bmatrix}$, giving $\mathbf{Q}^{-1}\mathbf{AQ}$

$= \begin{bmatrix} 2 & 1 \\ 0 & 2 \end{bmatrix}$, showing that **A** is similar to $\mathbf{J}_3$.

23. Matrix **A** has the eigenvalues $\lambda_1 = 1$ and $\lambda_2 = 5$, with the eigenvectors $[1, -1]^{\mathrm{T}}$ and $[1, 1]^{\mathrm{T}}$. The matrix is diagonalizable by the matrix $\mathbf{Q} = \begin{bmatrix} 1 & 1 \\ -1 & 1 \end{bmatrix}$ when $\mathbf{Q}^{-1}\mathbf{AQ} = \begin{bmatrix} 1 & 0 \\ 0 & 5 \end{bmatrix}$, so the matrix is similar to $\mathbf{J}_1$.

24. Matrix **A** has a repeated eigenvalue $\lambda = 3$, and the single eigenvector $[-1, 1]^{\mathrm{T}}$.

Set $\mathbf{C} = \begin{bmatrix} -1 & 1 \\ 1 & 0 \end{bmatrix}$ when $\mathbf{C}^{-1} = \begin{bmatrix} 0 & 1 \\ 1 & 1 \end{bmatrix}$, so $\mathbf{C}^{-1}\mathbf{AC} = \begin{bmatrix} 3 & -2 \\ 0 & 3 \end{bmatrix}$.

$\mathbf{M} = \begin{bmatrix} 1 & 0 \\ 0 & -\frac{1}{2} \end{bmatrix}$, so $\mathbf{Q} = \mathbf{CM} = \begin{bmatrix} -1 & -\frac{1}{2} \\ 1 & 0 \end{bmatrix}$ and $\mathbf{Q}^{-1} = \begin{bmatrix} 0 & 1 \\ -2 & -2 \end{bmatrix}$,

giving $\mathbf{Q}^{-1}\mathbf{AQ} = \begin{bmatrix} 3 & 1 \\ 0 & 3 \end{bmatrix}$, showing that **A** is similar to $\mathbf{J}_3$.

25. Matrix **A** has the complex conjugate eigenvalues $\lambda_\pm = 1 \pm i\sqrt{6}$, so it is similar to $\mathbf{J}_4$ with $\alpha = 1,\ \beta = \sqrt{6}$, so it is similar to $\begin{bmatrix} 1 & -\sqrt{6} \\ \sqrt{6} & 1 \end{bmatrix}$.

26. Matrix **A** has a repeated eigenvalue $\lambda = 4$, and the single eigenvector $[1, 0]^{\mathrm{T}}$. Set $\mathbf{C} = \begin{bmatrix} 1 & 1 \\ 0 & 1 \end{bmatrix}$ when $\mathbf{C}^{-1} = \begin{bmatrix} 1 & -1 \\ 0 & 1 \end{bmatrix}$, so $\mathbf{C}^{-1}\mathbf{AC} = \begin{bmatrix} 4 & -2 \\ 0 & 4 \end{bmatrix}$.

$$\mathbf{M}=\begin{bmatrix}1 & 0\\ 0 & -\frac{1}{2}\end{bmatrix},\quad \text{so } \mathbf{Q}{=}\mathbf{CM}=\begin{bmatrix}-1 & -\frac{1}{2}\\ 0 & -\frac{1}{2}\end{bmatrix} \text{ and } \mathbf{Q}^{-1}=\begin{bmatrix}1 & -1\\ 0 & -2\end{bmatrix},$$

giving $\mathbf{Q}^{-1}\mathbf{AQ}=\begin{bmatrix}4 & 1\\ 0 & 4\end{bmatrix}$, showing that $\mathbf{A}$ is similar to $\mathbf{J}_3$.

27. Matrix $\mathbf{A}$ has the complex conjugate eigenvalues $\lambda_\pm = -1 \pm 2i$, so it is similar to $\mathbf{J}_4$ with $\alpha = -1$, $\beta = 2$, so it is similar to $\begin{bmatrix}-1 & -2\\ 2 & -1\end{bmatrix}$.
28. Self checking.
29. Let ζ be a zero of the nth degree polynomial $P_n(z) = z^n + a_1 z^{n-1} + \cdots + a_n$, then $\zeta^n + a_1\zeta^{n-1} + \cdots + a_n = 0$. Taking the complex conjugate of this equation in which the coefficients $a_1, a_2, \ldots, a_n$ are real numbers, so $\bar{a}_r = a_r$, and using the fact that $\overline{\zeta^r} = \bar{\zeta}^r$, shows that $\overline{a_r\zeta^r} = a_r\overline{\zeta^r} = a_r\bar{\zeta}^r$, so $\bar{\zeta}^n + a_1\bar{\zeta}^{n-1} + \cdots + a_n = 0$, and hence $\bar{\zeta}$ must also be a zero of $P_n(z)$. So if there are complex zeros they must occur in complex conjugate pairs. Thus $(z-\zeta)$ and $(z-\bar{\zeta})$ are both factors of $P_n(z)$. Setting $\zeta = \alpha + i\beta$ the product of the two factors

$$(z-\zeta)(z-\bar{\zeta}) = (z-\alpha-i\beta)(z-\alpha+i\beta) = z^2 - 2\alpha z + \alpha^2 + \beta^2.$$

Thus the complex conjugate zeros are seen to produce a real quadratic factor of $P_n(z)$. Consequently, if the degree of $P_n(z)$ is odd, either all of its zeros are real or, if some are complex conjugate pairs, there must be at least one real zero of $P_n(z)$.

30. Q is positive definite and $\mathbf{A} = \begin{bmatrix}4 & -1 & 0\\ -1 & 4 & 0\\ 0 & 0 & 1\end{bmatrix}$. $|\det \mathbf{A}| = 15$ and the Hadamard inequality gives 16.
31. Q is positive definite and $\mathbf{A} = \begin{bmatrix}\frac{5}{2} & 0 & \frac{1}{2}\\ 0 & 1 & 0\\ \frac{1}{2} & 0 & \frac{5}{2}\end{bmatrix}$. $|\det \mathbf{A}| = 6$ and the Hadamard inequality gives $\frac{25}{4} = 6.25$.
32. Q is positive definite and $\mathbf{A} = \begin{bmatrix}2 & 2 & -1 & -1\\ 2 & 7 & -2 & 1\\ -1 & -2 & 12 & 0\\ -1 & 1 & 0 & 3\end{bmatrix}$. $|\det \mathbf{A}| = 192$ and the Hadamard inequality gives 504. Poor though this estimate is, it is better than the estimate provided by the inequality in Exercise 32, of Chapter 3, which gives 974.997. One reason for this is that the result in Exercise 32, of Chapter 3, applies to an arbitrary determinant, whereas the result in this case is restricted to the class of determinants associated with positive definite matrices, so this constraint leads to a better estimate.
33. (a) $\rho(\mathbf{A}) = 0$ so $\lim\limits_{n\to\infty} \mathbf{A} = \mathbf{0}$, and in fact $\mathbf{A}^4 = \mathbf{0}$.

(b) $|\mathbf{A} - \lambda\mathbf{I}| = 0$ gives the characteristic equation $(\lambda - 1)\left(\lambda^2 - \frac{13}{12}\lambda + \frac{1}{12}\right) = 0$, so the eigenvalues of $\mathbf{A}$ (the spectrum of $\mathbf{A}$) are$\lambda = 1$ (twice) and $\lambda = \frac{1}{12}$. Thus $\rho(\mathbf{A}) = 1$ so $\mathbf{L} = \lim_{n\to\infty} \mathbf{A}^n$ is bounded, and rounded to four figures

$$\mathbf{L} = \begin{bmatrix} 1 & 0 & 0 \\ 0 & 0.2727 & 0.7273 \\ 0 & 0.2727 & 0.7273 \end{bmatrix}.$$

34. Self checking.
35. Self checking.
36. Use $\mathbf{U}_{n+1} = \mathbf{A}\mathbf{U}_n$ with $\mathbf{U}_n = \begin{bmatrix} u_{n+1} \\ u_n \end{bmatrix}$ and $\mathbf{A} = \begin{bmatrix} 1 & 2 \\ 1 & 0 \end{bmatrix}$. Reason as in Section 5.11 with $\lambda_1 = -1$, $\mathbf{x}_1 = \begin{bmatrix} 1 \\ -1 \end{bmatrix}$, $\lambda_2 = 2$, $\mathbf{x}_2 = \begin{bmatrix} 2 \\ 1 \end{bmatrix}$.
37. Exact solution $u(x) = \left(\frac{1}{2\pi^3}\right)(-5\sin(2\pi x) + 5\pi x\cos(2\pi x) - 5\pi x)$.

x	0.2	0.4	0.6	0.8
Exact	−0.1158	−0.2563	−0.2673	−0.0842
Approx	−0.1117	−0.2307	−0.2275	−0.0633

38.
$$\lambda_1 = -2,\ \mathbf{x}_1 = \begin{bmatrix} 1 \\ -i \end{bmatrix},\quad \lambda_2 = 4,\ \mathbf{x}_2 = \begin{bmatrix} -i \\ 1 \end{bmatrix}.$$

39.
$$\lambda_1 = 0,\ \mathbf{x}_1 = \begin{bmatrix} -i \\ 1 \end{bmatrix},\quad \lambda_2 = 2,\ \mathbf{x}_2 = \begin{bmatrix} 1 \\ -i \end{bmatrix}.$$

40.
$$\lambda_1 = \lambda_2 = -1 \text{ (twice)},\ \mathbf{x}_1 = \begin{bmatrix} -1 \\ 0 \\ 1 \end{bmatrix},\ \mathbf{x}_2 = \begin{bmatrix} -i \\ 1 \\ 0 \end{bmatrix},\ \lambda_3 = 2,\ \mathbf{x}_3 = \begin{bmatrix} 1 \\ -i \\ 1 \end{bmatrix}.$$

41.
$$\lambda_1 = -\sqrt{2},\ \mathbf{x}_1 = \begin{bmatrix} 1 \\ -i\sqrt{2} \\ -i \end{bmatrix},\ \lambda_2 = \sqrt{2},\ \mathbf{x}_2 = \begin{bmatrix} 1 \\ i\sqrt{2} \\ -i \end{bmatrix},\ \lambda_3 = 0,$$

$$\mathbf{x}_3 = \begin{bmatrix} 1 \\ 0 \\ i \end{bmatrix}.$$

42. By definition, the purely real parts of each element form a real symmetric matrix $\mathbf{A}_1$, while the purely imaginary parts of each entry form a real skew-symmetric matrix $\mathbf{A}_2$, so $\mathbf{A} = \mathbf{A}_1 + i\mathbf{A}_2$.

43. The reasoning follows that in the text for an Hermitian matrix except that now when $\bar{\mathbf{A}}^{\mathrm{T}}$ occurs it must be replaced by $-\mathbf{A}$. This leads to the conclusion that $\lambda + \bar{\lambda} = 0$, which is only possible if λ is purely imaginary.
44. Take the transpose of $\mathbf{U}^{\mathrm{T}} = \mathbf{U}^{-1}$ and use the fact that the transpose operation and the inverse of a matrix commute.
45. The characteristic equation is $\lambda^3 - 2\lambda^2 + 2\lambda - 1 = 0$, and inspection shows one eigenvalue is 1. After removing the factor $(\lambda - 1)$ from the characteristic equation and finding the roots of the remaining quadratic equation the other two eigenvalues are seen to be $\frac{1}{2}(1 \pm i\sqrt{3})$, and all three have modulus 1.

Solutions 6

1. Self checking.
2. Self checking.
3.

$$[\mathbf{GH}]^{-1} = \mathbf{H}^{-1}\mathbf{G}^{-1} \text{ so } \frac{d}{dt}[\mathbf{GH}]^{-1} = \frac{d}{dt}[\mathbf{H}^{-1}\mathbf{G}^{-1}]$$

$$= \left[\frac{d\mathbf{H}^{-1}}{dt}\right]\mathbf{G}^{-1} + \mathbf{H}^{-1}\left[\frac{d\mathbf{G}^{-1}}{dt}\right].$$

Hence $\frac{d}{dt}[\mathbf{G}(t)\mathbf{H}(t)]^{-1} = -\mathbf{H}^{-1}\frac{d\mathbf{H}}{dt}\mathbf{H}^{-1}\mathbf{G}^{-1} - \mathbf{H}^{-1}\mathbf{G}^{-1}\frac{d\mathbf{G}}{dt}\mathbf{G}^{-1}$.

4. Self checking.
5.

$$\frac{d\mathbf{G}^{-1}(t)}{dt} = \begin{bmatrix} -\frac{1}{10}\sin t + \frac{3}{10}\cos t & \frac{1}{5}\sin t + \frac{2}{5}\cos t \\ -\frac{1}{10}\cos t - \frac{3}{10}\sin t & \frac{1}{5}\cos t - \frac{2}{5}\sin t \end{bmatrix}.$$

6. The result is almost immediate from the definition of the sum of two matrices and the fact that α and β are scalar constants.
7. The result follows in the same way as the usual formula for integration by parts by differentiating $\mathbf{AB}$, rearranging terms, and then integrating the result. We have $\frac{d[\mathbf{AB}]}{dt} = \frac{d\mathbf{A}}{dt}\mathbf{B} + \mathbf{A}\frac{d\mathbf{B}}{dt}$, so integrating this gives $\mathbf{AB} = \int \frac{d\mathbf{A}}{dt}\mathbf{B}dt + \int \mathbf{A}\frac{d\mathbf{B}}{dt}dt$, from which the result follows after rearranging terms.
8.

$$\lambda_1 = -1,\ \mathbf{x}_1 = \begin{bmatrix} 1 \\ 2 \\ 1 \end{bmatrix},\ \lambda_2 = 1,\ \mathbf{x}_2 = \begin{bmatrix} -1 \\ 0 \\ 1 \end{bmatrix},\ \lambda_3 = -2,\ \mathbf{x}_3 = \begin{bmatrix} 0 \\ 1 \\ 0 \end{bmatrix}.$$

$$x_1(t) = \tfrac{1}{2}C_1(e^t + e^{-t}) + \tfrac{1}{2}C_3(e^{-t} - e^t),\ x_2(t) = (C_1 + C_3)e^{-t} + (C_2 - C_1 - C_3)e^{-2t},$$

$$x_3(t) = \tfrac{1}{2}C_1(e^{-t} - e^t) + \tfrac{1}{2}C_3(e^t + e^{-t}).$$

$$x_1(t) = \tfrac{1}{2}e^{-t} + \tfrac{3}{2}e^t,\ x_2(t) = e^{-t} + e^{-2t},\ x_3(t) = \tfrac{1}{2}e^{-t} - \tfrac{3}{2}e^t.$$

9.
$$\lambda_1 = -1,\ \mathbf{x}_1 = \begin{bmatrix} 1 \\ 0 \\ 1 \end{bmatrix},\quad \lambda_2 = -1,\ \mathbf{x}_2 = \begin{bmatrix} 0 \\ 1 \\ 0 \end{bmatrix},\quad \lambda_3 = 3,\ \mathbf{x}_3 = \begin{bmatrix} 0 \\ 1 \\ 2 \end{bmatrix}.$$
$$x_1(t) = C_3 e^{-t},\ x_2(t) = C_2 e^{-t} + C_1 e^{3t},\ z(t) = C_3 e^{-t} + 2C_1 e^{3t}.$$
$$x_1(t) = 2e^{-t},\ x_2(t) = \tfrac{5}{2}e^{-t} - \tfrac{3}{2}e^{3t},\ x_3(t) = 2e^{-t} - 3e^{3t}.$$

10.
$$\lambda_1 = -1 + 2i,\ \mathbf{x} = \begin{bmatrix} 1 \\ -\frac{1}{2}i \end{bmatrix},\quad \lambda_2 = -1 - 2i,\ \mathbf{x} = \begin{bmatrix} 1 \\ \frac{1}{2}i \end{bmatrix}.$$

11.
$$\lambda_1 = 2 + 2i,\ \mathbf{x} = \begin{bmatrix} 1 \\ \frac{1}{2}i \end{bmatrix},\quad \lambda_2 = 2 - 2i,\ \mathbf{x} = \begin{bmatrix} 1 \\ -\frac{1}{2}i \end{bmatrix}.$$
$$x_1(t) = e^{2t}(C_1 \cos 2t + 2C_2 \sin 2t),\ x_2(t) = \tfrac{1}{2}e^{2t}(2C_2 \cos 2t - C_1 \sin 2t).$$
$$x_1(t) = 2e^{2t}(\sin 2t - \cos 2t),\ x_2(t) = e^{2t}(\sin 2t + \cos 2t).$$

12.
$$\lambda_1 = 1,\ \mathbf{x}_1 = \begin{bmatrix} -4 \\ 1 \\ 2 \end{bmatrix},\quad \lambda_2 = 2,\ \mathbf{x}_2 = \begin{bmatrix} -1 \\ 0 \\ 1 \end{bmatrix},\quad \lambda_3 = -1,\ \mathbf{x}_3 = \begin{bmatrix} 0 \\ 1 \\ 0 \end{bmatrix}.$$
$$x_1(t) = C_1(2e^t - e^{2t}) + 2C_3(e^t - e^{2t}),\ x_2(t) = \tfrac{1}{2}C_1(e^{-t} - e^t) + C_2 e^{-t} + \tfrac{1}{2}C_3(e^{-t} - e^t),$$
$$x_3(t) = C_1(e^{2t} - e^t) + C_3(2e^{2t} - e^t).$$
$$x_1(t) = 6e^t - 5e^{2t},\ x_2(t) = \tfrac{1}{2}e^{-t} - \tfrac{3}{2}e^t,\ x_3(t) = 5e^{2t} - 3e^t.$$

13.
$$\lambda_1 = 1,\ \mathbf{x}_1 = \begin{bmatrix} 0 \\ 1 \\ 1 \end{bmatrix},\quad \lambda_2 = -1,\ \mathbf{x}_2 = \begin{bmatrix} 1 \\ 0 \\ -1 \end{bmatrix},\quad \lambda_3 = 2,\ \mathbf{x}_3 = \begin{bmatrix} -1 \\ 1 \\ 1 \end{bmatrix}.$$
$$x_1(t) = C_1 e^{2t} + C_2(e^{-t} - e^{2t}) + C_3(e^{2t} - e^t),\ x_2(t) = C_1(e^t - e^{2t}) + C_2 e^{2t} + C_3(e^t - e^{2t}),$$
$$x_3(t) = C_1(e^t - e^{2t}) + C_2(e^{2t} - e^t) + C_3(e^{-t} + e^t - e^{2t}).$$
$$x_1(t) = 2e^{-t} - e^{2t},\ x_2(t) = e^{2t} - 2e^t,\ x_3(t) = e^{2t} - 2e^t + e^{-t}.$$

14. $\lambda_1 = i,\ \mathbf{x}_1 = \begin{bmatrix} 1 \\ 1+i \end{bmatrix},\quad \lambda_2 = -i,\ \mathbf{x}_2 = \begin{bmatrix} 1 \\ 1-i \end{bmatrix}$. General solution:

$$x_1(t) = C_1(\cos t - \sin t) + C_2 \sin t - t^2 + 2 + 2t - e^{-t},$$
$$x_2(t) = -2C_1 \sin t + C_2(\cos t + \sin t) - 2t^2 + 4.$$

(a) $x_1(t) = 2 - 5\sin t - t^2 + 2t - e^{-t}, \; x_2(t) = 4 - 5\cos t - 5\sin t - 2t^2.$

(b) $x_1(t) = -2\cos(t-1) + 4\sin(t-1) - 2e^{-1}\sin(t-1) - 2t^2 + 4,$

$$x_2(t) = \sin(t-1) - 3\cos(t-1) + e^{-1}(\cos(t-1) - \sin(t-1)) + 2 + 2t - t^2 - e^{-t}.$$

15. $\lambda_1 = -2, \; \mathbf{x}_1 = \begin{bmatrix} -1 \\ 1 \end{bmatrix}, \quad \lambda_2 = 2, \; \mathbf{x}_2 = \begin{bmatrix} 1 \\ 3 \end{bmatrix}$. General solution:

$$x_1(t) = \tfrac{1}{4}C_1(3e^{-2t} + e^{2t}) + \tfrac{1}{4}C_2(e^{2t} - e^{-2t}) + \tfrac{3}{4}t - \tfrac{3}{4} + \tfrac{1}{5}\sin t,$$
$$x_2(t) = \tfrac{3}{4}C_1(e^{2t} - e^{-2t}) + \tfrac{1}{4}C_2(3e^{2t} + e^{-2t}) - \tfrac{9}{4}t + \tfrac{1}{5}(\cos t + \sin t).$$

Solution of IVP:
$$x_1(t) = -\tfrac{11}{80}e^{-2t} - \tfrac{9}{80}e^{2t} + \tfrac{3}{4}t - \tfrac{3}{4} + \tfrac{1}{5}\sin t,$$
$$x_2(t) = -\tfrac{27}{80}e^{2t} + \tfrac{11}{80}e^{-2t} - \tfrac{9}{4}t + \tfrac{1}{5}(\cos t + \sin t).$$

16. $\lambda_1 = 1, \; \mathbf{x}_1 = \begin{bmatrix} 2 \\ 1 \end{bmatrix}, \quad \lambda_2 = 0, \; \mathbf{x}_2 = \begin{bmatrix} 1 \\ 1 \end{bmatrix}$. General solution:

$$x_1(t) = C_1(2e^t - 1) + 2C_2(1 - e^t) - 8t + \tfrac{1}{2}t^2 - 10,$$
$$x_2(t) = C_1(e^t - 1) + C_2(2 - e^t) - 5 - 9t + \tfrac{1}{2}t^2.$$

Solution of IVP:

(a)
$$x_1(t) = 4e^t - 6 - 8t + \tfrac{1}{2}t^2, \quad x_2(t) = 2e^t - 1 - 9t + \tfrac{1}{2}t^2.$$

(b)
$$x_1(t) = 10e^{t-1} + \tfrac{17}{2} - 8t + \tfrac{1}{2}t^2 - 10, \quad x_2(t) = 5e^{t-1} + \tfrac{17}{2} - 5 - 9t + \tfrac{1}{2}t^2.$$

17. $\lambda_1 = 2, \; \mathbf{x} = \begin{bmatrix} 1 \\ 1 \end{bmatrix}, \quad \lambda_2 = -4, \; \mathbf{x}_2 = \begin{bmatrix} -1 \\ 1 \end{bmatrix}$. General solution:

$$x_1(t) = \tfrac{1}{2}C_1(e^{2t} + e^{-4t}) + \tfrac{1}{2}C_2(e^{2t} - e^{-4t}) - \tfrac{1}{8}e^{-2t} - \tfrac{54}{85}\cos t + \tfrac{12}{85}\sin t,$$
$$x_2(t) = \tfrac{1}{2}C_1(e^{2t} - e^{-4t}) + \tfrac{1}{2}C_2(e^{2t} + e^{-4t}) + \tfrac{3}{8}e^{-2t} - \tfrac{14}{85}\cos t + \tfrac{22}{85}\sin t.$$

Solution of IVP:
$$x_1(t) = \tfrac{101}{68}e^{-4t} + \tfrac{11}{40}e^{2t} - \tfrac{1}{8}e^{-2t} - \tfrac{54}{85}\cos t + \tfrac{12}{85}\sin t,$$
$$x_2(t) = \tfrac{11}{40}e^{2t} - \tfrac{101}{68}e^{-4t} + \tfrac{3}{8}e^{-2t} - \tfrac{14}{85}\cos t + \tfrac{22}{85}\sin t.$$

18. $\lambda_1 = 1,\ \mathbf{x}_1 = \begin{bmatrix}1\\0\\0\end{bmatrix},\quad \lambda_2 = 0,\ \mathbf{x}_2 = \begin{bmatrix}1\\-1\\1\end{bmatrix},\quad \lambda_3 = 2,\ \mathbf{x}_3 = \begin{bmatrix}1\\1\\1\end{bmatrix}$. General solution:

$$\begin{aligned}
x_1(t) &= C_1e^t + \tfrac{1}{2}C_2\left(e^{2t}-1\right) + C_3\left(\tfrac{1}{2}e^{2t} - e^t + \tfrac{1}{2}\right) + \tfrac{7}{2}t + \tfrac{11}{4} + \tfrac{7}{130}\cos 3t + \tfrac{9}{130}\sin 3t,\\
x_2(t) &= \tfrac{1}{2}C_2\left(1+e^{2t}\right) + \tfrac{1}{2}C_3\left(e^{2t}-1\right) + \tfrac{3}{4} - \tfrac{3}{2}t + \tfrac{2}{13}\cos 3t - \tfrac{3}{13}\sin 3t,\\
x_3(t) &= \tfrac{1}{2}C_2\left(e^{2t}-1\right) + \tfrac{1}{2}C_3\left(e^{2t}+1\right) + \tfrac{3}{4} - \tfrac{11}{13}\cos 3t - \tfrac{3}{13}\sin 3t + \tfrac{3}{2}t.
\end{aligned}$$

Solution of IVP: $x_1(t) = -\tfrac{19}{10}e^t - \tfrac{99}{22}e^{2t} + \tfrac{15}{4} + \tfrac{7}{2}t + \tfrac{7}{130}\cos 3t + \tfrac{9}{130}\sin 3t,$

$$\begin{aligned}
x_2(t) &= -\tfrac{99}{52}e^{2t} - \tfrac{1}{4} + \tfrac{2}{13}\cos 3t - \tfrac{3}{13}\sin 3t - \tfrac{3}{2}t,\\
x_3(t) &= -\tfrac{99}{52}e^{2t} + \tfrac{7}{4} - \tfrac{11}{13}\cos 3t - \tfrac{3}{13}\sin 3t + \tfrac{3}{2}t.
\end{aligned}$$

19. $\lambda_1 = 1+i,\ \mathbf{x}_1 = \begin{bmatrix}1\\-i\\1\end{bmatrix},\quad \lambda_2 = 1-i,\ \mathbf{x}_2 = \begin{bmatrix}1\\i\\1\end{bmatrix},\quad \lambda_3 = 1,\ \mathbf{x}_3 = \begin{bmatrix}1\\0\\0\end{bmatrix}.$
General solution:

$$\begin{aligned}
x_1(t) &= C_1e^t - C_2e^t\sin t + C_3e^t(\cos t - 1) + \tfrac{1}{30}e^{-2t} + \tfrac{1}{5}\sin t - \tfrac{3}{2} + \tfrac{2}{5}\cos t,\\
x_2(t) &= C_2e^t\cos t + C_3e^t\sin t + \tfrac{3}{5}\sin t + \tfrac{1}{5}\cos t + \tfrac{1}{2} + \tfrac{1}{10}e^{-2t},\\
x_3(t) &= -C_2e^t\sin t + C_3e^t\cos t + \tfrac{2}{5}\cos t + \tfrac{1}{5}\sin t - \tfrac{1}{2} - \tfrac{3}{10}e^{-2t}.
\end{aligned}$$

Solution of IVP:

$$\begin{aligned}
x_1(t) &= \tfrac{8}{5}e^t - \tfrac{6}{5}e^t\sin t - \tfrac{3}{5}e^t\cos t + \tfrac{1}{30}e^{-2t} + \tfrac{1}{5}\sin t - \tfrac{3}{2} + \tfrac{2}{5}\cos t,\\
x_2(t) &= \tfrac{6}{5}e^t\cos t - \tfrac{3}{5}e^t\sin t + \tfrac{3}{5}\sin t + \tfrac{1}{5}\cos t + \tfrac{1}{2} + \tfrac{1}{10}e^{-2t},\\
x_3(t) &= -\tfrac{6}{5}e^t\sin t - \tfrac{3}{5}e^t\cos t + \tfrac{2}{5}\cos t + \tfrac{1}{5}\sin t - \tfrac{1}{2} - \tfrac{3}{10}e^{-2t}.
\end{aligned}$$

20. The system is in the form $\mathbf{B}d\mathbf{x}/dt = \mathbf{A}_1\mathbf{x} = \mathbf{f}_1(t)$, with $\mathbf{B} = \begin{bmatrix}1 & 2\\2 & 1\end{bmatrix}$, $\mathbf{A}_1 = \begin{bmatrix}7 & 1\\8 & -1\end{bmatrix}$, $\mathbf{f}_1(t) = \begin{bmatrix}-5+4t\\-1+2t\end{bmatrix}$. Pre-multiplication by $\mathbf{B}^{-1} = \begin{bmatrix}-\frac{1}{3} & \frac{2}{3}\\ \frac{2}{3} & 1\frac{1}{3}\end{bmatrix}$ brings it to the form $\mathbf{dx}/dt = \mathbf{Ax} + \mathbf{f}(t)$ with $\mathbf{A} = \begin{bmatrix}3 & -1\\2 & 1\end{bmatrix}$, $\mathbf{f}(t) = \begin{bmatrix}1\\2t-3\end{bmatrix}$.
The eigenvalues and eigenvectors of $\mathbf{A}$ are

$$\lambda_1 = 2+i,\ \mathbf{x}_1 = \begin{bmatrix}\frac{1}{2}(1+i)\\1\end{bmatrix},\quad \lambda_2 = 2-i,$$

$$\mathbf{x}_2 = \begin{bmatrix}\frac{1}{2}(1-i)\\1\end{bmatrix},\ \text{so}\ \mathbf{P} = \begin{bmatrix}\frac{1}{2}(1+i) & \frac{1}{2}(1-i)\\1 & 1\end{bmatrix}$$

and

$\mathbf{D} = \begin{bmatrix}2+i & 0\\0 & 2-i\end{bmatrix}$. General solution:

$$x_1(t) = C_1e^{2t}\cos t + (C_1 - C_2)e^{2t}\sin t + \tfrac{2}{25} - \tfrac{2}{5}t,$$
$$x_2(t) = 2C_1e^{2t}\sin t + C_2e^{2t}(\cos t - \sin t) + \tfrac{41}{25} - \tfrac{6}{5}t.$$

Solution of IVP:

$$x_1(t) = -\tfrac{52}{25}e^{2t}\cos t - \tfrac{36}{25}e^{2t}\sin t + \tfrac{2}{25} - \tfrac{2}{5}t,$$
$$x_2(t) = -\tfrac{88}{25}e^{2t}\sin t - \tfrac{16}{25}e^{2t}\cos t + \tfrac{41}{25} - \tfrac{6}{5}t.$$

21. The system is in the form $\mathbf{B}d\mathbf{x}/dt = \mathbf{A}_1\mathbf{x} = \mathbf{f}_1(t)$, with

$\mathbf{B} = \begin{bmatrix} 1 & -2 \\ -1 & 1 \end{bmatrix}$, $\mathbf{A}_1 = \begin{bmatrix} 1 & -2 \\ -1 & 1 \end{bmatrix}$, $\mathbf{f}_1(t) = \begin{bmatrix} 2-6t^2 \\ -2+3t^2 \end{bmatrix}$. Pre-multiplication by $\mathbf{B}^{-1} = \begin{bmatrix} -1 & -2 \\ -1 & -1 \end{bmatrix}$ brings it to the form $\mathbf{dx}/dt = \mathbf{Ax} + \mathbf{f}(t)$ with $\mathbf{A} = \begin{bmatrix} 0 & 1 \\ 1 & 0 \end{bmatrix}$, $\mathbf{f}(t) = \begin{bmatrix} 2 \\ 3t^2 \end{bmatrix}$. The eigenvalues and eigenvectors of $\mathbf{A}$ are

$$\lambda_1 = -1,\ \mathbf{x}_1 = \begin{bmatrix} 1 \\ -1 \end{bmatrix},\ \lambda_2 =,\ \mathbf{x}_2 = \begin{bmatrix} 1 \\ 1 \end{bmatrix},\ \text{so } \mathbf{P} = \begin{bmatrix} 1 & 1 \\ -1 & 1 \end{bmatrix}$$

$\mathbf{D} = \begin{bmatrix} -1 & 0 \\ 0 & 1 \end{bmatrix}$. General solution:

$$x_1(t) = \tfrac{1}{2}C_1(e^t + e^{-t}) + \tfrac{1}{2}C_2(e^t - e^{-t}) - 6 - 3t^2\,,$$
$$x_2(t) = \tfrac{1}{2}C_1(e^t - e^{-t}) + \tfrac{1}{2}(e^t + e^{-t}) - 2 - 6t\,.$$

Solution of IVP: $x_1(t) = \tfrac{7}{2}e^{-t} + \tfrac{9}{2}e^t - 6 - 3t^2\,,\ x_2(t) = \tfrac{9}{2}e^t - \tfrac{7}{2}e^{-t} - 2 - 6t\,.$

22. $$x_1(t) = \tfrac{5}{2}e^{-2t} - e^{-t} - \tfrac{3}{2} + 2t\,,\ x_2(t) = \tfrac{1}{2} - \tfrac{5}{2}e^{-2t} - te^{-t} + 2t\,,\ t \geqslant 0\,.$$

23. $$x_1(t) = \tfrac{1}{3} + \tfrac{7}{15}e^{3t} + \tfrac{1}{5}(\cos t - 2\sin t)\,,\ x_2(t)$$
$$= \tfrac{7}{15}e^{3t} - \tfrac{2}{3} + \tfrac{4}{5}(2\sin t - \cos t)\,, \qquad t \geq 0\,.$$

24. $$x_1(t) = -3t - \tfrac{5}{4}e^{-t} + \tfrac{9}{4}e^t + \tfrac{1}{2}\sin t\,,\ x_2(t) = -3 + \tfrac{9}{4}e^t + \tfrac{5}{4}e^{-t} - \tfrac{1}{2}\cos t\,,\ t \geq 0\,.$$

25. $$x_1(t) = -\tfrac{1}{18} + \tfrac{1}{3}t - \tfrac{3}{10}e^{-2t} + \tfrac{19}{180}e^{3t} + \tfrac{1}{4}e^{-t}\,,$$
$$x_2(t) = \tfrac{7}{18} - \tfrac{1}{3}t + \tfrac{19}{90}e^{3t} + \tfrac{9}{10}e^{-2t} - \tfrac{3}{2}e^{-t}\,,\ t \geq 0\,.$$

26.
$$x_1(t) = \tfrac{7}{10}e^t - \tfrac{3}{10}e^{-t} - \tfrac{2}{5}\cos t\ ,\ x_2(t) = 1 - 2t + \tfrac{7}{10}e^t - \tfrac{3}{10}e^{-t} - \tfrac{2}{5}\cos 2t\ ,$$
$$x_3(t) = -2 + \tfrac{7}{10}e^t + \tfrac{3}{10}e^{-t} + \tfrac{4}{5}\sin 2t\ ,\ t \geq 0\,.$$

27.
$$x_1(t) = -1 + \tfrac{11}{4}e^t - \tfrac{3}{4}\cos t + \sin t + \tfrac{1}{4}t(\sin t + \cos t)\ ,$$
$$x_2(t) = \tfrac{1}{2}(3\sin t - t\cos t)\ ,\ x_3(t) = 1 - \cos t - \tfrac{1}{2}t\sin t\ ,\ t \geq 0\,.$$

28.
$$x_1(t) = -\tfrac{5}{4} - \tfrac{3}{4}te^{2t} + \tfrac{3}{16}(11e^{2t} + e^{-2t})\ ,\ x_2(t) = \tfrac{1}{2}t + \tfrac{3}{8}(e^{-2t} - e^{2t}),$$
$$x_3(t) = -\tfrac{1}{4} - \tfrac{3}{8}(e^{2t} + e^{-2t})\ ,\ t \geq 0.$$

29.
$$x_1(t) = -1 - t + \tfrac{5}{4}e^t - \tfrac{1}{4}(t\sin t + t\cos t + \cos t),$$
$$x_2(t) = 1 + \tfrac{1}{2}(\sin t - t\cos t)\ ,\ x_3(t) = t - \tfrac{1}{2}t\sin t\ ,\ t \geq 0\,.$$

30.
$$x_1(t) = \tfrac{3}{8}e^t + \tfrac{1}{8}e^{-t} + \tfrac{1}{2}(\cos t - \sin t) + \tfrac{1}{4}t\cos t\ ,$$
$$x_2(t) = -t + \tfrac{3}{8}e^t + \tfrac{1}{8}e^{-t} - \tfrac{1}{2}\cos t - \tfrac{1}{4}t\cos t\ ,\ t \geq 0\,.$$

31.
$$x_1(t) = \tfrac{1}{4}e^{-t}(5\cos 2t + 2\sin 2t - 1)\ ,\ x_2(t)$$
$$= 3 + \tfrac{1}{4}e^{-t}(5 + 8\sin 2t - 9\cos 2t)\ ,\ t \geq 0\,.$$

32. $u = 2dy/dt$, $v = 4du/dt$, so $v = 8d^{\,2y}/dt^2$ and the equation is replaced by the system
$$\tfrac{1}{8}\frac{dv}{dt} + \tfrac{1}{4}v - \tfrac{1}{2}u - 2y = 1 + \sin t,\quad \frac{dy}{dt} = \tfrac{1}{2}u\ ,\quad \frac{du}{dt} = \tfrac{1}{4}v.$$
The initial conditions for the system are
$y(0) = 0$, $u(0) = 0$, $v(0) = 0$. The solution $y(t)$ is
$$y(t) = \tfrac{1}{4}e^{-t} + \tfrac{1}{4}e^t - \tfrac{1}{10}e^{-2t} - \tfrac{1}{2} + \tfrac{1}{10}(\cos t - 2\sin t) - \tfrac{1}{2},\quad t \geqslant 0.$$
Although not required, for reference purposes:
$$u(t) = \tfrac{1}{2}e^t - \tfrac{1}{2}e^{-t} + \tfrac{2}{5}e^{-2t} - \tfrac{2}{5}\cos t - \tfrac{1}{5}\sin t,$$
$$v(t) = 2e^{-t} - \tfrac{16}{5}e^{-2t} + 2e^t - \tfrac{4}{5}\cos t + \tfrac{8}{5}\sin t.$$

33. Nilpotent index $x = 5$.
$$e^{t\mathbf{A}} = \begin{bmatrix} 1 & 3t & \tfrac{3}{2}t^2 + t & t^3 + \tfrac{11}{2}t^2 + 2t \\ 0 & 1 & t & t^2 + 3t \\ 0 & 0 & 1 & 2t \\ 0 & 0 & 0 & 1 \end{bmatrix}.$$

34.

$$e^{t\mathbf{A}} = \begin{bmatrix} e^{-1} & 0 & 0 & 0 \\ 0 & e^{-2} & 0 & 0 \\ 0 & 0 & e & 0 \\ 0 & 0 & 0 & e^2 \end{bmatrix},$$

$$e^{t\mathbf{B}} = \begin{bmatrix} e & 0 & 0 & 0 \\ 0 & e^2 & 0 & 0 \\ 0 & 0 & e^{-1} & 0 \\ 0 & 0 & 0 & e \end{bmatrix} \quad \text{and} \quad e^{t(\mathbf{A}+\mathbf{B})} = \begin{bmatrix} 1 & 0 & 0 & 0 \\ 0 & 1 & 0 & 0 \\ 0 & 0 & 1 & 0 \\ 0 & 0 & 0 & e^3 \end{bmatrix}.$$

The result is true because the product of two upper diagonal 4×4 matrices is always commutative.

35.

$$e^{t\mathbf{A}} = \begin{bmatrix} \cosh t & \sinh t \\ \sinh t & \cosh t \end{bmatrix}.$$

36.

$$\lambda_1 = 3,\ \mathbf{x}_1 = [1,\ 1]^{\mathrm{T}},\ \ \lambda_2 = -1,\ \mathbf{x}_2 = [-1,\ 1],\ \mathbf{P} = \begin{bmatrix} 1 & -1 \\ 1 & 1 \end{bmatrix},$$

$$\mathbf{P}^{-1} = \begin{bmatrix} \frac{1}{2} & \frac{1}{2} \\ -\frac{1}{2} & \frac{1}{2} \end{bmatrix},$$

$$t\mathbf{D} = \begin{bmatrix} e^{3t} & 0 \\ 0 & e^{-t} \end{bmatrix},\ \ e^{t\mathbf{A}} = \mathbf{P}(t\mathbf{D})\mathbf{P}^{-1} = \begin{bmatrix} \frac{1}{2}(e^{-t} + e^{3t}) & \frac{1}{2}(e^{3t} - e^{-t}) \\ \frac{1}{2}(e^{3t} - e^{-t}) & \frac{1}{2}(e^{-t} + e^{3t}) \end{bmatrix}.$$

37.

$$\lambda_1 = 1 + 2i,\ \mathbf{x}_1 = [2i,\ \ 1]^{\mathrm{T}},\ \ \lambda_2 = -2i,\ \mathbf{x}_2 = [-2i,\ 1]^{\mathrm{T}},\ \ \mathbf{P} = \begin{bmatrix} 2i & -2i \\ 1 & 1 \end{bmatrix},$$

$$\mathbf{P}^{-1} = \begin{bmatrix} -\frac{1}{4}i & \frac{1}{2} \\ \frac{1}{4}i & \frac{1}{2} \end{bmatrix},$$

$$t\mathbf{D} = \begin{bmatrix} e^{\lambda_1 t} & 0 \\ 0 & e^{\lambda_2 t} \end{bmatrix},\ \ e^{t\mathbf{A}} = \mathbf{P}(t\mathbf{D})\mathbf{P}^{-1} = \begin{bmatrix} \frac{1}{2}(e^{\lambda_1 t} + e^{\lambda_2 t}) & i\left(e^{\lambda_1 t} - e^{\lambda_2 t}\right) \\ -\frac{1}{4}i\left(e^{\lambda_1 t} - e^{\lambda_2 t}\right) & \frac{1}{2}(e^{\lambda_1 t} + e^{\lambda_2 t}) \end{bmatrix}$$

so after simplification

$$e^{t\mathbf{A}} = \begin{bmatrix} e^t \cos(2t) & -2e^t \sin(2t) \\ \frac{1}{2}e^t \sin(2t) & e^t \cos(2t) \end{bmatrix}.$$

38.

$$e^{t\mathbf{A}} = \begin{bmatrix} \cos t & -\sin t \\ \sin t & \cos t \end{bmatrix},\ \ x_1(t) = \cos t - \sin t,\ \ x_2(t) = -\cos t - \sin t.$$

39.
$$e^{t\mathbf{A}} = \begin{bmatrix} \cosh t & \sinh t \\ \sinh t & \cosh t \end{bmatrix}, \quad x_1(t) = e^t - 2, \quad x_2(t) = e^t - t.$$

40. (a) $e^{t\mathbf{A}} = \begin{bmatrix} e^t \cos t & e^t \sin t \\ -e^t \sin t & e^t \cos t \end{bmatrix}$, (b) $e^{t\mathbf{A}} = \begin{bmatrix} e^{2t} \cos t & e^{2t} \\ -e^{2t} \sin t & e^{2t} \cos t \end{bmatrix}$.

41.
$$\lambda_1 = 2,\ \mathbf{x}_1 = [1,\ 1,\ 2]^{\mathrm{T}},\ \ \lambda_2 = 0,\ \mathbf{x}_2 = [1, -1,\ 2]^{\mathrm{T}},\ \ \lambda_3 = 1,\ \mathbf{x}_3 = [1,\ 0,\ 1]^{\mathrm{T}}$$

$$\mathbf{P} = \begin{bmatrix} 1 & 1 & 1 \\ 1 & -1 & 0 \\ 2 & 2 & 1 \end{bmatrix}, \quad \mathbf{P}^{-1} = \begin{bmatrix} -\frac{1}{2} & \frac{1}{2} & \frac{1}{2} \\ -\frac{1}{2} & -\frac{1}{2} & \frac{1}{2} \\ 2 & 0 & -1 \end{bmatrix}, \quad t\mathbf{D} = \begin{bmatrix} 2t & 0 & 0 \\ 0 & 0 & 0 \\ 0 & 0 & t \end{bmatrix},$$

$$e^{t\mathbf{A}} = \mathbf{P}(t\mathbf{D})\mathbf{P}^{-1} = \begin{bmatrix} -\frac{1}{2}e^{2t} + 2e^t - \frac{1}{2} & \frac{1}{2}e^{2t} - \frac{1}{2} & \frac{1}{2}e^{2t} - e^t + \frac{1}{2} \\ -\frac{1}{2}e^{2t} + \frac{1}{2} & \frac{1}{2}e^{2t} + \frac{1}{2} & \frac{1}{2}e^{2t} - \frac{1}{2} \\ -e^{2t} + 2e^t - 1 & e^{2t} - 1 & e^{2t} - e^t + 1 \end{bmatrix}.$$

42.
$$e^{t\mathbf{A}} = \begin{bmatrix} \frac{1}{2}(e^{3t} + e^{-t}) & \frac{1}{2}(e^{3t} - e^{-t}) & 0 \\ \frac{1}{2}(e^{3t} - e^{-t}) & \frac{1}{2}(e^{3t} + e^{-t}) & 0 \\ -\frac{1}{2}e^{-t} + \frac{1}{6}e^{3t} + \frac{1}{3} & \frac{1}{6}e^{3t} + \frac{1}{2}e^{-t} - \frac{2}{3} & 1 \end{bmatrix},$$

$$x_1(t) = C_1 e^{-t} + C_2 e^{3t},\ x_2(t) = -C_1 e^{-t} + C_2 e^{3t},\ \ x_3(t) = -C_1 e^{-t} + C_2 e^{3t} + C_3.$$

Solutions 7

1. The result is straightforward, because the axioms of a vector space are taken directly from the rules governing the manipulation of geometrical vectors.
2. Yes, because when vectors are added or multiplied by a scalar the zeros on the leading diagonal remain unchanged so this set of matrices forms a vector space.
3. No, because when vectors are added or multiplied by a scalar the first entry in the leading diagonal is altered so that the resulting vector does not belong to V. Thus this set of vectors does not form a vector space.
4. Yes. It is easily verified that the axioms of a vector space are satisfied when the matrices contain complex elements and the scalars λ and μ in Definition 7.1.1 are complex numbers. Thus this set of complex matrices forms a complex vector space.
5. Yes, When a vector is any member of the set of all cubic polynomials denoted by P_3, performing all of the operations in Definition 7.1.1 will produce another cubic polynomial. Thus the set of all cubic polynomials P_3 form a vector space.
6. The set of all quadratic polynomials forms a subspace of the vector space P_3. The set of all cubic polynomials in which $a_1 = 0$ form a subspace of the vector space P_3. The set of all cubic polynomials in which $a_0 = 1$ does not form a subspace of P_3, because when polynomials are added or scaled the coefficient a_0 is altered so the resulting polynomial is no longer of the required type.

7. The result follows because the sum of continuous functions is a continuous function and scaling a continuous function yields another continuous function. All of the other requirements of Definition 7.1.1 are satisfied.
8. Yes, because the vectors in the vector space P_3 of cubic polynomials are all differentiable functions.
9. The result follows from the fact that the addition and scaling of real continuous integrable functions defined over $a \le x \le b$is continuous and integrable, and the other properties of Definition 7.1.1 are all satisfied.
10. The discontinuities must be bounded and finite in number so the functions are integrable. If improper integrals are included, the integrals must be such that their Cauchy principal value is defined.
11. No. Multiplication of a convex function $f(x)$ by a negative scalar causes the points on the chord PQ that were *above* the graph of $y = f(x)$ to lie *below* it. Thus the resulting function is *not* convex, and in fact it is called a *concave* function.
12. It is a routine matter to show that $f(x)$ satisfies all the requirements of Definition 7.1.1 and so forms a vector space we can call V. The functions $f'(x) = 2b\cos x - 2c\sin x$, but b and c are arbitrary constants so $f'(x)$ has the same form as $f(x)$, though without the arbitrary additive constant. The functions $f'(x)$ also satisfy all of the requirements of Definition 7.1.1 and so form a vector space W. The vector space W is a subspace of the vector space V because the functions $f'(x)$ have the same form as $f(x)$, but without the arbitrary additive constant.
13. The determinant test in Chapter 4 should be used to test for linear independence. One set of basis vectors is $\{\mathbf{v}_1, \mathbf{v}_3, \mathbf{v}_4, \mathbf{v}_5\}$. This set of basis vectors is not unique, because another set of basis vectors is $\{\mathbf{v}_1, \mathbf{v}_2, \mathbf{v}_4, \mathbf{v}_5\}$. In fact $\mathbf{v}_3 = \mathbf{v}_1 + 2\mathbf{v}_2$, so yet another choice could be $\{\mathbf{v}_2, \mathbf{v}_3, \mathbf{v}_4, \mathbf{v}_5\}$.
14. [2, 1, 2, 0].
15. [1, 2, 0, −2,1].
16. The 1-norm. Clearly N1 and N2 are satisfied. N3 is satisfied because $\|\lambda\mathbf{u}\| = |\lambda u_1| + |\lambda u_2| + \cdots + |\lambda u_n| = |\lambda|\{|u_1| + |u_2| + \cdots |u_n|\}$. N4 is satisfied because

$$\begin{aligned}\|\mathbf{u}+\mathbf{v}\|_1 &= |u_1+v_1| + |u_2+v_2| + \cdots + |u_n+v_n| \\ &\le |u_1| + |v_1| + |u_2| + |v_2| + \cdots + |u_n| + |v_n|.\end{aligned}$$

The infinity norm. Clearly N1 and N2 are satisfied. N3 is satisfied because $\|\lambda\mathbf{u}\|_\infty = \max\{\lambda u_1, \lambda u_2, \ldots, \lambda u_n\} = \lambda\max\{\ u_1, u_2, \ldots, u_n\}$. N4 is satisfied because

$$\begin{aligned}\|\mathbf{u}+\mathbf{v}\|_\infty &= \max\{u_1+v_1, u_2+v_2, \ldots, u_n+v_n\} \\ &\le \max\{u_1, u_2, \ldots, u_n\} + \max\{v_1, v_2, \ldots, v_n\} = \|\mathbf{u}\|_\infty + \|\mathbf{v}\|_\infty.\end{aligned}$$

17. (i) Write axiom P2 as $\langle \mathbf{a}+\mathbf{b}, \mathbf{c}\rangle = \langle \mathbf{a}, \mathbf{c}\rangle + \langle \mathbf{b}, \mathbf{c}\rangle$, and set $\mathbf{a} = \mathbf{0}$, $\mathbf{b} = \mathbf{u}$ and $\mathbf{c} = \mathbf{v}$.

Then $\langle \mathbf{0}+\mathbf{u},\mathbf{v}\rangle = \langle \mathbf{0},\mathbf{v}\rangle + \langle \mathbf{u},\mathbf{v}\rangle$, but $\mathbf{0} + \mathbf{u} = \mathbf{u}$, so$\langle \mathbf{u},\mathbf{v}\rangle = \langle \mathbf{0},\mathbf{v}\rangle +$ $\langle \mathbf{u},\mathbf{v}\rangle$. Hence $\langle \mathbf{0},\mathbf{v}\rangle = 0$, but $\langle \mathbf{0},\mathbf{v}\rangle = \langle \mathbf{v},\mathbf{0}\rangle$, so $\langle \mathbf{v},\mathbf{0}\rangle = 0$.

(ii) From axiom P1 $\langle \mathbf{u},\mathbf{v}+\mathbf{w}\rangle = \langle \mathbf{v}+\mathbf{w},\mathbf{u}\rangle = \langle \mathbf{v},\mathbf{u}\rangle + \langle \mathbf{w},\mathbf{u}\rangle$. However from axiom P1 $\langle \mathbf{v},\mathbf{u}\rangle=\langle \mathbf{u},\mathbf{v}\rangle$ and $\langle \mathbf{w},\mathbf{u}\rangle = \langle \mathbf{u},\mathbf{w}\rangle$, so the result is proved.

(iii) From axioms P1 and P3$\langle \mathbf{u},k\mathbf{v}\rangle = \langle k\mathbf{v},\mathbf{u}\rangle = k\langle \mathbf{v},\mathbf{u}\rangle$. An application of axiom P1 shows that $k\langle \mathbf{v},\mathbf{u}\rangle = k\langle \mathbf{u},\mathbf{v}\rangle$, and the result is proved.

18. Examination of the definition of a projection shows that in $\text{proj}_{\mathbf{u}}\mathbf{v}$ the roles of $\mathbf{u}$ and $\mathbf{v}$ are reversed with respect to those in $\text{proj}_{\mathbf{v}}\mathbf{u}$, so in general the results will be different, because although $\langle \mathbf{u},\mathbf{v}\rangle = \langle \mathbf{v},\mathbf{u}\rangle$, the two denominators will be different. However, the two projections will be the same if $\|\mathbf{u}\| = \|\mathbf{v}\|$. We have $\langle \mathbf{u},\mathbf{v}\rangle = \langle \mathbf{v},\mathbf{u}\rangle = 8, \|\mathbf{u}\| = \sqrt{14}$ and $\|\mathbf{v}\| = \sqrt{6}$, so by definition $\text{proj}_{\mathbf{v}}\mathbf{u} = 8/\sqrt{6}$ and $\text{proj}_{\mathbf{u}}\mathbf{v} = 8/\sqrt{14}$, showing that $\text{proj}_{\mathbf{v}}\mathbf{u} \neq \text{proj}_{\mathbf{u}}\mathbf{v}$, while $\cos\theta = \frac{8}{\sqrt{14}\sqrt{6}} = 0.5097$, corresponding to $\theta = 29.2°$.

19.
$$\mathbf{w}_1 = \left(\frac{1}{\sqrt{2}},\frac{1}{\sqrt{2}},0\right),\quad \mathbf{w}_2 = \left(\frac{1}{\sqrt{3}},-\frac{1}{\sqrt{3}},\frac{1}{\sqrt{3}}\right),\quad \mathbf{w}_3 = \left(\frac{1}{\sqrt{6}},-\frac{1}{\sqrt{6}},-\frac{2}{\sqrt{6}}\right).$$

20. The vectors are linearly independent because the determinant with their entries as its rows equals 6, and so is not zero.

$$\mathbf{w}_1 = \left(\frac{1}{\sqrt{3}},\frac{1}{\sqrt{3}},\frac{1}{\sqrt{3}}\right),\ \mathbf{w}_2 = \left(\frac{1}{\sqrt{6}},-\frac{2}{\sqrt{6}},\frac{1}{\sqrt{6}}\right),\ \mathbf{w}_3 = \left(-\frac{1}{\sqrt{2}},0,\frac{1}{\sqrt{2}}\right).$$

21. The vectors are linearly independent because the determinant with their entries as its rows equals -1, and so is not zero.

$$\mathbf{w}_1 = \left(\tfrac{1}{2},\tfrac{1}{2},\tfrac{1}{2},\tfrac{1}{2}\right),\ \mathbf{w}_2 = \left(\frac{1}{\sqrt{12}},-\frac{3}{\sqrt{12}},\frac{1}{\sqrt{12}},\frac{1}{\sqrt{12}}\right),\ \mathbf{w}_3 = \left(\frac{1}{\sqrt{2}},0,-\frac{1}{\sqrt{2}},0\right),$$
$$\mathbf{w}_4 = \left(\frac{1}{\sqrt{6}},0,\frac{1}{\sqrt{6}},-\frac{2}{\sqrt{6}}\right).$$

22. Steps 1 and 2 of the Gram–Schmidt orthogonalization process do not involve $\mathbf{u}_3$, so they proceed as before and generate the orthogonal vectors $\mathbf{v}_1$ and $\mathbf{v}_2$. Step 3 tells us that

$$\mathbf{v}_3 = \mathbf{u}_3 - \frac{\langle \mathbf{v}_1,\mathbf{u}_3\rangle}{\langle \mathbf{v}_1,\mathbf{v}_1\rangle}\mathbf{v}_1 - \frac{\langle \mathbf{v}_2,\mathbf{u}_3\rangle}{\langle \mathbf{v}_2,\mathbf{v}_2\rangle}\mathbf{v}_2,$$

so substituting for $\mathbf{u}_3$ this becomes

$$\mathbf{v}_3 = \alpha\mathbf{u}_1 + \beta\mathbf{u}_2 - \frac{\langle \mathbf{v}_1,\alpha\mathbf{u}_1+\beta\mathbf{u}_2\rangle}{\langle \mathbf{v}_1,\mathbf{v}_1\rangle}\mathbf{v}_1 - \frac{\langle \mathbf{v}_2,\alpha\mathbf{u}_1+\beta\mathbf{u}_2\rangle}{\langle \mathbf{v}_2,\mathbf{v}_2\rangle}\mathbf{v}_2.$$

Now $\mathbf{u}_1 = \mathbf{v}_1$, so

$$\mathbf{v}_3=\alpha\mathbf{v}_1+\beta\mathbf{u}_2-\alpha\frac{\langle\mathbf{v}_1,\mathbf{v}_1\rangle}{\langle\mathbf{v}_1,\mathbf{v}_1\rangle}\mathbf{v}_1-\beta\frac{\langle\mathbf{v}_1,\mathbf{u}_2\rangle}{\langle\mathbf{v}_1,\mathbf{v}_1\rangle}-\alpha\frac{\langle\mathbf{v}_2,\mathbf{v}_1\rangle}{\langle\mathbf{v}_2,\mathbf{v}_2\rangle}\mathbf{v}_2-\beta\frac{\langle\mathbf{v}_2,\mathbf{u}_2\rangle}{\langle\mathbf{v}_2,\mathbf{v}_2\rangle}\mathbf{v}_2.$$

The first and third terms on the right cancel, while the fourth term on the right vanishes because of the orthogonality of $\mathbf{v}_1$ and $\mathbf{v}_2$. Consequently the result reduces to

$$\mathbf{v}_3=\beta\mathbf{u}_2-\beta\frac{\langle\mathbf{v}_1,\mathbf{u}_2\rangle}{\langle\mathbf{v}_1,\mathbf{v}_1\rangle}\mathbf{v}_1-\beta\frac{\langle\mathbf{v}_2,\mathbf{u}_2\rangle}{\langle\mathbf{v}_2,\mathbf{v}_2\rangle}\mathbf{v}_2.$$

The last two terms on the right are $-\beta\{(\mathrm{proj}_{\mathbf{v}_1}\mathbf{u}_2)\mathbf{v}_1+(\mathrm{proj}_{\mathbf{v}_2}\mathbf{u}_2)\mathbf{v}_2\}=-\beta\mathbf{u}_2$, showing that $\mathbf{v}_3=\mathbf{0}$.

23. The vectors proposed for a basis for the subspace W are suitable because they are linearly independent (they are not proportional). An application of the Gram–Schmidt orthogonalization procedure to these vectors yields the two orthonormal basis vectors for W $\mathbf{w}_1=\left(\frac{1}{\sqrt{6}},-\frac{2}{\sqrt{6}},\frac{1}{\sqrt{6}}\right)$ and $\mathbf{w}_2=\left(\frac{1}{\sqrt{3}},\frac{1}{\sqrt{3}},\frac{1}{\sqrt{3}}\right)$. As $\mathbf{p}=(2,\ 1,\ 4)$ it follows that $\langle\mathbf{p},\mathbf{w}_1\rangle=4/\sqrt{6}$ and $\langle\mathbf{p},\mathbf{w}_2\rangle=7/\sqrt{3}$. Thus adding the two vector contributions gives $\mathbf{q}=\frac{4}{\sqrt{6}}\left(\frac{1}{\sqrt{6}},-\frac{2}{\sqrt{6}},\frac{1}{\sqrt{6}}\right)+\frac{7}{\sqrt{3}}\left(\frac{1}{\sqrt{3}},\frac{1}{\sqrt{3}},\frac{1}{\sqrt{3}}\right)$, and so

$$\mathbf{q}=(3,\ 1,\ 3)\ \text{and}\ \|\mathbf{q}\|=\sqrt{19}.$$

24. The vectors proposed for a basis for the subspace W are suitable because they are linearly independent. This can be seen from the fact that when they are arranged as the rows of a matrix, its rank is found to be 3. An application of the Gram–Schmidt orthogonalization procedure applied to these vectors shows that an orthonormal basis for W is $\mathbf{w}_1=\left(\frac{1}{\sqrt{3}},0,\frac{1}{\sqrt{3}},-\frac{1}{\sqrt{3}}\right)$, $\mathbf{w}_2=\left(\frac{1}{\sqrt{15}},\frac{1}{\sqrt{15}},\frac{1}{\sqrt{15}},\frac{2}{\sqrt{15}}\right)$ and $\mathbf{w}_3=\left(\frac{1}{\sqrt{2}},0,-\frac{1}{\sqrt{2}},0\right)$.
Thus $\langle\mathbf{p},\mathbf{w}_1\rangle=-2/\sqrt{3}$, $\langle\mathbf{p},\mathbf{w}_2\rangle=1/\sqrt{15}$ and $\langle\mathbf{p},\mathbf{w}_3\rangle=\sqrt{2}$, and adding the three vector contributions gives
$\mathbf{q}=-\frac{2}{\sqrt{3}}\mathbf{w}_1+\frac{1}{\sqrt{15}}\mathbf{w}_2+\sqrt{2}\mathbf{w}_3=(\frac{2}{5},\ \frac{3}{15},-\frac{8}{5},\ \frac{4}{5})$, and $\|\mathbf{p}\|=\sqrt{17/5}$.

25. Let the x_1-axis in the (x_1,x_2)-plane be the axis about which the plane is rotated through an angle α, with $-\pi/2\le\alpha\le\pi/2$, to form the plane Π. When rotated, let the x_2-axis become the x_2'-axis in the plane Π, so the included angle between these two axes is α. Unit vectors along the x_1 and x_2'-axis are $\mathbf{u}_1=[1,0,0]$, and $\mathbf{u}_2=[0,\ \cos\alpha,\ \sin\alpha]$.

The vector $\mathbf{q}_1$ in the x_1-direction with magnitude equal to the component of $\mathbf{p}$ in that direction is

$$\mathbf{q}_1=\frac{\langle\mathbf{p},\mathbf{u}_1\rangle}{\langle\mathbf{u}_1,\mathbf{u}_1\rangle}\mathbf{u}_1=\frac{\langle[a\cos t,\ b\sin t,\ 0],[1,\ 0,\ 0]\rangle}{\langle[1,\ 0,\ 0],[1,\ 0,\ 0]\rangle}[1,0,0]=[a\cos t,\ 0,\ 0].$$

The vector $\mathbf{q}_2$ along the plane Π in the $\mathbf{u}_2$ direction with magnitude equal to the component of $\mathbf{p}$ in that direction is

$$\mathbf{q}_2 = \frac{\langle \mathbf{p}, \mathbf{u}_2 \rangle}{\langle \mathbf{u}_2, \mathbf{u}_2 \rangle}\mathbf{u}_2 = \frac{\langle [a\cos t,\ b\sin t,\ 0], [0,\ \cos\alpha,\ \sin\alpha] \rangle}{\langle [0,\ \cos\alpha,\ \sin\alpha], [0,\ \cos\alpha,\ \sin\alpha] \rangle}[0, \cos\alpha, \sin\alpha]$$
$$= [0,\ b\cos^2\alpha\sin t,\ b\cos\alpha\sin\alpha\sin t].$$

The position vector $\mathbf{q}$ of a point on the projection of the ellipse onto the plane Π is thus

$$\mathbf{q} = \mathbf{q}_1 + \mathbf{q}_2 = [a\cos t,\ b\cos^2\alpha\sin t,\ b\cos\alpha\sin\alpha\sin t].$$

This is an equation of an ellipse on plane Π, with a semi-axis of length a along the x_1-axis, but with its other semi-axis no longer equal to b. To find the length of the second semi-axis in the plane Π, we use the fact that the tip of the axis occurs when $t = \pi/2$, so as one end of the semi-axis lies at the origin, the other end will lie at the point $(0,\ b\cos^2\alpha,\ b\cos\alpha\sin\alpha)$. An application of Pythagoras' theorem shows the length of this semi-axis to be

$$b\left[\cos^4\alpha + \cos^2\alpha\sin^2\alpha\right]^{1/2} = b\cos\alpha, \quad -\pi/2 \le \alpha \le \pi/2.$$

So the projection of the ellipse in the (x_1, x_2)-plane onto plane Π is an ellipse with semi-axes a and $b\cos\alpha$. The algebraic equation of this ellipse relative to the x_1 and x_2'-axes is thus

$$\frac{x_1^2}{a^2} + \frac{x_2'2}{b^2\cos^2\alpha} = 1.$$

This result could, of course, have been derived directly, and much more simply, by using purely geometrical arguments.

26. Proceeding as in Example 7.8, vector $\mathbf{q}_1$, the vector in the direction of unit vector $\mathbf{u}_1$ in the x_1 direction with magnitude equal to the projection of $\mathbf{p}$ in that direction, is

$$\mathbf{q}_1 = \frac{\langle \mathbf{p}, \mathbf{u}_1 \rangle}{\langle \mathbf{u}_1, \mathbf{u}_1 \rangle}\mathbf{u}_1 = \frac{\langle [a\cos t,\ b\sin t, t^2], [1,\ 0,\ 0] \rangle}{\langle [1,\ 0,\ 0], [1,\ 0,\ 0] \rangle}[1,\ 0,\ 0] = [a\cos t,\ 0,\ 0].$$

The corresponding vectors $\mathbf{q}_2$ and $\mathbf{q}_3$ in the direction of the unit vectors $\mathbf{u}_2$ and $\mathbf{u}_3$ with magnitudes equal to the projection of $\mathbf{p}$ in these directions are

$$\mathbf{q}_2 = \frac{\langle \mathbf{p}, \mathbf{u}_2 \rangle}{\langle \mathbf{u}_2, \mathbf{u}_2 \rangle}\mathbf{u}_2 = \frac{\langle [a\cos t,\ b\sin t, t^2], [0,\ 1,\ 0] \rangle}{\langle [0,\ 1,\ 0], [0,\ 1,\ 0] \rangle}[0,\ 1,\ 0] = [0,\ b\sin t,\ 0],$$

and

$$\mathbf{q}_3 = \frac{\langle \mathbf{p}, \mathbf{u}_3 \rangle}{\langle \mathbf{u}_3, \mathbf{u}_3 \rangle}\mathbf{u}_1 = \frac{\langle [a\cos t,\ b\sin t, t^2], [0,\ 0,\ 1] \rangle}{\langle [0,\ 0,\ 1], [0,\ 0,\ 1] \rangle}[0,\ 0,\ 1] = [0,\ 0,\ t^2].$$

(a) The position vector of a point on the projection of **p** onto the (x_1, x_2)-plane is $\mathbf{q} = \mathbf{q}_1 + \mathbf{q}_2 = [a\cos t,\ b\sin t,\ 0]$, which is the parametric equation of an ellipse in the (x_1, x_2)-plane with semi-axes a and b.

(b) The position vector of a point on the projection of **p** onto the (x_2, x_3)-plane is $\mathbf{q} = \mathbf{q}_2 + \mathbf{q}_3 = [0,\ b\sin t,\ t^2]$, with $0 \le t \le 2\pi$. This is the parametric equation of a sinusoid stretched unevenly along the x_3-axis in the (x_2, x_3)-plane.

27. In terms of the unit vectors **i**, **j** and **k**, vectors $\mathbf{b}_1 = \mathbf{i} - \mathbf{j}$, $\mathbf{b}_2 = \mathbf{i} - \mathbf{k}$, $\mathbf{p} = 3\mathbf{i} + j + 3\mathbf{k}$.
A vector **n** normal to the plane defined by vectors $\mathbf{b}_1$ and $\mathbf{b}_2$ is $\mathbf{n} = \mathbf{b}_1 \times \mathbf{b}_2$, so $\mathbf{n} = \mathbf{b}_1 \times \mathbf{b}_2 = \begin{vmatrix} \mathbf{i} & \mathbf{j} & \mathbf{k} \\ 1 & -1 & 0 \\ 1 & 0 & -1 \end{vmatrix} = \mathbf{i} + \mathbf{j} + \mathbf{k}$. The unit vectors $\hat{\mathbf{n}}$ and $\hat{\mathbf{p}}$ in the directions **n** and **p** are $\hat{\mathbf{n}} = (1/\sqrt{3})(\mathbf{i} + \mathbf{j} + \mathbf{k})$ and $\hat{\mathbf{p}} = (1/\sqrt{19})(3\mathbf{i} + \mathbf{j} + 3\mathbf{k})$. So if θ is the angle between $\hat{\mathbf{n}}$ and $\hat{\mathbf{p}}$, taking the scalar product of $\hat{\mathbf{n}}$ and $\hat{\mathbf{p}}$ gives $\cos\theta = 7/(\sqrt{3}\sqrt{19}) = 0.927173$, so $\theta = 0.384002$ rad. The length of vector **p** is $\sqrt{19}$. So, if **q** is the vector projection of **p** on the plane defined by vectors $\mathbf{b}_1$ and $\mathbf{b}_2$, and the length of **q** is l, then the geometry of the problem shows that $l = \sqrt{19}\sin\theta = 1.632993 = \sqrt{8/3}$. This confirms the result in Example 7.7 which was obtained automatically, without the need for the geometrical intuition required by the elementary vector analysis approach used here. The only intermediate calculations that were necessary involved using the Gramm–Schmidt method to find an orthogonal set of basis vectors.
28. Orthonormal system: $\frac{1}{\sqrt{\pi}}\sin x,\ \frac{1}{\sqrt{\pi}}\sin 2x, \ldots,\ \frac{1}{\sqrt{2\pi}},\ \frac{1}{\sqrt{\pi}}\cos x,\ \frac{1}{\sqrt{\pi}}\cos 2x,\ \ldots$.
29. Self-Checking.
30. $P_0(x) = 1$, $P_1(x) = x$, $P_2(x) = \frac{1}{2}(3x^2 - 1)$, $P_3(x) = \frac{1}{2}(5x^3 - 3x)$.

Solutions 8

1. Yes, because $\mathbf{A}\mathbf{A}^{\mathrm{T}} = \mathbf{I}$.
2. Self-checking.
3. (a) Rotation with $\theta = \frac{2}{3}\pi$ (b) Rotation with $\theta = \frac{3}{4}\pi$ (c) Rotation with $\theta = -\frac{2}{3}\pi$ (d) Rotation with $\theta = -\frac{1}{4}\pi$.
4. Proceed as in Example 8.7.
5. Proceed as in Example 8.7.
6. (a) $\begin{bmatrix} 1 & -1 & 3 \\ 2 & 1 & -1 \\ -1 & -1 & 2 \end{bmatrix}$ (b) $\begin{bmatrix} 0 & 1 & 3 & -1 \\ 1 & -1 & 1 & -1 \\ 0 & 0 & 1 & 1 \\ 1 & -2 & 3 & 2 \end{bmatrix}$.

7. The linearity of the definite integral ensures that conditions (i) and (ii), sufficient and necessary for a linear transformation, are satisfied, because
(i)
$$\int_{-\infty}^{\infty} K(s,t)\{f(t)+g(t)\}dt = \int_{-\infty}^{\infty} K(s,t)f(t)dt + \int_{-\infty}^{\infty} K(s,t)g(t)dt,$$
(ii)
$$\int_{-\infty}^{\infty} \lambda K(s,t)f(t)dt = \lambda \int_{-\infty}^{\infty} K(s,t)f(t)dt.$$

8. Each satisfies the definition of a linear integral transformation that precedes Exercise 7.
   (a) The kernel of the Mellin transform is $K(p,t) = \begin{cases} 0, & t<0 \\ t^{p-1}, & t>0, \end{cases}$ with $f(t)$ defined for $0 < t < \infty$.
   (b) The kernel of the Fourier transform is $K(\omega,t) = e^{-i\omega t}$, with $f(t)$ defined for $-\infty<t<\infty$.
   (c) The kernel of the Hankel transform of order ν is $K_\nu(\sigma,t) = \begin{cases} 0, & t<0 \\ tJ_\nu(\sigma,t), & t>0, \end{cases}$ with $f(t)$ defined for $0 < t < \infty$.

9.
$$\mathbf{T}(\mathbf{e}_1) = \begin{bmatrix} 1 \\ -1 \\ -\frac{1}{2} \end{bmatrix}, \quad \mathbf{T}(\mathbf{e}_2) = \begin{bmatrix} 1 \\ 0 \\ 0 \end{bmatrix}, \quad \mathbf{T}(\mathbf{e}_3) = \begin{bmatrix} -1 \\ 1 \\ 1 \end{bmatrix} \quad \text{and so}$$
$$\mathbf{A} = \begin{bmatrix} 1 & 1 & -1 \\ -1 & 0 & 1 \\ -\frac{1}{2} & 0 & 1 \end{bmatrix}.$$

10.
$$\mathbf{T}(\mathbf{e}_1) = \begin{bmatrix} -4 \\ 4 \\ 0 \end{bmatrix}, \quad \mathbf{T}(\mathbf{e}_2) = \begin{bmatrix} 2 \\ -3 \\ 1 \end{bmatrix}, \quad \mathbf{T}(\mathbf{e}_3) = \begin{bmatrix} -4 \\ 4 \\ 0 \end{bmatrix},$$
$$\mathbf{T}(\mathbf{e}_4) = \begin{bmatrix} 0 \\ -1 \\ 1 \end{bmatrix} \quad \text{and so } \mathbf{A} = \begin{bmatrix} -4 & 2 & -4 & 0 \\ 4 & -3 & 4 & -1 \\ 0 & 1 & 0 & 1 \end{bmatrix}.$$

11. (a) $\det \mathbf{A} = 0$, so there is linear dependence between the rows of $\mathbf{A}$. Inspection (or Gaussian elimination) shows the first two rows are linearly independent so $\text{rank}(\mathbf{A}) = 2$. When written out in full, the equation $\mathbf{Ax} = \mathbf{y}$ becomes
$$\begin{aligned} 2x_1 + x_2 + 3x_3 &= y_1, \\ -x_1 + x_2 + 2x_3 &= y_2, \\ -6x_1 - 6x_2 - 16x_3 &= y_3, \end{aligned}$$

but as rank(**A**) $= 2$, for consistency y_3 must be determined in terms of y_1 and y_2 from the first two equations. Solving the first two equations for x_1 to x_3 gives

$$x_1 = k, \quad x_2 = 7k - 2y_1, \quad x_3 = -3k + y_1,$$

where k is arbitrary. Substituting these results into the third equation shows that $y_3 = -4y_1$. So the vector $\mathbf{y} = [y_1,\ y_2,\ -4y_1]^{\mathrm{T}}$ is the image of all vectors $\mathbf{x}$ of the form $\mathbf{x} = [k,\ 7k - 2y_1, -3k + y_1]^{\mathrm{T}}$, for arbitrary k. Thus a vector $\mathbf{y}$ of the given form is the image of arbitrarily many vectors $\mathbf{x} = [k,\ 7k - 2y_1, -3k + y_1]^{\mathrm{T}}$, where k is arbitrary. Any vector $\mathbf{y}$ that is not of the form $\mathbf{y} = [y_1,\ y_2,\ -4y_1]^{\mathrm{T}}$ will not be the image of a vector $\mathbf{x}$.

(b) $\det \mathbf{A} \neq 0$, so $\mathbf{A}^{-1}$exists and is unique. So from $\mathbf{A}^{-1}\mathbf{A}\mathbf{x} = \mathbf{A}^{-1}\mathbf{y}$, we have $\mathbf{x} = \mathbf{A}^{-1}\mathbf{y}$, showing the transformation to be one-to-one.

(c) Gaussian elimination shows rank (**A**) $= 3$, so writing out in full the first three equations gives

$$\begin{aligned} x_1 - x_2 + x_3 &= y_1, \\ x_1 + x_2 + x_3 &= y_2, \\ x_1 - x_3 &= y_3, \end{aligned}$$

and solving these for x_1 to x_3 gives $x_1 = \frac{1}{4}(y_1 + y_2 + 2y_3), x_2 = \frac{1}{2}(-y_1 + y_2)$, $x_3 = \frac{1}{4}(y_1 + y_2 - 2y_3)$. For the fourth equation $x_1 - 2x_2 + x_3 = y_4$ to be compatible with the first three equations, y_4 must be found by substituting these values into the fourth equation, when we find $y_4 = \frac{1}{2}(3y_1 - y_2)$. Thus for any given y_1, y_2, y_3, the vector $\mathbf{y} = [y_1,\ y_2,\ y_3,\ \frac{1}{2}(3y_1 - y_2)]^{\mathrm{T}}$ will be the image of the unique vector $\mathbf{x} = [\frac{1}{4}(y_1 + y_2 + 2y_3),\ \frac{1}{2}(-y_1 + y_2),\ \frac{1}{4}(y_1 + y_2 - 2y_3]^{\mathrm{T}}$. Any other vector $\mathbf{y}$ for which the fourth component is not $y_4 = \frac{1}{2}(3y_1 - y_2)$ will not be the image of a vector $\mathbf{x}$.

12. Writing out the scalar form of $\mathbf{Ax} = \mathbf{0}$ and solving for x_1 , x_2 and x_3 gives $x_1 = k$, $x_2 = -4k$ and $x_3 = -3k$, where k is arbitrary. Thus the vector $\mathbf{x}$ is proportional to k, so setting $k = 1$ we find that the vector in the nullspace is $\mathbf{x} = [1, -4, -3]^{\mathrm{T}}$, and so $\nu_{\mathbf{A}} = 1$. As **A** is a 3×3 matrix $n = 3$, but the form of **x** shows that rank (**A**) $= 2$, so rank(**A**) + $\nu_{\mathbf{A}} = 2 + 1 = 3 = n$, confirming the result of Theorem 8.2.
13. $\det \mathbf{A} \neq 0$, so $\mathbf{A}^{-1}$ exists, and so $\mathbf{A}^{-1}\mathbf{Ax} = \mathbf{Ix} = \mathbf{x} = \mathbf{0}$, showing there is no vector in the nullspace of **A**, with the result that $\nu_{\mathbf{A}} = 0$. As **A** is a 3×3 matrix $n = 3$, so as rank(**A**) $= 3$ we have rank (**A**) + $\nu_{\mathbf{A}} = 3 + 0 = 3$, confirming the result of Theorem 8.2.
14. Writing out the scalar form of $\mathbf{Ax} = \mathbf{0}$ and solving for x_1, x_2, x_3 and x_4 gives $x_1 = -2\alpha - 5\beta$, $x_2 = \alpha + 2\beta$, $x_3 = \alpha, x_4 = \beta$, where α and β are arbitrary, showing that rank(**A**) $= 2$, because from the four variables x_1 to x_4, two may be expressed in terms of the remaining two. Making the arbitrary assignment $\alpha = 1$, $\beta = 0$ shows one vector in the null space to be $\mathbf{x} = [-2,\ 1,\ 1,\ 0]^{\mathrm{T}}$, while making the assignment $\alpha = 0$, $\beta = 1$ shows another vector in the null space to be $\mathbf{x} = [-5,\ 2,\ 0,\ 1]^{\mathrm{T}}$, so that $\nu_{\mathbf{A}} = 2$. Different choices for α and β will give different vectors **x** in the nullspace, though any pair will span the nullspace.

Inspection of $\mathbf{A}$ shows that $n = 4$, so that rank $(\mathbf{A}) + \nu_{\mathbf{A}} = 2+2 = 4 = n$, confirming the result of Theorem 8.2.

15. Writing out the scalar form of $\mathbf{Ax} = \mathbf{0}$ and solving for the four variables x_1, x_2, x_3 and x_4 gives $x_1 = -\alpha,\ x_2 = 0,\ x_3 = \alpha,\ x_4 = \alpha$, where α is arbitrary, and $n = 4$. As the three variables x_1 to x_3 are expressible in terms of x_4 shows that rank$(\mathbf{A}) = 3$. Making the assignment $\alpha = 1$ shows a vector in the nullspace to be $\mathbf{x} = [-1,\ 0,\ 0,\ 1]^{\mathrm{T}}$, and so $\nu_{\mathbf{A}} = 1$. Thus rank$(\mathbf{A}) + \nu_{\mathbf{A}} = 3+1 = 4 = n$, confirming the result of Theorem 8.2.

16. (a) A reflection in a line L through the origin inclined at an angle $\theta = \pi/6$ to the x_1-axis.
    (b) A counterclockwise rotation about the origin through an angle $\theta = \pi/3$.
17. (a) A reflection in a line L through the origin inclined at an angle $\theta = \pi/4$ to the x_1-axis.
    (b) A projection onto a line L inclined at an angle $\theta = \pi/4$ to the x_1-axis.
18. (a) A clockwise rotation about the origin through an angle $\theta = -\pi/4$.
    (b) A reflection in a line L through the origin inclined at an angle $\theta = -\pi/2$ to the x_1-axis.
19. (a) A projection onto a line L inclined at an angle $\theta = -\pi/4$ to the x_1-axis.
    (b) A counterclockwise rotation about the origin through an angle $\theta = 3\pi/4$.
20. $A'\,(-1.366, -0.366)$, $B'\,(-3.098,\ 0.634)$, $C'\,(-2.598,\ 1.5)$, $D'\,(-0.866,\ 0.500)$.

21. The projection matrix $\mathbf{A} = \begin{bmatrix} \frac{1}{2} & \frac{1}{2} \\ \frac{1}{2} & \frac{1}{2} \end{bmatrix}$. If P is the point (3, 5), the coordinates of its orthogonal projection P'onto line L are given by $\begin{bmatrix} y_1 \\ y_2 \end{bmatrix} = \begin{bmatrix} \frac{1}{2} & \frac{1}{2} \\ \frac{1}{2} & \frac{1}{2} \end{bmatrix} \begin{bmatrix} 3 \\ 5 \end{bmatrix} = \begin{bmatrix} 4 \\ 4 \end{bmatrix}$, so P' is at the point (4, 4).

22. The projection matrix $\mathbf{A} = \begin{bmatrix} \frac{3}{4} & \frac{\sqrt{3}}{4} \\ \frac{\sqrt{3}}{4} & \frac{1}{4} \end{bmatrix}$. If P is the point (2, 6), the coordinates of its orthogonal projection P'onto line L are given by $\begin{bmatrix} y_1 \\ y_2 \end{bmatrix} = \begin{bmatrix} \frac{3}{4} & \frac{\sqrt{3}}{4} \\ \frac{\sqrt{3}}{4} & \frac{1}{4} \end{bmatrix} \begin{bmatrix} 2 \\ 6 \end{bmatrix} = \begin{bmatrix} 4.098 \\ 2.366 \end{bmatrix}$, so P' is at the point (4.098, 2.366).

23. This involves a composite transformation, because repeated reflection is described by the matrix product $\mathbf{A}^2$. We have

$$\mathbf{A}^2 = \begin{bmatrix} \cos 2\alpha & \sin 2\alpha \\ \sin 2\alpha & -\cos 2\alpha \end{bmatrix}^2$$

$$= \begin{bmatrix} \cos^2 2\alpha + \sin^2 2\alpha & 0 \\ 0 & \cos^2 2\alpha + \sin^2 2\alpha \end{bmatrix} = \begin{bmatrix} 1 & 0 \\ 0 & 1 \end{bmatrix}.$$

So $\mathbf{A}^2 = \mathbf{I}$, which is the identity transformation, showing that whatever the original geometrical shape, it is reconstructed after two repeated reflections.

24. The first part of this exercise involves a reflection first in the x_2- axis and then in the x_1-axis. The two reflection matrices involved are $\begin{bmatrix} -1 & 0 \\ 0 & 1 \end{bmatrix}$ and $\begin{bmatrix} 1 & 0 \\ 0 & -1 \end{bmatrix}$ so the composite effect is given by the product $\begin{bmatrix} -1 & 0 \\ 0 & 1 \end{bmatrix}\begin{bmatrix} 1 & 0 \\ 0 & -1 \end{bmatrix} = \begin{bmatrix} -1 & 0 \\ 0 & -1 \end{bmatrix}$.
The second part of this exercise involves a reflection first in a line inclined to the x_1-axis at an angle $\alpha = \pi/4$, and then a reflection in a line inclined to the x_1-axis at an angle $\alpha = -\pi/4$. The two reflection matrices involved are $\begin{bmatrix} 0 & 1 \\ 1 & 0 \end{bmatrix}$ and $\begin{bmatrix} 0 & -1 \\ -1 & 0 \end{bmatrix}$ so the composite effect is given by the product

$$\begin{bmatrix} 0 & 1 \\ 1 & 0 \end{bmatrix}\begin{bmatrix} 0 & -1 \\ -1 & 0 \end{bmatrix} = \begin{bmatrix} -1 & 0 \\ 0 & -1 \end{bmatrix}.$$

As the composite transformation is identical in each case, each set of reflections produces the same result.

25. This is a composite transformation with the first reflection described by the matrix $\mathbf{A} = \begin{bmatrix} \cos 2\alpha & \sin 2\alpha \\ \sin 2\alpha & -\cos 2\alpha \end{bmatrix}$, while the second reflection is described by the matrix $\mathbf{B} = \begin{bmatrix} \cos 2\beta & \sin 2\beta \\ \sin 2\beta & -\cos 2\beta \end{bmatrix}$. After simplification, it is found that the product $\mathbf{BA} = \begin{bmatrix} \cos 2(\beta - \alpha) & -\sin 2(\beta - \alpha) \\ \sin 2(\beta - \alpha) & \cos 2(\beta - \alpha) \end{bmatrix}$, which corresponds to a rotation through the angle $2(\beta - \alpha)$.

When $\alpha = \pi/6$ and $\beta = \pi/3$, a point P at (1, 1) in the (x_1, x_2)-plane is mapped to the point Q at $(-0.366,\ 1.336)$. To interpret correctly the rotation through the angle $2(\beta - \alpha)$, it should be remembered that the first reflection of the point P about the line L inclined at an angle $\pi/6$ to the x_1-axis, is in the *clockwise* direction, so $\alpha = -\pi/6$, whereas the second reflection is in the *counterclockwise* direction, so $\beta = \pi/3$, with the result that $2(\beta - \alpha) = \pi/3$. This is in agreement with the result that after two reflections the point (1, 1) at P is mapped to the point Q at $(-0.366,\ 1.336)$. Because if the origin is O, the vector OP to the original point P is expanded and rotated *counterclockwise* through an angle $\pi/3$ to become the vector OQ. Check this on a sketch drawn to scale.

26. The composite transformation comprising a uniform magnification, a rotation and then a shear, in this order, takes the form

$$\begin{bmatrix} 2 & 0 \\ 0 & 2 \end{bmatrix}\begin{bmatrix} \cos \pi/6 & -\sin \pi/6 \\ \sin \pi/6 & \cos \pi/6 \end{bmatrix}\begin{bmatrix} 1 & 1 \\ 0 & 1 \end{bmatrix} = \begin{bmatrix} \sqrt{3} & \sqrt{3} - 1 \\ 1 & \sqrt{3} + 1 \end{bmatrix}.$$

$$A(1.5,\ 0) \to A'(2.598, 1.5),\ \ B(1.5,\ 1) \to B'(3.330,\ 4.232)$$

$$C(0,\ 1) \to C'(0,732,\ 2.732),\ D(0,\ 0) \to D'(0,\ 0).$$

The matrix representing a uniform magnification is symmetric, so it may appear in any position in the composite transformation without altering the matrix product. However, the matrices representing the rotation and shear are not symmetric, so interchanging their order in the composite transformation will change the matrix product, and so change the effect of the composite transformation.

27. Self-checking. Magnification factor is $|\det \mathbf{A}| = 4$.
28. To produce a magnification factor of 3 the elements in **A** must be multiplied by a factor $\sqrt{3}$.
29. The matrix **A** is such that $|\det \mathbf{A}| = 1$.
30. If

$$\mathbf{A} = \begin{bmatrix} a & 0 \\ 0 & b \end{bmatrix},\ \text{and } \mathbf{B} = \begin{bmatrix} 1 & k \\ 0 & 1 \end{bmatrix},\ \text{then } \mathbf{AB} = \begin{bmatrix} a & ka \\ 0 & b \end{bmatrix} \text{ and } \mathbf{BA} = \begin{bmatrix} a & kb \\ 0 & b \end{bmatrix}.$$

Thus the first effect of **AB** is to produce a shear, after which the result is scaled differently in the x_1 and x_2 directions. The product **BA** reverses the order in which these effects are produced.

31. If $\mathbf{A} = \begin{bmatrix} a & b \\ c & d \end{bmatrix}$, then **A** will be singular if $ad - bc = 0$. In the trivial case when $\mathbf{A} = \mathbf{0}$ the entire $(x_1,\ x_2)$-plane will be mapped to the origin. Suppose, instead, that $a \neq 0$. Then $d = bc/a$, so $\mathbf{A} = \begin{bmatrix} a & b \\ c & bc/a \end{bmatrix}$ with the result that $\mathbf{Ax} = \begin{bmatrix} ax_1 + bx_2 \\ (c/a)(ax_1 + bx_2) \end{bmatrix}$. This shows that every point in the (x_1, x_2)-plane is mapped onto the line $y_2 = (c/a)y_1$ in the $(\ y_1, y_2)$-plane.
32. If **A** produces a stretch, changing the sign of an element in **A** simply reverses the direction of stretch about the appropriate axis, resulting in a reflection about that axis.
33. Pre-multiplication by **P** interchanges the rows of **A** , so it leaves a rectangle or triangle in the $(x_1,\ x_2)$-plane unchanged, but it reverses the sense around the rectangle or triangle.

Index

M

N

O

P

Q

R

S